Springer Finance

Springer
Berlin
Heidelberg
New York
Hong Kong
London
Milan
Paris
Tokyo

Springer Finance

Springer Finance is a programme of books aimed at students, academics and practitioners working on increasingly technical approaches to the analysis of financial markets. It aims to cover a variety of topics, not only mathematical finance but foreign exchanges, term structure, risk management, portfolio theory, equity derivatives, and financial economics.

Matthias Gundlach
Frank Lehrbass (Eds.)

CreditRisk$^+$
in the Banking
Industry

Springer

Matthias Gundlach
Aareal Bank AG
Konzerncontrolling Kreditrisiko
Paulinenstraße 15
65189 Wiesbaden
Germany
e-mail: matthias.gundlach@aareal-bank.com

Frank Lehrbass
Deutsche Genossenschafts-
Hypothekenbank AG
Portfolio Management
& Structured Investments
Credit Treasury
Rosenstraße 2
20095 Hamburg
Germany
e-mail: Frank.Lehrbass@dghyp.de

Mathematics Subject Classification (2000): 91B28,91B30, 60-08, 60E99
JEL Classification: C5, G11, G22

CreditRisk$^+$ TM is a service mark owned by Credit Suisse Group and is registered in numerous jurisdictions

Library of Congress Control Number: 2004104244

ISBN 3-540-20738-4 Springer-Verlag Berlin Heidelberg New York

Springer-Verlag Berlin Heidelberg New York
Springer-Verlag is a part of Springer Science+Business Media

springeronline.com

© Springer-Verlag Berlin Heidelberg 2004
Printed in Germany

The use of general descriptive names, registered names, trademarks, etc. in this publication does not imply, even in the absence of a specific statement, that such names are exempt from the relevant protective laws and regulations and therefore free for general use.

Cover design: *design & production*, Heidelberg

Typesetting by the authors using a Springer LaTeX macro package
Printed on acid-free paper   41/3142db - 5 4 3 2 1 0

Preface

The object of this book is to offer an overview of and describe new developments in the credit risk model CreditRisk$^+$. In the selection of topics and in its organization, the book addresses both experts and practitioners, as well as anyone looking for an introduction to the subject.

We are happy to present a book that can be seen as the product of a community. This community has arisen over the past six years out of the desire to develop and perfect a credit risk model that is regarded by its supporters as the most promising and the most elegant of the current models.

Such a community would not have been possible without Tom Wilde and Credit Suisse Financial Products (CSFP), who originally created this model and decided to share it with the public. It would also not have been established without those senior figures in the banking industry who have had the foresight and confidence to support their employees working on credit risk models of their own, in particular using CreditRisk$^+$. Our special thanks go to CSFP, Tom Wilde, Fritz-Thomas Murray of WestLB and Markus Petry of Aareal Bank.

Our sincere thanks go to all the contributors for their work, their enthusiasm, their reliability and their cooperation. We know that most of the research had to be done in valuable spare time. We are glad that all of them were willing to make such sacrifices for the sake of this book.

Our thanks go to all those people in the banking industry who work with the model and try to extend it with such great enthusiasm.

We are grateful to Springer-Verlag, who immediately welcomed our idea for this book and encouraged us to work on it.

We would also like to thank all our colleagues and friends who agreed to work as referees, Carsten Binnenhei, Leif Boegelein, Götz Giese, Hermann Haaf, Hans-Joachim Hüskemann, Markus Klintworth, Daniel Kluge, Marc

Klünger, Michael Lesko, Oliver Reiß, Oliver Steinkamp and Dirk Tasche. Jeffrey Boys did a marvellous job of proofreading the manuscript and correcting the English.

Moreover, we are much obliged to Steffen Gräf of Aareal Bank and Frank Holzwarth of Springer-Verlag for their help with technical problems in preparing the manuscript.

Finally we would like to thank our families, in particular our wives Angelika and Jutta for their continued support and understanding.

Wiesbaden, Germany   VOLKER MATTHIAS GUNDLACH
Hamburg, Germany   FRANK BERTHOLD LEHRBASS
                                                     October 2003

Contents

Contributors

NESE AKKAYA
Credit Risk Control, UBS AG, 8050 Zürich, Switzerland
E-MAIL: nese.akkaya-neuenschwander@ubs.com

CARSTEN BINNENHEI
Landesbank Baden-Württemberg, Risikocontrolling, Fritz-Elsas-Str. 31,
D-70174 Stuttgart, Germany
E-MAIL: Carsten.Binnenhei@lbbw.de

LEIF BOEGELEIN
Siemens Financial Services, Credit (CR), Seidlstr. 24 a,
D-80335 München, Germany
E-MAIL: Leif.Boeglein@sfs.siemens.de

FRANK BRÖKER
SCHUFA HOLDING AG, Hagenauer Str. 44, D-65203 Wiesbaden, Germany
E-MAIL: Frank.Broeker@Schufa.de

GÖTZ GIESE
Commerzbank AG, Kaiserplatz 1, D-60261 Frankfurt, Germany
E-MAIL: Goetz.Giese@commerzbank.com

MICHAEL GORDY
Division of Research & Statistics, Board of Governors of the Federal Reserve
System, Washington, DC 20551, USA
E-MAIL: michael.gordy@frb.gov

MATTHIAS GUNDLACH
Aareal Bank AG, Konzerncontrolling Kreditrisiko, Paulinenstr. 15,
D-65189 Wiesbaden, Germany
E-MAIL: matthias.gundlach@aareal-bank.com

HERMANN HAAF
Commerzbank AG, Risk Control (ZRC), 60261 Frankfurt/Main, Germany
E-MAIL: Hermann.Haaf@commerzbank.com

ALFRED HAMERLE
Lehrstuhl für Statistik, Wirtschaftswissenschaftliche Fakultät,
Universität Regensburg, Germany
E-MAIL: Alfred.Hamerle@wiwi.uni-regensburg.de

MARTIN HELLMICH
Landesbank Baden-Württemberg, 8830 - Kapitalmarktinvestitionen/
Strukturierung, Am Hauptbahnhof 2, D-70173 Stuttgart, Germany
E-MAIL: Martin.Hellmich@lbbw.de

DANIEL KLUGE
Portfolio Management/Structured Investments, Credit Treasury,
Deutsche Genossenschafts-Hypothekenbank AG, Rosenstr. 2,
D-20095 Hamburg, Germany
E-MAIL: Daniel.Kluge@dghyp.de

MICHAEL KNAPP
Lehrstuhl für Statistik, Wirtschaftswissenschaftliche Fakultät,
Universität Regensburg, Germany
E-MAIL: Michael.Knapp@wiwi.uni-regensburg.de

ALEXANDRE KURTH
Credit Risk Control, UBS AG, 8050 Zürich, Switzerland
E-MAIL: alexandre.kurth@ubs.com

FRANK LEHRBASS
Portfolio Management/Structured Investments, Credit Treasury,
Deutsche Genossenschafts-Hypothekenbank AG, Rosenstr. 2,
D-20095 Hamburg, Germany
E-MAIL: Frank.Lehrbass@dghyp.de

MICHAEL LESKO
Research, GILLARDON AG financial software, Alte Wilhelmstr. 4,
D-75015 Bretten, Germany
E-MAIL: Michael.Lesko@gillardon.de

SANDRO MERINO
Credit Research, Wealth Management & Business Banking,
UBS AG, Switzerland
E-MAIL: sandro.merino@ubs.com

MARK NYFELER
Credit Portfolio Risk Control, Investment Bank, UBS AG, USA
E-MAIL: mark.nyfeler@ubs.com

OLIVER REIß
Weierstrass Institute for Applied Analysis and Stochastics, Mohrenstraße 39,
D-10117 Berlin, Germany
E-MAIL: reiss@wias-berlin.de

DANIEL RÖSCH
Lehrstuhl für Statistik, Wirtschaftswissenschaftliche Fakultät,
Universität Regensburg, Germany
E-MAIL: Daniel.Roesch@wiwi.uni-regensburg.de

FRANK SCHLOTTMANN
Research, GILLARDON AG financial software, Alte Wilhelmstr. 4,
D-75015 Bretten, Germany
E-MAIL: Frank.Schlottmann@gillardon.de

JOHN SCHOENMAKERS
Weierstrass Institute for Applied Analysis and Stochastics, Mohrenstraße 39,
D-10117 Berlin, Germany
E-MAIL: schoenma@wias-berlin.de

STEFAN SCHWEIZER
zeb/rolfes.schierenbeck.associates, Hammer Straße 165,
D-48153 Münster, Germany
E-MAIL: sschweizer@zeb.de

DETLEF SEESE
Institut AIFB, Fakultät für Wirtschaftswissenschaften,
Universität Karlsruhe (TH), D-76128 Karlsruhe, Germany
E-MAIL: seese@aifb.uni-karlsruhe.de

OLIVER STEINKAMP
Landesbank Baden-Württemberg, 8830 - Kapitalmarktinvestitionen/
Strukturierung,Am Hauptbahnhof 2, D-70173 Stuttgart, Germany
E-MAIL: Oliver.Steinkamp@web.de

DIRK TASCHE
Deutsche Bundesbank, Banken und Finanzaufsicht, Postfach 10 06 02,
D-60006 Frankfurt, Germany
E-MAIL: tasche@ma.tum.de

STEPHAN VORGRIMLER
Research, GILLARDON AG financial software, Alte Wilhelmstr. 4,
D-75015 Bretten, Germany
E-MAIL: Stephan.Vorgrimler@gillardon.de

ARMIN WAGNER
Credit Risk Control, UBS AG, 8050 Zürich, Switzerland
E-MAIL: armin.wagner@ubs.com

CHRISTIAN WIECZERKOWSKI
Portfolio Management – Business Development – Financial Engineering,
HSH-Nordbank, Martensdamm 6, D-24103 Kiel, Germany
E-MAIL: christian.wieczerkowski@hsh-nordbank.com

1

Introduction

Volker Matthias Gundlach and Frank Berthold Lehrbass

Summary. We give a brief description of the history, possibilities and applications of CreditRisk$^+$ as well as an overview of the book.

1.1 Birth of the Model

Shortly after J. P. Morgan published the portfolio model CreditMetrics^{TM} (henceforth CreditMetrics), Credit Suisse Financial Products (CSFP) introduced its own portfolio model CreditRisk^{+TM} (henceforth CreditRisk$^+$) in October 1997. The model was created by Tom Wilde of CSFP and is still unique among existing portfolio models with respect to the mathematics involved. CreditRisk$^+$ applies techniques from actuarial mathematics in an enhanced way to calculate the probabilities for portfolio loss levels (i.e. the portfolio loss distribution, henceforth loss distribution).

In contrast to JP Morgan, CSFP decided not to market its credit risk model but to publish adequate information on the basics of CreditRisk$^+$. Those banks used to buying complete solutions for their problems avoided the CSFP model. However, for other banks CreditRisk$^+$ became a popular choice, used in business units such as Treasury, Credit Treasury, Controlling and Portfolio Management. It was not surprising therefore that CreditRisk$^+$ – along with CreditMetrics – played a significant role in the development of the New Basel Capital Accord.

One of the main reasons for the success of CreditRisk$^+$ is that it is the only credit risk model that focuses on the event of default, which is the event of most interest in all credit risk analysis. Moreover, it can easily be adapted to different kinds of credit business. This makes it particularly popular amongst practitioners who need to work with a credit risk model and its developments fitted to their needs.

Among other things, this book gives an idea of the demonstrative versatility of CreditRisk$^+$: the professional backgrounds of the authors and the topics of their contributions underline how widely implemented, influential, and flexible CreditRisk$^+$ is. Moreover, the contributions suggest that CreditRisk$^+$ is not only a powerful tool for dealing with credit risk, but also an intellectual challenge.

The aim of this book is to present the current position of CreditRisk$^+$ and to provide an overview of those developments that are, or could be, used in the banking industry. Before drawing up a short outline of the book, we illustrate the main reasons why CreditRisk$^+$ is well suited to and popular in the banking industry.

1.2 Reasons for Adoption

Why is CreditRisk$^+$ attractive for practitioners/experts?

In CreditRisk$^+$ the development of a loan is understood as a Bernoulli random variable: either the obligor pays the amount due or the loan defaults. No profits or losses from rating migrations have to be considered. In contrast, CreditMetrics tries to model a mark-to-market approach of liquid loans, in which each cash flow is discounted at a rate that is appropriate for its overall rating. The uncertainty of the future rating presents the source of randomness. It is not simply a question of two possibilities (obligor pays or defaults) being of interest, as there are as many possibilities as there are rating classes. Experts in the lending business quickly realized that CreditRisk$^+$ is more intuitive than CreditMetrics and better suited to practical needs. There is no requirement to provide the complete cash flow vector for each loan and no requirement to update spread curves for each rating class on a regular basis.

One final remark with respect to CreditMetrics seems appropriate: the mark-to-market approach of CreditMetrics is in most cases merely a mark-to-model approach. But the model does not pay you cash if the rating of the loan has improved. Hence, CreditMetrics sometimes measures (pseudo-mark-to-market) profits that cannot be realized.

Since CreditRisk$^+$ is an analytic-based portfolio approach it allows the rapid and unambigous calculation of the loss distribution. The speed of the calculation is helpful when comparative statistics ("what if" analyses) are performed, e.g. for analysing a new deal with a significant influence on the loss distribution or for pricing asset-backed securities (ABS) in a wider sense (ABS, mortgage-backed-securities (MBS), collateralized debt obligations (CDO), etc.) including different scenarios for the parameters.

Since practitioners tend to base decisions on their calculations, the unambiguity of the distribution is important – especially as far as the tail of the loss distribution is concerned. However, simulation-based portfolio approaches usually fail to give a single answer because their results depend on the number of simulation runs, type of random numbers, etc.

In handling huge client portfolios with low probabilities of default, simulation-based approaches face two problems: firstly, the absolute number of asset values (abilities to pay or other variables) to be simulated will become too extensive (e.g. in the tens of thousands for a residential mortgage-backed security (RMBS) pool); secondly, most of the simulation runs will not produce any defaults. Therefore, only very few results provide the basis for approximating the loss distribution as relative frequencies.

For these reasons, the motivation for an analytical approach is clear. However, especially with regard to the case mentioned above, a question arises: Why does one not apply the Central Limit Theorem? One is interested in the sum of individual losses that occur randomly and therefore, if the individual losses are mutually independent and share a common distribution – that has first and second moments – then the loss distribution should be normal.

The answer to this proposal is the following: even in an RMBS or credit card ABS pool, which are favourable portfolios in light of the assumptions of the Central Limit Theorem, the assumptions tend not to be valid. Firstly, the assets in the pool do not share a common (Bernoulli) distribution because the probabilities of default vary. Even if the exposure sizes are more or less equal, no further benefit can be achieved. Secondly, and more seriously, the losses are not completely independent of each other because obligors from the same industry or country share a similar development. CreditRisk$^+$ accounts for such facts.

CreditRisk$^+$ shows even greater flexibility. In accounting for stochastic probabilities of default with different expected values and rich structures of default correlation among the obligors and different levels of (net) exposures, CreditRisk$^+$ analytically derives an algorithm for the calculation of the loss distribution. It thus avoids the need to manage simulations.

Disregarding its numerous advantages, there is a price to pay for the analytical nature of CreditRisk$^+$: for example the Poisson approximation in CreditRisk$^+$ requires the expected default probabilities to be small, typically a single digit percentage figure. This approximation leads to two astonishing phenomena: firstly, there is normally some probability of the calculated losses being larger than the sum of the net exposures, i.e. you can lose more money than you have lent! Secondly, the risk contributions (as calculated in the original version) may turn out higher than the net exposure itself. However, even if the model is operated with moderate levels of expected default probabilities, high levels may return through the back door by setting high levels of standard deviations for the default probabilities. Fortunately, the probability of losses larger than the sum of the net exposures can be quantified. This usually reveals that the price is lower than expected.

A further price component is the limited range of default correlation that can be produced by CreditRisk$^+$. Due to the mathematical approach chosen (the intensity approach), the original CreditRisk$^+$ produces only positive and usually moderate levels of default correlation among two obligors sharing a common risk driver. In a particular industry, default correlation may theoretically be negative because one company can take capacity away from a defaulting competitor and thus decrease its default probability. However, economy-wide systematic risk factors should outweigh this negative dependence to a large extent. Summing up default correlation should be positive. The only price paid is the limited range that CreditRisk$^+$ – in its original version – can produce. There are ways to deal with this. One example for achieving the extreme case of a perfect default correlation is an artificial merger of two clients into one net exposure figure.

Finally, one more price component is the assumption that recovery rates and defaults are independent of one another. This is a common assumption made by several other credit risk models. A cautious approach, to cope with this problem and to acknowledge the fact of negative correlation, is the use of conservatively estimated recovery rates.

Weighing the advantages of credit risk modelling against the price paid, one should never forget that there is no such thing as *the true model* and keep in mind where the value of modelling lies: it imposes discipline and consistency on an institution's lending practices and allocation of risk capital. Recent research has outlined the severe problems of backtesting credit risk models. This underlines the value of starting with a practical model – instead of waiting or searching for the *ultimate model*.

1.3 Outline of the Book

This book comprises four parts, each with a different purpose. The first part should be viewed as an introduction to the subject and its background, while the second part covers the optimal implementation of CreditRisk$^+$. We then go on to consider various developments of the model, before we finally address practical problems of applying CreditRisk$^+$.

PART I: The Basics

This book starts with an introduction of the original (or standard) version of CreditRisk$^+$ by Gundlach. The evolution of CreditRisk$^+$ in both theory and practice is considered in this contribution. Tasche then introduces measures of credit risk such as value-at-risk and expected shortfall, emphasizing the significance of the latter two for all modifications and implementations of CreditRisk$^+$. A comparison between CreditMetrics, the other outstanding credit risk model, and CreditRisk$^+$ is drawn by Wieczerkowski.

PART II: Implementations of CreditRisk$^+$

The execution of the basic algorithm for the caculation of the loss distribution may lead to numerical errors, as investigated by Haaf, Reiß and Schoenmakers. They propose a new numerical recursion avoiding such shortcomings. Further contributions focus on avoiding numerical problems by looking at new ways of obtaining the loss distribution. While the standard model uses probability-generating functions, the contribution by Giese presents an approach based on moment-generating functions, which offer further possibilities for modelling default correlations. Saddlepoint approximation, based on the cumulant generating function, offers another alternative. Gordy shows how this technique can be applied to the estimation of the loss distribution in CreditRisk$^+$ as well as in an extended version of the model that allows for severity risk. A completely different option is presented by Reiß who focuses on the characteristic function for obtaining the loss distribution. He recommends using Fourier inversion techniques when dealing with characteristic functions.

PART III: Refining the working of risk drivers in CreditRisk$^+$

The original CreditRisk$^+$ assumes fixed net exposures and independent risk drivers. These deviations from reality are tackled in this part of the book. While Akkaya, Kurth, and Wagner make suggestions for both of these problems, Giese shows how his framework, using moment-generating functions, can be used to work with dependent risk drivers. The contributions by Bröker and Schweizer as well as by Binnenhei present how the two-state approach (obligor pays or defaults) can be enriched by *still* using ideas of CreditRisk$^+$. The possibility to integrate possible losses due to rating downgrades has been an important argument in favour of CreditRisk$^+$ – especially in the early years when, to some people, the competitor CreditMetrics seemed superior for this reason. Finally, Reiß tries to close the gap to the modelling of market risk by introducing dependent risk drivers as well as random default times. This leads to a version of CreditRisk$^+$ in continuous time at the cost of losing the analytical tractability.

PART IV: Applications of CreditRisk$^+$

Using CreditRisk$^+$ requires determining the probabilities of default and their volatilities, and performing a so-called sector allocation in order to establish the most suitable default correlation structure. Contributions by two groups of authors examine these problems from different angles: one group consists of Boegelein, Hamerle, Knapp and Rösch, who tackle the problem on a global basis, whereas Lesko, Schlottmann and Vorgrimler present empirical results for Germany. Besides the regional priorities, the two groups also differ in their methodological approach.

Knowing the loss distribution of your current portfolio is one thing. Another is finding the optimal portfolio. Lesko, Schlottmann, Seese and Vorgrimler show one possible route for risk-return analysis using CreditRisk$^+$ and evolutionary computation. Assessing the impact of a CDO on a portfolio in a realistic setting and bringing together techniques of fast Fourier transform and numerical integration using Monte Carlo methods is the focus of the analysis by Merino and Nyfeler. Pricing tranches of ABS is the subject of two contributions. While Kluge and Lehrbass make some general remarks on the applicability of CreditRisk$^+$ to the analysis of ABS, Hellmich and Steinkamp provide pricing formulae for structured credit derivatives such as first- and k-to-default baskets as well as for all kinds of tranches of CDO.

In order to avoid merely a collection of single papers on the subject we tried to use a common notation and to encourage the exchange of ideas and links between the papers as well as a common index. We therefore recommend reading Chapter 2 in order to grasp the notions and notations. In order to distinguish the citations within the book from outside sources, we cite the contributions in the book by reference to the corresponding chapters.

Basics of CreditRisk$^+$

Volker Matthias Gundlach

Summary. We present the fundamental ideas of CreditRisk$^+$ and give an introduction to the main notions of this credit risk model. In particular we set up the notations for the book and describe the original version of CSFP.

Introduction

Unlike most of the other credit risk models, CreditRisk$^+$ is not a commercial solution. Instead it is implemented by various users in their own way and style, using several extensions. Consequently there exist nowadays many versions of the model, nearly all of which are called CreditRisk$^+$. In this contribution we would like to present what can be seen as the common ground of the CreditRisk$^+$ versions; in particular we describe the original version of CSFP.

Basically, CreditRisk$^+$ is a default mode model that mimics acturial mathematics in order to describe credit risk in a portfolio. There are only a few fundamental ideas for the model, which can be summarized as follows:

- **No model for default event:** The reason for a default of an obligor is not the subject of the model. Instead the default is described as a purely random event, characterized by a probability of default.
- **Stochastic probability of default:** The probability of default of an obligor is not seen as a constant, but a randomly varying quantity, driven by one or more (systematic) risk factors, the distribution of which is usually assumed to be a gamma distribution.
- **Linear dependence on risk factors:** A linear relationship between the systematic risk factors and the probabilities of default is assumed.
- **Conditional independence:** Given the risk factors the defaults of obligors are independent.
- **Only implicit correlations via risk drivers:** Correlations between obligors are not explicit, but arise only implicitly due to common risk factors which drive the probability of defaults.

- **Discretization of losses:** In order to aggregate losses in a portfolio in a comfortable way, they are represented as multiples of a common loss unit.
- **Use of probability-generating functions:** The distribution of losses is derived from probability-generating functions, which can be used if the losses take discrete values.
- **Approximation of loss distribution:** In order to obtain the loss distribution for the portfolio with acceptable effort, an approximation for the probability-generating functions is used, which usually corresponds to a Poisson approximation for the loss of an obligor.

These ideas result in the following characteristic features of CreditRisk$^+$:

- The resulting portfolio loss distribution is described by a sum of independent compound negative binomial random variables.
- The distribution and its characteristic quantities can be calculated analytically in a fast and comfortable way; no simulations are necessary.

While most of these ideas and features can be found in any of the chapters in this book, some of them will be replaced or extended in some places. Here we want to explain the main topics in detail.

This chapter is organized as follows. In the first section we present some important general notions and notations from probability theory, which will be used throughout the book. Notations which describe the input for the credit risk model are fixed in Section 2.2. There we also introduce the idea of sectors and the discretization of losses. In Section 2.3 we explain how CreditRisk$^+$ deals with the distribution of losses due to credit defaults, before we deduce the basic output of the model in the following section. Section 2.5 is devoted to the details of the model as originally introduced by CSFP [3]. There a recursion scheme from actuarial mathematics was suggested in order to calculate the loss distribution. This specification is given in an appendix. In the final section we discuss the problem of verification of the model.

Let us mention that there exist various notations in the literature for describing CreditRisk$^+$. We try to follow those of [2].

2.1 Some Basic Notions from Probability Theory

Though stochastics plays a major role in the modelling of default risk by CreditRisk$^+$, the specification of a suitable probability space is of minor importance. We denote such a probability space by $(\Omega, \mathcal{F}, \mathbb{P})$. It is our standard assumption that this space is chosen in a way that keeps all variables measurable.

We denote by $\mathbb{E}$ the expectation, by var and by cov the variance and covariance, respectively, with respect to $\mathbb{P}$, while $\mathbb{E}[.|.]$, var$[.|.]$ and cov$[.|.]$ denote respective conditional expectations, variances and covariances. For a random variable Y on our probability space we will denote the probability density by f_Y, if it exists or should exist (which could mean another assumption on the probability space). If a random variable Y is distributed according to a well-known distribution f, we express this by $Y \sim f$. In CreditRisk$^+$ we ocasionally have to consider for random variables Y gamma distributions, the probability density function of which we denote by

$$f_Y^{(\alpha,\beta)}(y) = \frac{y^{\alpha-1}}{\beta^\alpha \Gamma(\alpha)} e^{-y/\beta} \quad \text{for } y \geq 0 \tag{2.1}$$

for parameters $\alpha, \beta > 0$ and the gamma function Γ. If the parameters are fixed, we also omit the superscript (α, β). We recall that the expectation and the variance of such a random variable Y are given by

$$\mathbb{E}[Y] = \alpha\beta, \qquad \text{var}[Y] = \alpha\beta^2. \tag{2.2}$$

If a random variable Y attains only values in the non-negative integers, its distribution can be derived easily from its probability-generating function (PGF) G_Y defined as

$$G_Y(z) = \mathbb{E}[z^Y] = \sum_{n \geq 0} \pi(n) z^n \quad \text{with } \pi(n) = \mathbb{P}[Y = n].$$

Note that this expectation exists for $|z| \leq 1$. The PGF contains all the information to calculate the distribution, namely we obtain by n-th differentiation

$$\pi(n) = \mathbb{P}[Y = n] = \frac{1}{n!} G_Y^{(n)}(0).$$

Analogously, one can deal with conditional PGFs

$$G_Y(z|.) = \mathbb{E}[z^Y |.]. \tag{2.3}$$

G_Y can be obtained from $G_Y(.|.)$ via expectation with respect to the condition. Moreover, we have the following rules for the calculation of PGFs:

$$G_{Y_1 + Y_2} = G_{Y_1} G_{Y_2} \text{ for independent random variables } Y_1, Y_2, \tag{2.4}$$

$$G_{kY}(z) = G_Y(z^k) \quad \text{for any natural number } k. \tag{2.5}$$

As examples we mention the PGF for a Bernoulli variable Y with probability p,

$$G_Y(z) = (1-p)z^0 + pz^1 = 1 + p(z-1) \tag{2.6}$$

and the one for a Poisson-distributed random variable Z with parameter μ,

$$G_Z(z) = \exp(\mu(z-1)). \tag{2.7}$$

The PGF cannot only be used to determine the distribution of a random variable, but also for calculating its variance. Note that

$$G_Y'(z) = \frac{1}{z}\mathbb{E}[Y z^Y], \qquad G_Y''(z) = \frac{1}{z^2}\mathbb{E}[(Y^2 - Y)z^Y],$$

in particular

$$G_Y'(1) = \mathbb{E}[Y], \qquad G_Y''(1) = \mathbb{E}[Y^2] - \mathbb{E}[Y].$$

Hence one has for the variance σ_Y^2 of the random variable Y

$$\sigma_Y^2 = \mathbb{E}[Y^2] - E[Y]^2 = G_Y''(1) + G_Y'(1) - G_Y'(1)^2.$$

If $G_Y(1) = \mathbb{E}[1] = 1$, then we can rewrite the equality using

$$G_Y' = G_Y(\ln G_Y)', \qquad G_Y'' = G_Y'(\ln G_Y)' + G_Y(\ln G_Y)''$$

as

$$\sigma_Y^2 = (\ln G_Y)''(1) + (\ln G_Y)'(1). \tag{2.8}$$

We denote the sets of natural, integer and real numbers by $\mathbb{N}$, $\mathbb{Z}$ and $\mathbb{R}$, respectively. A lower index $+$ denotes the restriction to non-negative numbers, an upper index the dimension of a corresponding vector space.

2.2 Input Data for the Model

We consider a loan portfolio with respect to K different obligors. We use a lower index A to indicate the dependence on one of the K obligors. We denote by $\tilde{p}_A$ the expected probability of default of obligor A. In general this quantity is the output of a rating process. Furthermore we denote by $\mathcal{E}_A$ the outstanding exposure of obligor A. This is assumed to be constant. In the case of a default, parts of $\mathcal{E}_A$ can be recovered and we only have to consider the potential loss $\tilde{\nu}_A$ for obligor A. Thus the expected loss for obligor A is

$$EL_A = \tilde{p}_A \tilde{\nu}_A.$$

It is one of the features of CreditRisk$^+$ to work with discretized losses. For this purpose one fixes a loss unit L_0 and chooses a positive integer ν_A as a rounded version of $\tilde{\nu}_A/L_0$. In order to compensate for the error due to the rounding process one adjusts the expected probability of default to keep the expected loss unchanged, i.e. instead of $\tilde{p}_A$ one considers

$$p_A = \frac{EL_A}{\nu_A L_0} = \frac{\tilde{\nu}_A}{\nu_A L_0}\tilde{p}_A.$$

Therefore we consider from now on the expected probability of defaults p_A and integer-valued losses ν_A.

As indicated by the notion, p_A describes an expectation rather than an actual value. The expectation usually represents a mean over time and economic situations in different countries or industries for obligors with similar risk profiles. In order to compensate for the influence of the expectation and to take into account the economy of a specific industry or country the obligor is active in, one can try to adjust the probability of default via positive random scaling factors to the respective current risk situation. The latter can be seen to be determined by certain risk drivers according to countries or industries, for example. This corresponds to the introduction of segments – sectors in the terminology of CreditRisk$^+$ – for the consideration of systematic default risk.

We consider N different and independent sectors $S_1, \ldots, S_N$ with corresponding positive random scalar factors S_i, $i = 1, \ldots, N$, which are also called risk factors. Each obligor A can be active in more than one sector, therefore we introduce factor loadings, also called factor weights, to fix shares w_{Ak} in the sectors. Such factor loadings have to satisfy

$$0 \leq w_{Ak} \leq 1, \qquad \sum_{k=1}^{N} w_{Ak} \leq 1.$$

Thus $\sum_{k=1}^{N} w_{Ak}$ specifies the systematic default risk share for obligor A, while $w_{A0} := 1 - \sum_{k=1}^{N} w_{Ak}$ represents the share of the obligor specific or idiosyncratic risk.

The risk factors S_i are usually assumed to be independent gamma-distributed random variables. (The cases of other distributions and dependent distributions are considered in Chapters 6 by Giese, 12 by Binnenhei and 13 by Reiß; correlated sectors are also a topic in Chapter 9 by Akkaya, Kurth and Wagner.) This assumption means for our probability space that the σ-algebra generated by the factors S_i, $i = 1, \ldots, N$ should form a sub-σ-algebra of $\mathcal{F}$, conditional on which we have to require the measurability of all appearing quantities. Conditional on $S = (S_0, S_1, \ldots, S_N)$ with $S_0 \equiv 1$ we now have the probability of default given by

$$p_A^S = p_A \left(1 + \sum_{k=1}^{N} w_{Ak}(S_k/\mathbb{E}[S_k] - 1) \right) = p_A \sum_{k=0}^{N} w_{Ak} S_k/\mathbb{E}[S_k] \qquad (2.9)$$

with

$$\mathbb{E}[p_A^S] = p_A, \qquad \mathrm{var}[p_A^S] = p_A^2 \sum_{k=1}^{N} \frac{w_{Ak}^2}{\mathbb{E}[S_k]^2} \mathrm{var}[S_k].$$

It is one of our main assumptions that, conditional on S, all obligors are independent. This feature is also known as the conditional independence framework and is common to other credit risk models. In Chapter 4 by Wieczerkowski it plays an important role in relating CreditRisk$^+$ and CreditMetrics, the other widely used credit risk model.

Let us also remark that the losses ν_A represent expected values and could be turned into random variables by positive random scalar factors. This feature is not contained in the orignial version of CreditRisk$^+$. Stochastic losses are considered in several chapters (cf. e.g. Chapters 3 by Tasche and 7 by Gordy), in particular an extension involving systematic severity scaling factors is presented in Chapter 9 by Akkaya, Kurth and Wagner.

2.3 Determining the Loss Distribution

The object of interest is the distribution of the loss in the portfolio. We denote this loss by the random variable X. Using the default indicator given by the indicator function,

$$\mathbf{1}_A := \begin{cases} 1 \text{ if } A \text{ defaults} \\ 0 \text{ otherwise} \end{cases}$$

we can write this variable as

$$X = \sum_A \mathbf{1}_A \nu_A.$$

Therefore the expected portfolio loss conditional on S can be presented as

$$EL^S = \mathbb{E}[X|S] = \sum_A p_A^S \nu_A = \sum_{k=0}^{N} S_k / \mathbb{E}[S_k] \sum_A w_{Ak} p_A \nu_A,$$

while

$$\mathrm{var}[X|S] = \sum_A \nu_A^2 \mathrm{var}[\mathbf{1}_A].$$

An efficient way to calculate the loss distribution for X is via PGFs, as X attains only values in the non-negative integers. (If one prefers to work with the true potential loss $\tilde{\nu}_A$ and dispenses with the use of PGFs, there is the option to use characteristic functions. This is shown in the Chapters 8 and 13 by Reiß.) For this purpose we denote the PGF $G_{\mathbf{1}_A}$ for the loss of obligor A by G_A. According to (2.6) we can determine G_A conditional on S using (2.3) and (2.5) by

$$G_A(z|S) = (1 - p_A^S) + p_A^S z^{\nu_A} = 1 + p_A^S(z^{\nu_A} - 1). \tag{2.10}$$

Since we assume the independence of obligors conditional on S we deduce via (2.4) that

$$G_X(z|S) = \prod_A G_A(z|S).$$

Knowing the probability density function f_S for the risk factors S_k we can determine the generating function for X as

$$G_X(z) = \int G_X(z|S = s) f_S(s) ds, \qquad (2.11)$$

where f_S can be written as a product of $f_{S_i}(s_i)$ for realizations s_i of S_i due to the independence of the risk factors.

For favourable choices of f_S and suitable approximations of G_A the generating function G_X can be obtained analytically in a closed form. Usual choices in CreditRisk$^+$ are gamma distributions for the risk factor S_k and a Poisson approximation for the default events decribed by the default indicators $\mathbf{1}_A$. This combination guarantees the advantageous representation of G_X. We will give the details in Section 2.5. Other approaches to determining the loss contribution rely on cumulant-generating functions (see Chapter 7 by Gordy), moment-generating functions (see Chapters 6 and 10 by Giese) and characteristic functions (see Chapters 8 and 13 by Reiß). In Chapters 9, 10 and 13 the assumption of the independence of sectors is dropped. This allows for a stronger correlation of obligors. Note that in the current set-up the correlation of any two obligors A, B is given by

$$\rho_{AB} = \frac{\mathbb{E}[\mathbf{1}_A \mathbf{1}_B] - \mathbb{E}[\mathbf{1}_A]\mathbb{E}[\mathbf{1}_B]}{\sqrt{\mathbb{E}[\mathbf{1}_A]\mathbb{E}[1 - \mathbf{1}_A]\mathbb{E}[\mathbf{1}_B]\mathbb{E}[1 - \mathbf{1}_B]}}.$$

As

$$\mathbb{E}[\mathbf{1}_A \mathbf{1}_B] = \mathbb{E}[\mathbb{E}[\mathbf{1}_A|S]\mathbb{E}[\mathbf{1}_B|S]] = \mathbb{E}[p_A^S p_B^S] = p_A p_B \sum_{k,l=0}^{N} w_{Ak} w_{Bl} \frac{\mathbb{E}[S_k S_l]}{\mathbb{E}[S_k]\mathbb{E}[S_l]}$$

we obtain, due to the independence of the risk factors,

$$\mathbb{E}[\mathbf{1}_A \mathbf{1}_B] = p_A p_B \left(\sum_{k \neq l}^{N} w_{Ak} w_{Bl} + \sum_{k=0}^{N} w_{Ak} w_{Bk} \frac{\mathbb{E}[S_k^2]}{\mathbb{E}[S_k]^2} \right)$$

$$= p_A p_B \left(1 - \sum_{k=0}^{N} w_{Ak} w_{Bk} + \sum_{k=0}^{N} w_{Ak} w_{Bk} \frac{\mathrm{var}[S_k] + \mathbb{E}[S_k]^2}{\mathbb{E}[S_k]^2} \right),$$

which finally yields

$$\rho_{AB} = \frac{\sqrt{p_A p_B}}{\sqrt{(1 - p_A)(1 - p_B)}} \sum_{k=1}^{N} w_{Ak} w_{Bk} \frac{\mathrm{var}[S_k]}{\mathbb{E}[S_k]^2}. \qquad (2.12)$$

2.4 Basic Output

The central output of credit risk models is the distribution of portfolio loss. CreditRisk$^+$ also offers the possibility to derive many of the characteristics of this distribution analytically in a rather efficient way. It is interesting to discuss which quantities should be used to describe the risk of a portfolio. While it is clear to record the expected loss by

$$EL = \mathbb{E}[EL^S] = \sum_A EL_A = \sum_A p_A \nu_A L_0 = \sum_A \tilde{p}_A \tilde{\nu}_A \qquad (2.13)$$

which is an additive quantity that can be trivially split with respect to sub-portfolios, in particular sectors, the notion of unexpected loss is not uniquely determined. It is sometimes used to mean the standard deviation of loss, and sometimes to mean the gap between the value-at-risk and the expected loss. We will use the notion in the latter sense.

In the following we are going to describe the quantities standard deviation and unexpected loss. We leave further discussion of suitable quantities to characterize risk to other contributors. In particular we refer to Chapter 3 by Tasche for a general description of important risk measures, especially expected shortfalls.

In addition to the expected loss the variance and the standard deviation are natural quantities to look at. We denote them by

$$\sigma_X^2 = \text{var}[X] \quad \text{and} \quad \text{SD}_X = \sigma_X.$$

The variance can trivially be split into

$$\text{var}[X] = \mathbb{E}[\text{var}[X|S]] + \text{var}[\mathbb{E}[X|S]]. \qquad (2.14)$$

Here $\text{var}[\mathbb{E}[X|S]]$ describes the systematic risk, which is caused by the risk drivers and hence cannot be reduced by diversification – in contrast to $\mathbb{E}[\text{var}[X|S]]$, which hence is also called diversifiable risk. Therefore the splitting in (2.14) is also presented as

$$\sigma_X^2 = \sigma_{\text{div}}^2 + \sigma_{\text{syst}}^2, \qquad \sigma_{\text{div}}^2 = \mathbb{E}[\text{var}[X|S]], \qquad \sigma_{\text{syst}}^2 = \text{var}[\mathbb{E}[X|S]].$$

Thus if we use the notation

$$\nu^{(k)} = \sum_A w_{Ak} p_A \nu_A \quad \text{for } k = 1, \ldots, N,$$

we have

$$\sigma_X^2 = \mathbb{E}\left[\sum_A \nu_A^2 \text{var}[\mathbf{1}_A|S]\right] + \text{var}\left[\sum_{k=1}^N \nu^{(k)} S_k / \mathbb{E}[S_k]\right]$$

$$= \sum_A \nu_A^2 \text{var}[\mathbf{1}_A] + \sum_{k=1}^N \left(\frac{\nu^{(k)}}{\mathbb{E}[S_k]}\right)^2 \text{var}[S_k], \qquad (2.15)$$

in particular

$$\sigma_{\text{div}}^2 = \sum_A \nu_A^2 \text{var}[\mathbf{1}_A], \qquad \sigma_{\text{syst}}^2 = \sum_{k=1}^N \left(\frac{\nu^{(k)}}{\mathbb{E}[S_k]}\right)^2 \text{var}[S_k].$$

In the next section we will further specify the variance using moments for $\mathbf{1}_A$ and S_k, which are obtained under assumptions of certain distributions for these random variables.

While the standard deviation is quite instrumental in characterizing distributions that are similar to the normal distribution, it plays a less helpful role for skew distributions, which are common for losses. Therefore the value-at-risk is usually a quantity that is preferred to the standard deviation for quantifying risk. For a confidence level, also known as solvency level or solvency probability, $1 - \epsilon \in (0,1)$, we denote the value-at-risk (VaR) for a random variable Y by

$$\mathrm{VaR}_\epsilon[Y] = \inf\{c \in \mathbb{R} : \mathbb{P}[Y \leq c] \geq \epsilon\}.$$

Since the loss distributions produced by CreditRisk$^+$ are discrete, one either has to use for the portfolio loss X the approximation

$$\mathrm{VaR}_\epsilon[X] \approx \bar{\rho}_\epsilon(X) = \inf\{n \in \mathbb{N} : \mathbb{P}[X \leq n] \geq \epsilon\}$$

or one uses a linear approximation for a noninteger fraction of the loss:

$$\mathrm{VaR}_\epsilon[X] \approx \rho_\epsilon(X) = \bar{\rho}_\epsilon(X) - 1 + \frac{\epsilon - \mathbb{P}[X \leq \bar{\rho}_\epsilon(X) - 1]}{\mathbb{P}[X = \bar{\rho}_\epsilon(X)]}.$$

In any case, having determined the value-at-risk, we calculate the unexpected loss (UL), which sometimes is also called credit-value-at-risk,

$$\mathrm{UL}_\epsilon[X] = \mathrm{VaR}_\epsilon[X] - \mathrm{EL}.$$

If the solvency level $1 - \epsilon$ is fixed, we might omit the index ϵ for the UL.

How to assign to each obligor a respective part of the risk (for example in form of the standard deviation or the credit-value-at-risk) is a problem in itself, to which Chapter 3 by Tasche is dedicated. The original suggestion for risk contributions considers the standard deviation σ_X and is given by

$$r_A(\sigma) = \nu_A \frac{\partial \sigma_X}{\partial \nu_A} = \frac{\nu_A}{2\sigma_X} \frac{\partial \sigma_X^2}{\partial \nu_A}. \tag{2.16}$$

According to [3] this contribution can be carried over to the contribution to the value-at-risk and the unexpected loss, though there is no theoretical basis for this apporach:

$$r_A(\mathrm{VaR}_\epsilon) = \mathrm{EL}_A + \frac{\mathrm{UL}_\epsilon}{\sigma_X} r_A(\sigma), \qquad r_A(\mathrm{UL}_\epsilon) = \frac{\mathrm{UL}_\epsilon}{\sigma_X} r_A(\sigma).$$

2.5 The Original Model

In the following we will present the basic version of CreditRisk$^+$ as developed by Wilde for CSFP [3], to which we refer in the following as standard CreditRisk$^+$. We adopt the notations of Gordy [4].

Let us return to the PGF (2.10) for the loss of obligor A and rewrite it as

$$G_A(z|S) = \exp \log(1 + p_A^S(z^{\nu_A} - 1)).$$

For small probability of defaults we can use the approximation

$$\log(1 + p_A^S(z^{\nu_A} - 1)) \approx p_A^S(z^{\nu_A} - 1)$$

which yields

$$G_A(z|S) \approx \exp(p_A^S(z^{\nu_A} - 1))$$

and hence the PGF for a random variable $\nu_A D_A$, where D_A is a random variable, which is Poisson-distributed with parameter p_A^S and replaces $\mathbf{1}_A$. If we assume that this Poisson approximation is valid – as we do from now on – and if we think of $\nu_A \mathbf{1}_A$ as a Poisson random variable $\nu_A D_A$, then we obtain for the probability-generating function of X under our standard assumption of independency of obligors conditional on S that

$$G_X(z|S) = \prod_A G_A(z|S) = \exp\left(\sum_A p_A^S(z^{\nu_A} - 1)\right).$$

Now we assume without loss of generality that the distributions of the random scaling factors S_i are such that $\mathbb{E}[S_i] = 1$ – a natural condition that ensures $\mathbb{E}[p_A^S] = p_A$. Hence we obtain for the PGF via (2.9)

$$G_X(z|S) = \exp\left(\sum_A \sum_{k=0}^N p_A w_{Ak} S_k (z^{\nu_A} - 1)\right).$$

Rearranging the terms and setting

$$\mu_k = \sum_A w_{Ak} p_A, \qquad \mathcal{P}_k(z) = \frac{1}{\mu_k} \sum_A w_{Ak} p_A z^{\nu_A}$$

yields

$$G_X(z|S) = \exp\left(\sum_{k=0}^N S_k \left(\sum_A p_A w_{Ak}(z^{\nu_A} - 1)\right)\right)$$

$$= \exp\left(\sum_{k=0}^N S_k \mu_k (\mathcal{P}_k(z) - 1)\right). \tag{2.17}$$

The $\mathcal{P}_k$ are also called sector polynomials.

A further advantage of the Poisson approximation is that it allows a closed analytical form for the PGF when a gamma distribution is assumed for the random scalar factors. Namely, we get in this case according to (2.11)

$$G_X(z) = \int G_X(z) = \int G_X(z|S = s) f_S^{(\alpha,\beta)}(s) ds \tag{2.18}$$

using multiple integration, $s_0 \equiv 1$ and parameter vectors $\alpha = (\alpha_1,\ldots,\alpha_N)$ and $\beta = (\beta_1,\ldots,\beta_N)$. Note that the conditions $\mathbb{E}[S_i] = 1$ require $\alpha_i\beta_i = 1$ for all $i = 1,\ldots,N$. Thus we get for (2.18) from (2.1) and (2.17)

$$
G_X(z) = \int \exp\left(\sum_{k=0}^{N} s_k\mu_k(\mathcal{P}_k(z)-1)\right) \prod_{\ell=1}^{N} \frac{s_\ell^{\alpha_\ell-1}\alpha_\ell^{\alpha_\ell}}{\Gamma(\alpha_\ell)} e^{-\alpha_\ell s_\ell} ds_\ell
$$

$$
= e^{\mu_0(\mathcal{P}_0(z)-1)} \prod_{k=1}^{N} \frac{\alpha_k^{\alpha_k}}{\Gamma(\alpha_k)} \int_0^\infty e^{s_k[\mu_k(\mathcal{P}_k(z)-1)-\alpha_k]} s_k^{\alpha_k-1} ds_k.
$$

As $\Gamma(y) = \int_0^\infty x^{y-1}e^{-x}dx$, we obtain

$$
G_X(z) = e^{\mu_0(\mathcal{P}_0(z)-1)} \prod_{k=1}^{N} \frac{\alpha_k^{\alpha_k}}{\Gamma(\alpha_k)} \left(\frac{-1}{\mu_k(\mathcal{P}_k(z)-1)-\alpha_k}\right)^{\alpha_k} \int_0^\infty e^{-s_k} s_k^{\alpha_k-1} ds_k
$$

$$
= e^{\mu_0(\mathcal{P}_0(z)-1)} \prod_{k=1}^{N} \left(\frac{\alpha_k}{\alpha_k - \mu_k(\mathcal{P}_k(z)-1)}\right)^{\alpha_k}
$$

which can also be presented as

$$
G_X(z) = e^{\mu_0(\mathcal{P}_0(z)-1)} \prod_{k=1}^{N} \left(\frac{1-\delta_k}{1-\delta_k\mathcal{P}_k(z)}\right)^{\alpha_k} \tag{2.19}
$$

with $\delta_k = \mu_k/(\mu_k + \alpha_k)$ or equivalently as

$$
G_X(z) = \exp\left(\sum_A w_{A0}p_A(z^{\nu_A}-1)\right)
$$

$$
\times \exp\left(-\sum_{k=1}^{N} \frac{1}{\sigma_k^2}\log[1-\sigma_k^2\sum_A w_{Ak}p_A(z^{\nu_A}-1)]\right). \tag{2.20}
$$

It remains to specify the α_k. Recall from (2.2) and $\alpha_k\beta_k = 1$ that $\alpha_k = 1/\mathrm{var}(S_k)$. In the optimal case there exists a way to estimate the variance of the random scalar factors S_k. Often this is not possible yet. In this situation one can pick up an idea from [3] and use the standard deviations τ_A for the probability of defaults p_A – if they are known. This could be done by choosing as the standard deviation σ_k for S_k, $k = 1,\ldots,N$

$$
\sigma_k = \sum_A w_{Ak}\tau_A \qquad \text{or} \qquad \sigma_k = \frac{1}{\mu_k}\sum_A w_{Ak}\tau_A \tag{2.21}
$$

and setting

$$
\alpha_k = \sigma_k^{-2}.
$$

It was suggested by Lehrbass et al. (cf. [5]; see also Chapter 18 by Kluge and Lehrbass, in particular (18.6)), to use instead of (2.21)

$$\sigma_k = \sum_A \sqrt{w_{Ak}} \tau_A.$$

In the following we will also use $\beta_k = 1/\alpha_k$ for $k = 1, \ldots, N$ and $\beta_0 = 0$. Other choices of parameters are discussed in Chapters 15 by Lesko, Schlottmann and Vorgrimler, and 14 by Boegelein, Hamerle, Knapp and Rösch. The current choices lead to the following pairwise correlation of obligors A, B according to (2.12):

$$\rho_{AB} = \frac{\sqrt{p_A p_B}}{\sqrt{(1 - p_A)(1 - p_B)}} \sum_{k=1}^{N} \frac{w_{Ak} w_{Bk}}{\alpha_k}, \qquad (2.22)$$

which can be approximated for small probabilities of default by

$$\rho_{AB} \approx \sqrt{p_A p_B} \sum_{k=1}^{N} \frac{w_{Ak} w_{Bk}}{\alpha_k}. \qquad (2.23)$$

In order to determine the distribution of losses in the portfolio from the PGF G_X in an efficient way, it was suggested in [3] to use a recurrence procedure known in the insurance literature (see also [6]) as Panjer recursion. This is based on the presentation of the log derivative of G_X as the quotient of two polynomials $A(z)$, $B(z)$ of degree d_A and d_B, respectively, i.e.

$$(\ln G_X)'(z) = \frac{A(z)}{B(z)} = \frac{\sum_{j=0}^{d_A} a_j z^j}{\sum_{j=0}^{d_B} b_j z^j}.$$

In an appendix we present a derivation of formulas for the coefficients a_j, b_j as well as for the loss probabilities

$$g_n = \mathbb{P}[X = n].$$

There it is also shown that the polynomial degrees are given by

$$d_B = \deg B = \sum_{j=1}^{N} \deg \mathcal{P}_j, \quad d_A = \deg A = \sum_{j=0}^{N} \deg \mathcal{P}_j - 1. \qquad (2.24)$$

Note that the degree of $\mathcal{P}_j$ corresponds to the largest loss that can occur for any obligor from sector j.

Knowing the coefficients a_j, b_j the following recursion is valid for $n \geq 0$:

$$g_0 = e^{-\mu_0} \prod_{k=1}^{N} (1 - \delta_k)^{\alpha_k},$$

$$g_{n+1} = \frac{1}{(n+1)b_0} \left(\sum_{j=0}^{\min(d_A, n)} a_j g_{n-j} - \sum_{j=0}^{\min(d_B, n)-1} (n-j) b_{j+1} g_{n-j} \right). \qquad (2.25)$$

On the basis of (2.8) we are now going to calculate the standard deviation for the portfolio loss. For this purpose we deduce from (2.29) in the appendix

$$(\ln G_X)'(1) = \sum_{k=0}^{N} \eta_k P_k'(1) \frac{1}{1 - \delta_k P_k(1)}$$

$$(\ln G_X)''(1) = \sum_{k=0}^{N} \frac{\eta_k}{1 - \delta_k P_k(1)} \left(P_k''(1) + \frac{\delta_k P_k'(1)^2}{1 - \delta_k P_k(1)} \right).$$

Using (2.30) from the appendix we thus obtain for (2.8)

$$\sigma_X^2 = \sum_{k=0}^{N} \frac{\eta_k}{1 - \delta_k} \left(P_k'(1) + P_k''(1) + \frac{\delta_k P_k'(1)^2}{1 - \delta_k} \right)$$

$$= \sum_{k=0}^{N} \frac{\eta_k}{(1 - \delta_k)\mu_k} \left(\sum_{A:\nu_A \geq 1} w_{Ak} p_A \nu_A + \sum_{A:\nu_A \geq 2} w_{Ak} p_A \nu_A (\nu_A - 1) \right. $$

$$\left. + \frac{\delta_k}{\mu_k(1 - \delta_k)} \left(\sum_{A:\nu_A \geq 1} w_{Ak} p_A \nu_A \right)^2 \right)$$

$$= \sum_{k=0}^{N} \sum_{A:\nu_A \geq 1} w_{Ak} p_A \nu_A^2 + \sum_{k=0}^{N} \beta_k \left(\sum_{A:\nu_A \geq 1} w_{Ak} p_A \nu_A \right)^2$$

$$= \sum_{A:\nu_A \geq 1} p_A \nu_A^2 + \sum_{k=1}^{N} \frac{1}{\alpha_k} \nu^{(k)^2}. \tag{2.26}$$

Note that this result is in accordance with (2.15) due to the assumption of Poisson and gamma distributions. In particular we have

$$\sigma_{\text{div}}^2 = \sum_{A} p_A \nu_A^2, \qquad \sigma_{\text{syst}}^2 = \sum_{k=1}^{N} \frac{1}{\alpha_k} \nu^{(k)^2}.$$

We finally obtain for the contributions to the standard deviations according to (2.16)

$$r_A(\sigma) = \frac{\nu_A}{2\sigma_X} \left(p_a 2\nu_A + \sum_{k=1}^{N} \frac{1}{\alpha_k} 2\nu^{(k)} w_{Ak} p_A \right)$$

$$= \frac{\text{EL}_A}{\sigma_X} \left(\nu_A + \sum_{k=1}^{N} \frac{1}{\alpha_k} \nu^{(k)} w_{Ak} \right). \tag{2.27}$$

2.6 Verification of the Output

A model like CreditRisk$^+$ does not only require a validation but also a verification of its results. While the first topic is mainly a problem of calibration and input, which is discussed in Chapter 14 by Boegelein, Hamerle, Knapp and Rösch, the second is connected with the implementation of the model, in particular with numerical questions. Some of these problems can be detected by a change of parameters. The impact of the choice of the loss unit L_0 on the accurancy of the outcome (as well as on the speed of calculation), for example, is evident. Other problems can only be detected using other implementations, in particular different (numerical) algorithms.

In the Chapters 6 by Giese, 7 by Gordy and 8 by Reiß, algorithms using nested evaluation, saddlepoint approximation or Fourier inversion, respectively, instead of the Panjer recursion, are presented. The use of such methods is quite recommendable, as it has turned out that the Panjer recursion is numerically fragile. This method stimulates the accumulation of roundoff errors that can cause drastic effects. This can be seen from (2.25) where $\min(d_A, n) + \min(d_B, n)$ coefficients for the polynomials A and B are required. While n might have to be chosen quite large in order to reach the value-at-risk one is interested in, the polynomial degrees d_A and d_B depend – according to (2.24) – on the number N of sectors as well as on the largest losses that can occur in the sectors, hence also on the size of the loss unit. Thus in applications several thousand coefficients a_j, b_j are quite common. These coefficients can obtain positive and negative values, become rather small and hence are a source for roundoff errors that accumulate in (2.25). This can become particulary dramatic, if summands occur in (2.25) that are of comparable size and opposite sign. As a result of such a possible accumulation of roundoff errors in (2.25), negative "probabilities" g_n have been observed by users, for example.

The roundoff effects need not be as drastic and obvious as in this example. Nevertheless it is of importance to identify the quality of the results obtained by the Panjer recursion. A diagnostic for this purpose was suggested in Gordy [4]. It is based on the fact from extreme value theory that the tail of the CreditRisk$^+$-loss distribution, i.e. $G_X(n)$ for large n, can be well approximated by the tail of an exponential distribution with parameter κ, which can be determined as follows: if a quantile q^* with a corresponding loss n^* is chosen so large that $q^* = G_X(n^*) \approx 1 - \exp(-n^*/\kappa)$, then one puts $\kappa = -n^*/\log(1 - q^*)$. Now the tail obtained by the Panjer recursion can be replaced by the tail of the exponential distribution such that expectations for the random variables X^ℓ can be calculated as

$$\mathbb{E}[X^\ell] = \mathbb{E}[X^\ell | X \leq n^*]\mathbb{P}[X \leq n^*] + \mathbb{E}[X^\ell | X > n^*]\mathbb{P}[X > n^*]$$

$$\approx \sum_{n \leq n^*} g_n n^\ell + (1 - q^*) \sum_{j=0}^{\ell} \binom{\ell}{j} \frac{j!}{(-\log(1 - q^*))^j}.$$

In particular, moments of the loss distribution can be obtained this way. They can be used to check the quality of the distribution of the Panjer recursion by comparing it with respective quantities obtained by other methods like the one based on cumulants and presented in Gordy [4].

Appendix: The Panjer Recursion

In the following we provide details for the Panjer recursion that was introduced in Section 2.5 in order to determine the loss distribution in an efficient way. For this purpose we use a representation suggested by Binnenhei [1]. We recall from (2.19) that

$$\ln G_X(z) = \mu_0(\mathcal{P}_0(z) - 1) + \sum_{k=1}^{N} \alpha_k \ln \frac{1 - \delta_k}{1 - \delta_k \mathcal{P}_k(z)}$$

and hence

$$(\ln G_X)'(z) = \mu_0 \mathcal{P}_0'(z) + \sum_{k=1}^{N} \alpha_k \frac{1 - \delta_k \mathcal{P}_k(z)}{1 - \delta_k} \frac{1 - \delta_k}{(1 - \delta_k \mathcal{P}_k(z))^2} \delta_k \mathcal{P}_k'(z)$$

$$= \mu_0 \mathcal{P}_0'(z) + \sum_{k=1}^{N} \alpha_k \delta_k \mathcal{P}_k'(z) \frac{1}{1 - \delta_k \mathcal{P}_k(z)}$$

$$= \eta_0 \frac{\mathcal{P}'(z)}{1 - \delta_0 \mathcal{P}_0(z)} + \sum_{k=1}^{N} \eta_k \frac{\mathcal{P}_k'(z)}{1 - \delta_k \mathcal{P}_k(z)} \qquad (2.28)$$

with

$$\delta_0 = 0, \qquad \eta_k = \frac{\mu_k}{1 + \mu_k \beta_k},$$

i.e. $\eta_0 = \mu_0$ and $\eta_k = \alpha_k \delta_k$ for $k = 1, \ldots, N$. The fundamental idea of the Panjer recursion is to rewrite (2.28) as

$$(\ln G_X)'(z) = \frac{\sum_{k=0}^{N} \eta_k \mathcal{P}_k'(z) \prod_{j=0, j \neq k}^{N} (1 - \delta_j \mathcal{P}_j(z))}{\prod_{\ell=0}^{N} (1 - \delta_\ell \mathcal{P}_\ell(z))} = \frac{A(z)}{B(z)} \qquad (2.29)$$

with polynomials $A(z) = \sum_{j=0}^{d_A} a_j z^j$ and $B(z) = \sum_{j=0}^{d_B} b_j z^j$, the degrees of which are determined by (2.24). One can find the coefficients a_n and b_n of the polynomials A and B respectively by differentiation

$$a_n = \frac{1}{n!} A^{(n)}(0), \qquad b_n = \frac{1}{n!} B^{(n)}(0)$$

with the help of Leibniz's rule

$$\frac{1}{n!} \left(\prod_{k=0}^{N} h_k \right)^{(n)} = \sum_{n_1 + \ldots + n_N = n} \prod_{k=0}^{N} \frac{h_k^{(n_k)}}{n_k!}$$

for any differentiable functions h_k, and using

$$\mathcal{P}_k(0) = 0,$$

$$\mathcal{P}_k^{(r)}(z) = \frac{1}{\mu_k} \sum_{A:\nu_A \geq r} w_{Ak} p_A \nu_A (\nu_A - 1) \dots (\nu_A - r + 1) z^{\nu_A - r}, \qquad (2.30)$$

$$\mathcal{P}_k^{(r)}(0) = \frac{1}{\mu_k} \sum_{A:\nu_A = r} w_{Ak} p_A r!.$$

With the notation

$$c_{r,k} = \sum_{A:\nu_A = r} w_{Ak} p_A$$

we thus obtain for the coefficients b_n

$$b_n = \frac{1}{n!} \left(\prod_{k=1}^{N} (1 - \delta_k \mathcal{P}_k) \right)^{(n)} (0) = \sum_{n_1 + \dots + n_N = n} \prod_{k=1}^{N} \left(\delta_{0n_k} - \chi(n_k) \delta_k \frac{\mathcal{P}_k^{(n_k)}}{n_k!} \right) (0)$$

$$= \sum_{n_1 + \dots + n_N = n} \prod_{k=1}^{N} \left(\delta_{0n_k} - \chi(n_k) \frac{\delta_k c_{n_k,k}}{\mu_k} \right)$$

where δ_{ij} denotes the Kronecker delta and χ is the characteristic function for the set of losses in the portfolio, i.e.

$$\delta_{ij} = \begin{cases} 1 \text{ if } i = j \\ 0 \text{ otherwise} \end{cases} \qquad \chi(n) = \begin{cases} 1 \text{ if } n = \nu_A \text{ for some } A \\ 0 \text{ otherwise} \end{cases}.$$

Analogously we get the following formula for the coefficients a_n:

$$a_n = \sum_{k=0}^{N} \sum_{n_0 + \dots + n_N = n} \frac{\eta_k c_{n_k + 1, k}}{\mu_k} (n_k + 1) \chi(n_k + 1) \prod_{j=0, j \neq k}^{N} \left(\delta_{0n_j} - \chi(n_j) \frac{\delta_j c_{n_j, j}}{\mu_j} \right).$$

Now we can finally calculate the distribution using

$$G_X'(z) = G_X(z)(\ln G_X)'(z) = G_X(z) \frac{A(z)}{B(z)},$$

$$G_X''(z) = \frac{1}{B(z)} [A'(z) G_X(z) + A(z) G_X'(z)] - \frac{B'(z)}{B(z)} G_X'(z)$$

$$= \frac{1}{B(z)} [G_X(z) A'(t) + G_X'(z) \{A(z) - B'(z)\}],$$

$$G_X^{(n)}(z) = \frac{1}{B(z)} \left[G_X(z) A^{(n-1)}(z) + \sum_{j=1}^{n-1} G_X^{(j)} \left\{ \binom{n-1}{j} A^{(n-1-j)}(z) \right. \right.$$

$$\left. \left. - \binom{n-1}{n-j} B^{(n-j)}(z) \right\} \right].$$

This formula can be checked via induction. Hence we get for $g_n = \mathbb{P}[X = n]$

$$g_0 = G_X(0) = e^{\ln G_X(0)} = \exp\left(\mu_0(\mathcal{P}_0(0) - 1) + \sum_{k=1}^{N} \alpha_k \ln \frac{1 - \delta_k}{1 - \delta_k \mathcal{P}_k(0)}\right)$$

$$= e^{-\mu_0} \prod_{k=1}^{N} (1 - \delta_k)^{\alpha_k},$$

$$g_1 = G'_X(0) = g_0 \frac{a_0}{b_0}.$$

For $n \geq 1$ we obtain

$$g_n = \frac{1}{n!} G_X^{(n)}(0)$$

$$= \frac{1}{n! \, b_0} \left\{ g_0(n-1)! a_{n-1} + \sum_{j=1}^{n-1} j! \, g_j \left[\binom{n-1}{j} (n-1-j)! \, a_{n-1-j} \right. \right.$$

$$\left. \left. - \binom{n-1}{n-j} (n-j)! \, b_{n-j} \right] \right\}$$

$$= \frac{1}{n! \, b_0} \left\{ g_0(n-1)! a_{n-1} + \sum_{j=\max(1, n-d_A)}^{n-1} g_j \frac{(n-1)! j! (n-1-j)!}{j! (n-1-j)!} a_{n-1-j} \right.$$

$$\left. - \sum_{j=\max(1, n-1-d_B)}^{n-1} g_j \frac{j! (n-1)! (n-j)!}{(n-j)! (n-1-n+j)!} b_{n-j} \right\}$$

$$= \frac{1}{n! \, b_0} \left\{ g_0(n-1)! a_{n-1} + \sum_{j=\max(1, n-d_A)}^{n-1} g_j (n-1)! \, a_{n-1-j} \right.$$

$$\left. - \sum_{j=\max(1, n-1-d_B)}^{n-1} g_j j (n-1)! \, b_{n-j} \right\}$$

$$= \frac{1}{n b_0} \left(g_0 a_{n-1} + \sum_{k=1}^{\min(d_A, n-1)} g_{n-k} a_{k-1} - \sum_{k=1}^{\min(d_B, n-1)} (n-k) g_{n-k} b_k \right)$$

which can be rewritten as

$$g_{n+1} = \frac{1}{(n+1) b_0} \left(\sum_{j=0}^{\min(d_A, n)} a_j g_{n-j} - \sum_{j=0}^{\min(d_B, n)-1} (n-j) b_{j+1} g_{n-j} \right),$$

the formula presented in Section 2.5.

References

1. C. Binnenhei. Anmerkungen zur Methodik von CreditRisk$^+$. Unpublished notes. Stuttgart, 2000.
2. P. Bürgisser, A. Kurth, and A. Wagner. Incorporating severity variations into credit risk. *Journal of Risk*, 3(4):5–31, 2001.
3. Credit Suisse Financial Products. CreditRisk$^+$: A credit risk management framework. London, 1997. Available at `http://www.csfb.com/creditrisk`.
4. M. B. Gordy. Saddlepoint approximation of CreditRisk$^+$. *Journal of Banking & Finance*, 26:1335–1353, 2002.
5. F. Lehrbass, I. Boland, and R. Thierbach. Versicherungsmathematische Risikomessung für ein Kreditportfolio. *Blätter der Deutschen Gesellschaft für Versicherungsmathematik*, XXV(2):285–308, 2001.
6. H. H. Panjer and G. E. Willmot. Insurance Risk Models. *Society of Actuaries*, Schaumberg, IL, 1992.

3

Capital Allocation with CreditRisk$^+$

Dirk Tasche*

Summary. Capital allocation for credit portfolios has two meanings. First, at portfolio level it means to determine capital as a buffer against an unexpected negative cash-flow resulting from credit losses. In this case, the allocation method can be specified by means of a risk measure. Its result is called economic capital of the portfolio. Second, at subportfolio or transaction level, capital allocation means breaking down the economic capital of the portfolio into its sub-units. The resulting capital assignments are called risk contributions. We discuss several current concepts for economic capital and risk contributions in a general setting. Then we derive formulas and algorithms for these concepts in the special case of the CreditRisk$^+$ methodology with individual independent potential exposure distributions.

Introduction

For the purpose of credit portfolio management, it is crucial to not only know the loss distribution but also to be able to identify the main sources of risk in the portfolio. In this context, risk is understood as economic capital, i.e. as a capital reserve held by the credit institution in order to guarantee its own solvency in case of high losses. Hence, the portfolio manager's first task is to determine the economic capital of the portfolio. She does so by applying a so-called risk measure.

The main sources of risk are those subportfolios or transactions that contribute most to the economic capital of the portfolio. Therefore, the portfolio manager's second task is to break down the economic capital to subportfolios or even to single transactions. The result of this process are the risk contributions of the subportfolios and transactions respectively.

Note that the risk contributions described so far are determined in a top-down approach, since the point of departure is the economic capital at portfolio level and the risk contributions are derived from the economic capital.

---

* The views expressed in this chapter do not necessarily reflect those of the Deutsche Bundesbank or of its staff.

This is in contrast to the method that is applied in order to compute the regulatory capital of the portfolio. For this purpose, at first capital charges of the transactions are determined. The capital charges are then added up to obtain the regulatory capital. This bottom-up approach may in general lead to counter-intuitive results when it is applied to portfolio management.[1] For this reason, in the following our interest will be focused on economic capital.

In the original version of CreditRisk$^+$, two ways to calculate the economic capital are provided – as the standard deviation of the loss distribution times some constant multiplier or as the value-at-risk (VaR) of the loss distribution. However, from the economic viewpoint both standard deviation and VaR show serious deficiencies when they are used as risk measures. Below, we will briefly discuss this issue and then argue that there is an alternative risk measure that should be preferred to standard deviation and VaR – namely the expected shortfall (ES).

As seen in Chapter 2, the original CreditRisk$^+$ also suggests a method to calculate risk contributions to standard deviation and VaR. Later in this chapter, we will show that this original approach with regard to VaR is not compatible with economic intuition as far as optimization of the portfolio return on risk-adjusted capital (RORAC) is concerned. Instead, we will propose an alternative that is both economically well-founded and easy to calculate in the original CreditRisk$^+$ framework.

The remainder of this chapter is organized as follows. In Section 3.1 we present the coherence axioms introduced by Artzner et al. in [3] in order to give an economic foundation to the notion of risk measure. We argue that ES is the cheapest coherent risk measure in a certain sense. The question of how to break down economic capital as specified by a risk measure to transaction level is tackled in Section 3.2. It turns out that the Euler allocation joins economic intuition to conceptual simplicity. In order to apply to CreditRisk$^+$ the concepts developed in the preceding two sections, in Section 3.3 we briefly review the methodology with an emphasis on its probabilistic aspects. In particular, we allow for a simple concept of stochastic exposures that can be introduced without any extra effort. With these preparations, we can present the CreditRisk$^+$-formulas for the contributions to VaR and ES in Section 3.4. These formulas are of particular interest because they can be evaluated within the original CreditRisk$^+$ framework.

3.1 Allocation at Portfolio Level

We consider the general problem of finding intuitive and operational methods for determining the economic capital of a portfolio. The assets in the portfolio need not necessarily be loans; they could also be stocks or even derivative

---

[1] Nevertheless, there are a few situations such that regulatory and economic capital of the portfolio coincide. See [7] for a detailed discussion of this topic.

instruments. The viewpoint that we adopt here is that of a banking supervisor or of a creditor who has funded the portfolio and wants the portfolio managers to deposit some equity capital as collateral.

Assumption I. *The portfolio is composed of assets $1, \ldots, n$. The i-th asset has weight u_i. Its expected return is r_i (interpreted as excess return over the risk-free rate). The economic capital of the portfolio is allocated according to a risk measure ρ, i.e. $\rho(u) = \rho(u_1, \ldots, u_n)$ specifies the economic capital required by the portfolio with asset weights $u = (u_1, \ldots, u_n)$.*

In Assumption I "weight" need not necessarily be understood as a percentage. In the context of this chapter "weight" equally means "number of shares" or "amount of invested money".

Choosing risk measures that guarantee some fixed probability of solvency appears to be the current standard in the financial industry. Of course, in this case, one has to rely on a stochastic portfolio model.

Assumption II. *The return of the i-th asset is stochastic and is denoted by the random variable R_i. We assume $\mathbb{E}[R_i] = r_i$. The total portfolio return is given by $R(u) = \sum_{i=1}^{n} u_i R_i$.*

Under Assumption II, the risk measure ρ can be described in terms of the joint distribution of the return vector $(R_1, \ldots, R_n)$. The standard deviation and VaR are the two most popular risk measures of this kind. We briefly introduce both of them in a form adapted to the purpose of ensuring some probability of solvency.

Definition 1 (Standard deviation as risk measure). *Fix a solvency probability $\theta \in (0,1)$ (close to 1). Let Assumption II be satisfied. Define the risk measure ρ_{Std} by*

$$
\begin{aligned}
\rho_{\text{Std}}(u) &= c_\theta \sqrt{\operatorname{var}(R(u))} - \mathbb{E}[R(u)] \\
&= c_\theta \sqrt{\operatorname{var}(R(u))} - \sum_{i=1}^{n} u_i r_i,
\end{aligned}
\tag{3.1a}
$$

where var *as usual denotes the variance. The constant $c_\theta > 0$ has to be chosen in such a way that*

$$
\mathbb{P}[R(u) + \rho_{\text{Std}}(u) \geq 0] \geq \theta
\tag{3.1b}
$$

is granted.

Note that the constant c_θ in Definition 1 is not uniquely determined by (3.1b). However, from an economic point of view it is clear that the smallest of all the numbers c_θ that satisfy (3.1b) should be chosen. For instance, c_θ can be determined by assuming that $(R_1, \ldots, R_n)$ follows some known distribution or by applying Chebyshev's inequality. In particular, the assumption that the return vector $(R_1, \ldots, R_n)$ is jointly normally distributed implies that c_θ is the θ-quantile of the standard normal distribution.

The application of Chebyshev's inequality is based on the observation that (3.1b) is equivalent to

$$\mathbb{P}\left[R(u) - \mathbb{E}[R(u)] < -c_\theta \sqrt{\operatorname{var}(R(u))}\right] \leq 1 - \theta. \tag{3.2}$$

By Chebyshev's inequality, the left-hand side of (3.2) is not greater than $1/(1 + c_\theta^2)$. Hence, a distribution-free c_θ that satisfies (3.2) can be found by solving the equation $1/(1 + c^2) = 1 - \theta$ for c.

However, the approach to solvency-related risk measures as in Definition 1 is not really satisfactory. If c_θ is determined according to some distributional assumption, the method may turn out to be unsafe in the sense of yielding too small solvency probabilities whenever the assumption is violated. In contrast, if c_θ is calculated using Chebyshev's inequality the resulting economic capital in general will be deterringly high. In order to avoid this dilemma, in the late 1980s the risk measure value-at-risk was introduced. For any real random variable Y define its θ-quantile $q_\theta(Y)$ by

$$q_\theta(Y) = \min\{y \in \mathbb{R} : \mathbb{P}[Y \leq y] \geq \theta\}. \tag{3.3}$$

Definition 2 (Value-at-risk, VaR). *Keep a solvency probability $\theta \in (0,1)$ (close to 1) fixed. Let Assumption II be satisfied. Define the risk measure VaR$_\theta$ by*

$$\operatorname{VaR}_\theta(u) = q_\theta(-R(u)). \tag{3.4}$$

From (3.3) follows that VaR satisfies (3.1b) with ρ_{Std} replaced by VaR and hence ensures the solvency probability θ. Unfortunately, even if VaR has found widespread acceptance as a risk measure, VaR also has deficiencies that make it appear questionable. In particular, VaR neglects risks with high impact but low frequency. This is a consequence of the fact that by its very definition VaR$_\theta$ does not take into account high losses that occur with probability less than $1 - \theta$.

Another problem with VaR is its (in general) missing sub-additivity, i.e.

$$\operatorname{VaR}_\theta(u) + \operatorname{VaR}_\theta(v) < \operatorname{VaR}_\theta(u + v)$$

may happen.[2] Thus, a merger of two portfolios may create a need for economic capital that is not covered by the sum of the economic capital values of the merged portfolios. These problems led Artzner et al. in [3] to define their famous coherence axioms that are assumed to describe an economically sound framework for risk measures.

Definition 3 (Coherence). *Let Assumptions I and II be satisfied. Then a risk measure ρ is called coherent if it satisfies the following conditions:*

---

[2] Equivalently, the risk contribution of a subportfolio may be larger than its stand-alone economic capital (cf. [10, 18]). This phenomenon may crop up when credit portfolio risk is measured with VaR.

1. *monotonicity: u, v such that $R(v) \leq R(u) \Rightarrow \rho(v) \geq \rho(u)$,*
2. *sub-additivity: $\rho(u + v) \leq \rho(u) + \rho(v)$,*
3. *positive homogeneity: $h > 0 \Rightarrow \rho(h\,u) = h\,\rho(u)$,*
4. *translation invariance: if the i-th asset is risk free (i.e. with constant return) then*

$$\rho(u_1, \ldots, u_i + a, \ldots, u_n) = \rho(u_1, \ldots, u_i, \ldots, u_n) - a.$$

See [3] for motivations of the axioms appearing in Definition 3. As practitioners tend to regard economic capital as a potential loss, the translation invariance axiom may seem a bit strange at first glance because it indicates that coherent risk measures can assume negative values. But this is a consequence of the fact that the authors of [3] considered risk measures as indicators of the additional amount of capital that has to be deposited in order to make acceptable the portfolio to the supervisors. From this perspective, negative economic capital indicates that capital can be released.

Both the standard deviation as well as VaR are not coherent risk measures. The standard deviation is not monotonous, whereas VaR – as noticed above – in general is not sub-additive. Therefore, it might be interesting to look for the smallest coherent risk measure to dominate VaR_θ. This risk measure would guarantee the solvency probability θ and would be the cheapest in the sense that it would cause less cost of capital than any other coherent risk measure that dominates VaR_θ. Unfortunately, no such smallest coherent risk measure exists (see [5, Theorem 14]). Nevertheless, if we require the dominating coherent risk measure to belong to the class of risk measures ρ that depend on their arguments only through the distribution, i.e.

$$\mathbb{P}[R(u) \leq t] = \mathbb{P}[R(v) \leq t], \ t \in \mathbb{R} \Rightarrow \rho(u) = \rho(v), \tag{3.5}$$

then existence of the dominating coherent risk measure will be granted. Indeed, it turns out that it is provided by the following example.

Definition 4 (Expected shortfall, ES). *Fix a solvency probability $\theta \in (0, 1)$. Let Assumption II be satisfied. Define the risk measure ES_θ by*

$$\text{ES}_\theta(u) = (1 - \theta)^{-1} \int_\theta^1 \text{VaR}_t(u)\, dt. \tag{3.6}$$

In the literature, ES is also called conditional value-at-risk (CVaR). For any fixed solvency probability $\theta \in (0, 1)$ expected shortfall is a coherent risk measure (cf. [2, 5]). Since VaR_t is non-decreasing in t, from (3.6) it is clear that ES_θ dominates VaR_θ. The following theorem clarifies in which sense ES_θ can be understood as the *smallest* coherent risk measure to dominate VaR_θ.

Theorem 1 ([5], Theorem 13). *Fix a solvency probability $\theta \in (0, 1)$. Let Assumption II be satisfied. If ρ is any coherent risk measure such that $\rho(u) \geq \text{VaR}_\theta(u)$ for all u and (3.5) is satisfied, then $\rho(u) \geq \text{ES}_\theta(u)$ for all u follows.*

Property (3.5) can be interpreted as being estimable from sample data. Besides (3.6) there are a lot of other equivalent definitions of ES. In particular, if the distribution of $R(u)$ is continuous, i.e. $\mathbb{P}[R(u) = t] = 0$ for all t, then one obtains

$$\mathrm{ES}_\theta(u) = -\mathbb{E}\big[R(u) \mid -R(u) \geq q_\theta(-R(u))\big]. \tag{3.7}$$

Hence, ES_θ can be considered the conditional expectation of the loss given that it exceeds VaR_θ. In general, ES_θ can only be represented as a weighted sum of the conditional expectation in (3.7) and VaR_θ. See [2, 15] for this and other properties of ES. In most cases of practical importance, the difference of the values calculated according to (3.6) and (3.7) respectively will be negligible. In this chapter, therefore, we will make use of (3.7) as a definition of ES.

As ES_θ dominates VaR_θ, a financial institution that replaces risk measurement with VaR_θ by risk measurement with ES_θ will observe a growth of economic capital if it keeps the solvency probability θ unchanged. If the level of economic capital is to be kept constant, the solvency probability τ underlying ES must be lower than the solvency probability θ for VaR. Under the restriction of keeping the economic capital constant, a very rough rule of thumb for the relationship between τ and θ can be derived from (3.7). This rule is based on the assumption that the conditional median of the loss $-R(u)$ given the event $\{-R(u) \geq q_\tau(-R(u))\}$ is a reasonable approximation of the conditional expectation $\mathbb{E}\big[-R(u) \mid -R(u) \geq q_\tau(-R(u))\big]$. But that conditional median just coincides with the $\frac{\tau+1}{2}$-quantile of $-R(u)$, i.e. with $\mathrm{VaR}_{(\tau+1)/2}$. Hence

$$\tau = 2\theta - 1 \tag{3.8a}$$

is a reasonable guess for the solvency probability τ such that

$$\mathrm{ES}_\tau(u) \approx \mathrm{VaR}_\theta(u). \tag{3.8b}$$

Observe that the choice of τ according to (3.8a) is conservative because in case of a heavy-tailed loss distribution the conditional expectation will exceed the conditional median.

The notion of ES can be generalized to the class of spectral risk measures which allows us to model stronger risk aversion than that expressed by ES (see [1, 18]).

3.2 Allocation at Transaction Level

In this section, we assume that some methodology – a risk measure ρ – for allocating economic capital at portfolio level has been fixed. Thus, the setting of this section is given by Assumption I. How can we meaningfully define the risk contribution $\rho_i(u)$ of the i-th asset to the economic capital $\rho(u)$?

As the goal of our allocation exercise is the ability to identify the important sources of risk, we need criteria in order to fix in which sense an asset can be particularly risky. Such criteria were suggested in [6, 10, 17]. Denault [6] proposed a game-theoretic approach to the problem. Kalkbrener [10] put emphasis on the issue of diversification. The results of both [6, 10] apply only to sub-additive risk measures, but in this case they coincide essentially with the suggestions by Tasche [17]. Tasche's approach is based on two simple economic requirements – RORAC-compatibility and full allocation.

Definition 5. *Given Assumption I and assuming that a family $\rho_1,\ldots,\rho_n$ of risk contributions corresponding to the assets $1,\ldots,n$ has been specified,[3]*

1. the portfolio RORAC (return on risk-adjusted capital) is defined by

$$\text{RORAC}(u) = \frac{\sum_i r_i u_i}{\rho(u)},\tag{3.9a}$$

2. the RORAC of the i-th asset is defined by

$$\text{RORAC}_i(u) = \frac{r_i u_i}{\rho_i(u)},\tag{3.9b}$$

3. the family $\rho_1,\ldots,\rho_n$ of risk contributions is called RORAC-compatible if for each u and each i there is a number $\epsilon > 0$ such that

$$\text{RORAC}_i(u) > \text{RORAC}(u)$$

implies

$$\text{RORAC}(u_1,\ldots,u_i+h,\ldots,u_n) > \text{RORAC}(u)$$

for all $0 < h < \epsilon$,
4. the family $\rho_1,\ldots,\rho_n$ of risk contributions is called fully allocating if

$$\sum_{i=1}^n \rho_i(u) = \rho(u).\tag{3.9c}$$

A result from [17, Theorem 4.4] shows that RORAC-compatibility in connection with differentiability suffices to determine uniquely a family of risk contributions.

Theorem 2. *Let Assumption I be satisfied. If the risk measure $\rho = \rho(u)$ is partially differentiable with respect to the components u_i of the weight vector $u = (u_1,\ldots,u_n)$ of the assets, then there is only one family $\rho_1,\ldots,\rho_n$ of risk contributions that is RORAC-compatible for all return vectors $(r_1,\ldots,r_n)$. This family is given by*

$$\rho_i(u) = u_i \frac{\partial\rho}{\partial u_i}(u), \quad i = 1,\ldots,n.\tag{3.10}$$

---

[3] Note that in Definition 5 we implicitly assume that the economic capital and the risk contributions are positive. However, this assumption is not essential for the results in this section. The definition of RORAC-compatibility can be adapted to the case of negative risk contributions, but in this case it is intuitive only at second glance.

Euler's theorem states that a partially differentiable function $\rho = \rho(u)$ is positively homogeneous in the sense of Definition 3 if and only if it satisfies the equation

$$\rho(u) = \sum_{i=1}^{n} u_i \frac{\partial \rho}{\partial u_i}(u). \qquad (3.11)$$

Hence the family of risk contributions from Theorem 2 is fully allocating if and only if the risk measure ρ is positively homogeneous. The allocation principle stated in Theorem 2 is therefore called Euler allocation. To the author's best knowledge, Litterman [13] was the first to suggest the Euler allocation principle and to apply it to the standard deviation.

Theorem 2 has provided a recipe that tells us how to obtain risk contributions when a risk measure is given. Of course, we have to check whether this recipe is really useful.[4] We do so by examining the examples from Section 3.1. In case of the standard deviation we arrive at the well-known result (cf. [4, Chapter 5])

$$\rho_{\mathrm{Std},i}(u) = u_i \left(c_\theta \frac{\mathrm{cov}(R_i, R(u))}{\sqrt{\mathrm{var}(R(u))}} - r_i \right), \quad i = 1, \ldots, n. \qquad (3.12a)$$

The corresponding result on quantile derivatives[5] that is needed for the VaR contributions was independently derived by several authors [8, 12, 17]:

$$\mathrm{VaR}_{\theta,i}(u) = -u_i \, \mathbb{E}[R_i \mid -R(u) = q_\theta(-R(u))], \quad i = 1, \ldots, n. \qquad (3.12b)$$

The contributions[6] to ES according to Theorem 2 were derived in [17]:

$$\mathrm{ES}_{\theta,i}(u) = -u_i \, \mathbb{E}[R_i \mid -R(u) \geq q_\theta(-R(u))], \quad i = 1, \ldots, n. \qquad (3.12c)$$

3.3 A Short Review of CreditRisk+

In the remainder of this chapter we will derive analytical expressions for the risk contributions according to (3.12b) and (3.12c) in the CreditRisk+ methodology. The approach to the loss distribution in Chapter 2 was driven by analytical considerations and – to some extent – hid the role the default indicators play for the composition of the loss distribution. Therefore, while preserving the notation from Chapter 2, we will in this section review the steps that

---

[4] In case that the derivatives involved in the Euler allocation principle cannot be computed, an approach based on the saddlepoint approximation can be applied (see [14]).

[5] Certain conditions must be assumed for the joint distribution of the returns in order to justify formula (3.12b). Cf. [8, 12, 17].

[6] The differentiability of ES is subject to the same conditions as the differentiability of VaR.

lead to the probability-generating function in (2.19). When doing so, we will slightly generalize the methodology to the case of stochastic exposures that are independent of the default events and the random factors expressing the dependence on sectors or industries. This generalization can be afforded at no extra cost, as the result is again a generating function in the form of (2.19), the only difference being that the polynomials $\mathcal{P}_k$ are composed in a different way.

In Chapter 2, an approximation was derived for the distribution of the portfolio loss variable $X = \sum_A 1_A \nu_A$ with the ν_A denoting deterministic potential losses. A careful inspection of the beginning of Section 2.5 reveals that the main step in the approximation procedure was to replace the indicators 1_A by random variables D_A with the same expected values. These variables D_A were conditionally Poisson distributed given the random factors $S_1, \ldots, S_N$.

Here, we want to study the distribution of the more general loss variable $X = \sum_A 1_A \mathcal{E}_A$, where $\mathcal{E}_A$ denotes the random outstanding exposure of obligor A. We assume that $\mathcal{E}_A$ takes on positive integer values. However, just replacing 1_A by D_A as in the case of deterministic potential losses does not yield a nice generating function – "nice" in the sense that the CreditRisk$^+$ algorithms for extracting the loss distribution can be applied. We will instead consider the approximate loss variable

$$ X = \sum_A \sum_{i=1}^{D_A} \mathcal{E}_{A,i}, \tag{3.13a} $$

where $\mathcal{E}_{A,1}, \mathcal{E}_{A,2}, \ldots$ are independent copies of $\mathcal{E}_A$. Thus, we approximate the terms $1_A \mathcal{E}_A$ by conditionally compound Poisson sums. For the sake of brevity, we write

$$ Y_A = \sum_{i=1}^{D_A} \mathcal{E}_{A,i} \tag{3.13b} $$

for the loss suffered due to obligor A. A careful inspection of the arguments presented to derive (2.19) now yields the following result on the generating function of X.

Theorem 3. *Assume the setting of Section 2.5. Let the loss variable X be specified by (3.13a) where*

1. *given the factors $S = (S_1, \ldots, S_N)$, the D_A are conditionally independent, and the conditional distribution of D_A given S is Poisson with conditional intensity $p_A^S = p_A \sum_{k=0}^{N} w_{Ak} S_k$ as in (2.9),*
2. *the factors $S = (S_1, \ldots, S_N)$ are gamma distributed with unit expectations $\mathbb{E}[S_i] = 1$, $i = 1, \ldots, N$ and parameters $(\alpha_k, \beta_k) = (\alpha_k, 1/\alpha_k)$, and*
3. *the random variables $\mathcal{E}_{A,1}, \mathcal{E}_{A,2}, \ldots$ are independent copies of $\mathcal{E}_A$ that are also independent of the D_A and S. The distribution of the positive integer-valued random variable $\mathcal{E}_A$ is given through its generating function*

$$ H_A(z) = \mathbb{E}\big[z^{\mathcal{E}_A}\big]. \tag{3.14a} $$

Define for $k = 0, 1, \ldots, N$ the portfolio polynomial $\mathcal{Q}_k$ by

$$\mathcal{Q}_k(z) = \frac{1}{\mu_k} \sum_A w_{Ak} \, p_A \, H_A(z), \qquad (3.14b)$$

with μ_k and w_{Ak} as in Section 2.5.

Then the PGF G_X of the loss variable X can be represented as

$$G_X(z) = e^{\mu_0 \, (\mathcal{Q}_0(z) - 1)} \prod_{k=1}^{N} \left(\frac{1 - \delta_k}{1 - \delta_k \, \mathcal{Q}_k(z)} \right)^{\alpha_k}, \qquad (3.14c)$$

where the constants δ_k are defined as in (2.19).

Remark 1.

1. The setting of Section 2.5 can be regained from Theorem 3 by choosing constant exposures, i.e. $\mathcal{E}_A = \nu_A$. Then the generating functions of the exposures turn out to be just monomials, namely $H_A(z) = z^{\nu_A}$.

2. Theorem 3 states duality of the "individual" actuarial model (3.13a) and the "collective" actuarial model (3.14c). See [16, Chapter 4] for a detailed discussion of this kind of duality. In particular, Theorem 3 generalizes Theorem 4.2.2 of [16] in the sense that negative binomial compounds instead of Poisson compounds are considered. The portfolio polynomial $\mathcal{Q}_k$ is the generating function of a random variable that may be interpreted as a kind of "typical" claim in the sector k.

By means of Theorem 3 the loss distribution of the generalized model (3.13a) can be calculated with the same algorithms as in the case of the original CreditRisk$^+$ model. Once the probabilities $\mathbb{P}[X = t]$, $t \in \mathbb{N}$, are known, it is easy to calculate the loss quantiles $q_\theta(X)$ as defined by (3.3).

When working with Theorem 3, one has to decide whether random exposures are to be taken into account, and in the case of a decision in favour of doing so, how the exposure distributions are to be modelled. The following example suggests a way to deal with these questions.

Example 1 (Distributions of random exposures). Let F denote the face value of a loan that has been granted to a fixed obligor. Then the exposure to this obligor can be represented by

$$\mathcal{E} = F L, \qquad (3.15)$$

where L is a random variable that takes on values in $(0, 1]$. L is called *loss given default* (LGD). Assume that the distribution of L is unknown but that estimates ℓ of its expectation and $v^2 > 0$ of its variance are available. We suggest approximating $\mathcal{E}$ as

$$\mathcal{E} = 1 + E, \qquad (3.16a)$$

where E obeys a negative binomial distribution such that

$$\mathbb{E}[E] + 1 = \mathbb{E}[\mathcal{E}] = F\ell \tag{3.16b}$$

and

$$\text{var}[E] = \text{var}[\mathcal{E}] = F^2 v^2. \tag{3.16c}$$

The distribution of E is specified by two parameters $r > 0$ and $q \in (0,1)$ by means of the relation

$$\mathbb{P}[E = n] = \binom{r + n - 1}{n} q^r (1-q)^n. \tag{3.17a}$$

In particular, we have

$$\mathbb{E}[E] = \frac{r(1-q)}{q} \quad \text{and} \quad \text{var}[E] = \frac{r(1-q)}{q^2} \tag{3.17b}$$

for the expectation and variance of E as well as

$$\mathbb{E}[z^E] = \left(\frac{q}{1 - (1-q)z}\right)^r \tag{3.17c}$$

for the generating function of E. Note that, as a consequence of (3.16a), the generating function of $\mathcal{E}$ will be given by

$$\mathbb{E}[z^{\mathcal{E}}] = z\,\mathbb{E}[z^E] = z\left(\frac{q}{1 - (1-q)z}\right)^r. \tag{3.17d}$$

With respect to the values of the parameters r and q, the moment-matching approach via (3.16b), (3.16c), and (3.17b) requires us to solve

$$\frac{r(1-q)}{q} = F\ell - 1 \quad \text{and} \quad \frac{r(1-q)}{q^2} = F^2 v^2 \tag{3.18a}$$

for r and q. The formal solution of (3.18a) reads

$$q = \frac{F\ell - 1}{F^2 v^2} \quad \text{and} \quad r = \frac{F\ell(F\ell - 1)}{F^2 v^2 - F\ell + 1}. \tag{3.18b}$$

Obviously, not all solutions specified by (3.18b) are admissible in the sense that $q \in (0,1)$ and $r > 0$. In order to find simple approximate conditions for admissibility, assume that $\mathbb{E}[\mathcal{E}] = F\ell$ is large compared with 1 so that we have $\frac{F\ell - 1}{F\ell} \approx 1$. Then (3.18b) changes to

$$q \approx \frac{\ell}{F v^2} \quad \text{and} \quad r \approx \frac{F\ell^2}{F v^2 - \ell}. \tag{3.18c}$$

The approximate values for q and r according to (3.18c) will be admissible if and only if

$$F v^2 > \ell \tag{3.19}$$

or equivalently var$[\mathcal{E}] > \mathbb{E}[\mathcal{E}]$, i.e. if $\mathcal{E}$ is over-dispersed. Hence, approximation of the random exposure with a negative binomial distribution is feasible as long as the random exposure exhibits *over-dispersion*. By (3.19), this will be the case if neither the face value of the loan nor the variance of the loss given default are too small. However, these two cases are not of primary interest as a small face value would indicate that the loan can be neglected in the portfolio context, whereas a small variance of the loss given default would imply the loss given default to be nearly constant. As a consequence, restricting the random exposure modelling to the case of over-dispersed exposures (as specified by (3.19)) appears to be a maintainable approach.   □

When dealing with risk measures and risk contributions, we have considered random returns rather than random losses as we do in this section. Hence, in order to apply the results of Sections 3.1 and 3.2 we have to define the return R_A that arises through the business with obligor A. The potential loss with obligor A is denoted by Y_A as defined in (3.13b). Observe that by the independence assumption on the variables $\mathcal{E}_A$ we have

$$\mathbb{E}[Y_A] = p_A \, \mathbb{E}[\mathcal{E}_A]. \tag{3.20a}$$

Since a credit portfolio manager will at least require the expected loss as return on a granted loan, we suggest describing the return realized with obligor A as

$$R_A = s_A + p_A \, \mathbb{E}[\mathcal{E}_A] - Y_A, \tag{3.20b}$$

where $s_A > 0$ denotes the expected spread through the business with A. Then the return R at portfolio level reads $R = \sum_A R_A$. As an obvious consequence, we see that var$(R) = $ var(X). For the risk measures[7] VaR and ES as given by (3.4) and (3.7), respectively, we obtain

$$\text{VaR}_\theta(R) = q_\theta(X) - \sum_A (s_A + p_A \, \mathbb{E}[\mathcal{E}_A]) \tag{3.21a}$$

and

$$\text{ES}_\theta(R) = \mathbb{E}[X \mid X \geq q_\theta(X)] - \sum_A (s_A + p_A \, \mathbb{E}[\mathcal{E}_A]). \tag{3.21b}$$

Similarly, we obtain for obligor A's contributions[7] to portfolio standard deviation, VaR, and ES according to (3.12a), (3.12b), and (3.12c), respectively,

---

[7] Compared with Sections 3.1 and 3.2, the notation here is slightly changed in so far as we implicitly assume $u = (1, \dots, 1)$ and replace the argument u by the underlying random variable R.

$$\rho_{\text{Std},A}(R) = c_\theta \frac{\text{cov}(Y_A, X)}{\sqrt{\text{var}(X)}} - s_A, \tag{3.22a}$$

$$\text{VaR}_{\theta,A}(R) = \mathbb{E}[Y_A \mid X = q_\theta(X)] - (s_A + p_A\, \mathbb{E}[\mathcal{E}_A]), \tag{3.22b}$$

and

$$\text{ES}_{\theta,A}(R) = \mathbb{E}[Y_A \mid X \geq q_\theta(X)] - (s_A + p_A\, \mathbb{E}[\mathcal{E}_A]). \tag{3.22c}$$

Note, however, that the contributions to VaR and ES as specified by (3.22b) and (3.22c) are definitions rather than inevitable given facts because due to its discrete nature there is no overall differentiability in the CreditRisk$^+$ model. Under these circumstances, Theorem 3 yields only the motivation for our choice of the contributions.

At first glance, the calculation of $\mathbb{E}[X \mid X \geq q_\theta(X)]$, which will be a necessary part of the calculation of portfolio ES_θ, appears to involve an infinite series, since by definition we have

$$\mathbb{E}[X \mid X \geq q_\theta(X)] = \mathbb{P}[X \geq q_\theta(X)]^{-1} \sum_{k=q_\theta(X)}^{\infty} k\, \mathbb{P}[X = k]. \tag{3.23a}$$

Observe, however, that the calculation can be carried out in finitely many steps when the decomposition

$$\sum_A p_A\, \mathbb{E}[\mathcal{E}_A] = \mathbb{E}[X] = \mathbb{P}[X \geq q_\theta(X)]\, \mathbb{E}[X \mid X \geq q_\theta(X)]$$
$$+ \mathbb{P}[X < q_\theta(X)]\, \mathbb{E}[X \mid X < q_\theta(X)] \tag{3.23b}$$

is used. This leads to the representation

$$\mathbb{E}[X \mid X \geq q_\theta(X)] = \frac{\sum_A p_A \mathbb{E}[\mathcal{E}_A] - \sum_{k=1}^{q_\theta(X)-1} k\, \mathbb{P}[X = k]}{1 - \sum_{k=0}^{q_\theta(X)-1} \mathbb{P}[X = k]}. \tag{3.23c}$$

3.4 Contributions to Value-at-Risk and Expected Shortfall with CreditRisk$^+$

The purpose of this section is to provide formulas for the contributions to VaR and ES according to (3.22b) and (3.22c) that can be evaluated with the known CreditRisk$^+$ algorithms. We will not deal with the contributions to the standard deviation, since they are considered in a more general form in Chapter 9. The following lemma – a generalization of [11, Eq. (20)] – yields the foundation of our results. Denote by $I(E)$ the indicator variable of the event E, i.e. $I(E; m) = 1$ if $m \in E$ and $I(E; m) = 0$ if $m \notin E$.

Lemma 1. *Define the individual loss variable Y_A as in (3.13b). Under the assumptions of Theorem 3, we have for any $x \in \mathbb{N}$*

$$\mathbb{E}[Y_A \, I(X = x)] = \mathbb{E}[p_A^S \, \mathcal{E}_A \, I(X = x - \mathcal{E}_A)].$$

Proof. Basically, the assertion is the result of a straightforward computation.

$$\mathbb{E}[Y_A \, I(X = x)] = \sum_{k=0}^{\infty} \mathbb{E}\Big[I(D_A = k, X = x) \sum_{i=1}^{k} \mathcal{E}_{A,i}\Big] \qquad (3.24)$$

$$= \sum_{k=0}^{\infty} \mathbb{E}\Big[I\Big(D_A = k, \sum_{B \neq A} Y_B = x - \sum_{i=1}^{k} \mathcal{E}_{A,i}\Big) \sum_{i=1}^{k} \mathcal{E}_{A,i}\Big].$$

Since the variables $\mathcal{E}_{A,1}, \mathcal{E}_{A,2}, \dots$ are independently identically distributed as $\mathcal{E}_A$ and independent of D_A and $(Y_B)_{B \neq A}$, (3.24) implies

$$\mathbb{E}[Y_A \, I(X = x)] = \sum_{k=1}^{\infty} k \, \mathbb{E}\Big[\mathcal{E}_A \, I\Big(D_A = k, \sum_{B \neq A} Y_B + \sum_{i=1}^{k-1} \mathcal{E}_{A,i} = x - \mathcal{E}_A\Big)\Big],$$

where $\mathcal{E}_A$ is independent of D_A, $(\mathcal{E}_{A,i})_{i=1,2,\dots}$, and $(Y_B)_{B \neq A}$. Since all involved variables are conditionally independent given the vector S of the scalar factors and D_A is conditionally Poisson distributed given S, we can continue:

$$\mathbb{E}[Y_A \, I(X = x)]$$

$$= \sum_{k=1}^{\infty} k \, \mathbb{E}\Big[\mathbb{P}[D_A = k | S] \, \mathbb{E}\Big[\mathcal{E}_A \, I\Big(\sum_{B \neq A} Y_B + \sum_{i=1}^{k-1} \mathcal{E}_{A,i} = x - \mathcal{E}_A\Big)\Big| S\Big]\Big]$$

$$= \sum_{k=1}^{\infty} k \, \mathbb{E}\Big[\frac{(p_A^S)^k}{k!} e^{-p_A^S} \, \mathbb{E}\Big[\mathcal{E}_A \, I\Big(\sum_{B \neq A} Y_B + \sum_{i=1}^{k-1} \mathcal{E}_{A,i} = x - \mathcal{E}_A\Big)\Big| S\Big]\Big]$$

$$= \sum_{k=0}^{\infty} \mathbb{E}\Big[p_A^S \, \mathbb{P}[D_A = k | S] \, \mathbb{E}\Big[\mathcal{E}_A \, I\Big(\sum_{B \neq A} Y_B + \sum_{i=1}^{k} \mathcal{E}_{A,i} = x - \mathcal{E}_A\Big)\Big| S\Big]\Big]$$

$$= \sum_{k=0}^{\infty} \mathbb{E}\Big[p_A^S \, \mathbb{E}\Big[\mathcal{E}_A \, I\Big(D_A = k, \sum_{B \neq A} Y_B + \sum_{i=1}^{k} \mathcal{E}_{A,i} = x - \mathcal{E}_A\Big)\Big| S\Big]\Big]$$

$$= \mathbb{E}[p_A^S \, \mathcal{E}_A \, I(X = x - \mathcal{E}_A)],$$

as stated in the assertion.   □

Recall that by (3.22b) we have to compute the conditional expectation $\mathbb{E}[Y_A \,|\, X = q_\theta(X)]$ in order to evaluate obligor A's VaR contribution. The following theorem provides the announced formula for doing so in the original CreditRisk$^+$ framework.

Theorem 4 (VaR contributions). *Define the individual loss variable Y_A as in (3.13b) and assume the setting of Theorem 3. Write $\mathbb{P}_\alpha[X \in .]$ in order to express the dependence[8] of the portfolio loss distribution upon the exponents $\alpha = (\alpha_1, \ldots, \alpha_N)$ in (3.14c). Define $\alpha(k)$, $k = 1, \ldots, N$ by*

$$\alpha(k) = (\alpha_1, \ldots, \alpha_{k-1}, \alpha_k + 1, \alpha_{k+1}, \ldots, \alpha_N). \tag{3.25a}$$

Let η_A denote a random variable with distribution specified by

$$\mathbb{P}[\eta_A = t] = \frac{t\,\mathbb{P}[\mathcal{E}_A = t]}{\mathbb{E}[\mathcal{E}_A]}, \tag{3.25b}$$

for any $t \in \mathbb{N}$. Assume that η_A and X are independent.
Then, for any $x \in \mathbb{N}$ with $\mathbb{P}_\alpha[X = x] > 0$, we have

$$\mathbb{E}[Y_A \mid X = x] = p_A\,\mathbb{E}[\mathcal{E}_A]\,\mathbb{P}_\alpha[X = x]^{-1}$$
$$\times \left(w_{A0}\,\mathbb{P}_\alpha[X = x - \eta_A] + \sum_{j=1}^{N} w_{Aj}\,\mathbb{P}_{\alpha(j)}[X = x - \eta_A] \right). \tag{3.25c}$$

Remark 2.

1. Here we define the loss quantile $q_\theta(X)$ by (3.3). As a consequence, $q_\theta(X)$ will always be a non-negative integer with $\mathbb{P}_\alpha[X = q_\theta(X)] > 0$.
2. The probabilities in the numerator of (3.25c) have to be computed by means of convolution, i.e.

$$\mathbb{P}_{\alpha(j)}[X = x - \eta_A] = \sum_{i=0}^{x} \mathbb{P}_{\alpha(j)}[X = i]\,\mathbb{P}[\eta_A = x - i]. \tag{3.26}$$

   In the case of $x = q_\theta(X)$ with θ close to one, most of the terms will disappear whenever $\max \mathcal{E}_A = \max \eta_A \ll q_\theta(X)$. In the case of a constant exposure to obligor A, i.e. if $\mathcal{E}_A = \nu_A = $ constant, by (3.25b) the transformed exposure η_A is also constant with $\eta_A = \nu_A$. The sum in (3.26) then collapses to the single term $\mathbb{P}_{\alpha(j)}[X = x - \nu_A]$.
3. If there exists an algorithm for extracting the probability distribution of the portfolio loss variable X from generating functions as in (3.14c), then Theorem 4 states that applying the same algorithm $(N + 1)$ times to slightly different generating functions also yields the corresponding VaR contributions (and, as we will see in Corollary 2, the corresponding ES contributions).
4. Due to the approximation procedure the VaR contributions may exceed the maximum exposures $\max \mathcal{E}_A$. This incident should be regarded as an indicator of low approximation quality when replacing indicator variables by conditionally Poisson-distributed variables.

---

[8] Of course, the distribution also depends on μ_0, $\mathcal{Q}_0, \ldots, \mathcal{Q}_N$, and $\delta_1, \ldots, \delta_N$. However, these input parameters are considered constant in Theorem 4.

5. Contributions equal to zero may occur in (3.25c). This is more likely in the case of constant exposures, i.e. if $\mathcal{E}_A = \nu_A = $ constant for all obligors A, but still limited to some exceptional situations. See [9] for a detailed discussion of this topic.

Proof (Proof of Theorem 4). We will derive (3.25c) by comparing the coefficients of two power series. The first one is $\mathbb{E}[Y_A\, z^X] = \sum_{k=0}^{\infty}\mathbb{E}[Y_A\, I(X = k)]\, z^k$, the second one is an expression that is equivalent to $\mathbb{E}[Y_A\, z^X]$ but involves generating functions similar to (3.14c).

Recall that we denote the generating function of $\mathcal{E}_A$ by $H_A(z)$. By means of Lemma 1, we can compute

$$\begin{aligned}
\mathbb{E}[Y_A\, z^X] &= \sum_{k=0}^{\infty}\mathbb{E}[p_A^S\, \mathcal{E}_A\, I(X + \mathcal{E}_A = k)]\, z^k \\
&= \mathbb{E}[p_A^S\, \mathcal{E}_A\, z^{X+\mathcal{E}_A}] = \mathbb{E}[p_A^S\, z^X]\,\mathbb{E}[\mathcal{E}_A\, z^{\mathcal{E}_A}] \\
&= \mathbb{E}[p_A^S\, z^X]\, z\, H_A'(z) = \mathbb{E}[p_A^S\, z^X]\,\mathbb{E}[\mathcal{E}_A]\, U_A(z), \quad (3.27a)
\end{aligned}$$

where $U_A(z) = \frac{z\, H_A'(z)}{\mathbb{E}[\mathcal{E}_A]} = \sum_{k=0}^{\infty}\mathbb{P}[\eta_A = k]\, z^k$ is just the generating function of the random variable η_A, which has been defined in the statement of the theorem.

By making use of the fact that the scalar factors $S = (S_1,\ldots,S_N)$ are gamma distributed with parameters $(\alpha_k, 1/\alpha_k)$, $k = 1,\ldots,N$, we obtain for $\mathbb{E}[p_A^S\, z^X]$ (cf. the proof of (2.19))

$$\begin{aligned}
\mathbb{E}[p_A^S\, z^X] &= \mathbb{E}[p_A^S\, \mathbb{E}[z^X \mid S]] = p_A \sum_{j=0}^{N} w_{Aj}\, \mathbb{E}\left[S_j \prod_{k=0}^{N} \exp\big(S_k\, (\mathcal{Q}_k(z) - 1)\big)\right] \\
&= p_A \sum_{j=0}^{N} w_{Aj}\, \mathbb{E}[S_j\, \exp\big(S_j\, (\mathcal{Q}_j(z) - 1)\big)]\, \mathbb{E}\left[\prod_{k=0,\,k\neq j}^{N} \exp\big(S_k\, (\mathcal{Q}_k(z) - 1)\big)\right] \\
&= p_A \left(w_{A0}\, G_X^{(\alpha)}(z) + \sum_{j=1}^{N} w_{Aj}\, G_X^{(\alpha(j))}(z)\right). \quad (3.27b)
\end{aligned}$$

In (3.27b),

$$G_X^{(\alpha)}(z) = \sum_{k=0}^{\infty} \mathbb{P}_\alpha[X = k]\, z^k \quad (3.28a)$$

and

$$G_X^{(\alpha(j))}(z) = \sum_{k=0}^{\infty} \mathbb{P}_{\alpha(j)}[X = k]\, z^k \quad (3.28b)$$

denote the generating functions of X according to (3.14c), as has been explained in the statement of the theorem.

Observe that $U_A(z)\,G_X^{(\alpha)}(z)$ and $U_A(z)\,G_X^{(\alpha(j))}(z)$ are the generating functions of the sequences $\mathbb{P}_\alpha[X+\eta_A=0], \mathbb{P}_\alpha[X+\eta_A=1], \ldots$ and $\mathbb{P}_{\alpha(j)}[X+\eta_A=0], \mathbb{P}_{\alpha(j)}[X+\eta_A=1], \ldots$, respectively. Now, combining (3.27a) and (3.27b) and comparing the coefficients of the corresponding power series yields (3.25c). □

Before turning to the case of ES contributions we consider the modification of Theorem 4 that arises when in (3.25c) the variable Y_A is replaced by D_A. Then the conditional expectation $\mathbb{E}[D_A\,|\,X=x]$ can be interpreted as the conditional probability of obligor A's default given that the portfolio loss X assumes the value x. A reasoning similar to that for Theorem 4 leads to the following result.

Corollary 1. *Adopt the setting and the notation of Theorem 4. Then, in the CreditRisk$^+$ framework, the conditional probability of obligor A's default given that the portfolio loss X assumes the value x may be expressed by*

$$\mathbb{E}[D_A\,|\,X=x] \;=\; p_A\,\mathbb{P}_\alpha[X=x]^{-1}$$

$$\times \left(w_{A0}\,\mathbb{P}_\alpha[X=x-\widetilde{\mathcal{E}}_A] + \sum_{j=1}^{N} w_{Aj}\,\mathbb{P}_{\alpha(j)}[X=x-\widetilde{\mathcal{E}}_A]\right), \quad (3.29)$$

where $\widetilde{\mathcal{E}}_A$ stands for a random variable that has the same distribution as $\mathcal{E}_A$ but is independent of X.

The probabilities in the numerator of (3.29) must also be calculated by convolution, cf. (3.26). Interestingly enough, Corollary 1 has an implication that can be used for constructing a certain kind of stress scenario for the portfolio under consideration. Observe that by the very definition of conditional probabilities we have

$$\mathbb{P}[A \text{ defaults}\,|\,X=x] \;=\; \frac{p_A\,\mathbb{P}[X=x\,|\,A \text{ defaults}]}{\mathbb{P}[X=x]}. \quad (3.30)$$

Since by Corollary 1 an approximation for $\mathbb{P}[A \text{ defaults}\,|\,X=x]$ is provided, the termwise comparison of (3.29) and (3.30) yields

$$\mathbb{P}_\alpha[X=x\,|\,A \text{ defaults}] \approx$$

$$w_{A0}\,\mathbb{P}_\alpha[X=x-\widetilde{\mathcal{E}}_A] + \sum_{j=1}^{N} w_{Aj}\,\mathbb{P}_{\alpha(j)}[X=x-\widetilde{\mathcal{E}}_A]. \quad (3.31)$$

Using (3.31), stressed portfolio loss distributions can be evaluated, conditional on the scenarios that single obligors have defaulted. If, for instance, the portfolio VaR changes dramatically when obligor A's default is assumed, then one may find that the portfolio depends too strongly upon A's condition.

According to (3.31), the conditional distribution $\mathbb{P}_\alpha[X=\cdot\,|\,A \text{ defaults}]$ of the portfolio loss X given that A defaults may be computed by means of a decomposition with respect to the economic factors that could have caused the default. A default due to factor j makes a high value of the realization of the factor appear more likely than under the unconditional distribution $\mathbb{P}_\alpha[X = \cdot]$. This fact is expressed by the exponent $\alpha_j + 1$ in the generating function of $\mathbb{P}_{\alpha(j)}[X = \cdot]$.

We conclude this chapter by providing evaluable analytical expressions for the contributions to ES according to (3.22c). The result follows immediately from Theorem 4 when

$$\mathbb{E}[Y_A \, I(X \geq x)] = \sum_{k=x}^{\infty} \mathbb{E}[Y_A \, I(X = k)] \qquad (3.32)$$

is taken into account.

Corollary 2. *Adopt the setting and the notation of Theorem 4. If x is any non-negative integer with $\mathbb{P}[X \geq x] > 0$ then the conditional expectation $\mathbb{E}[Y_A \mid X \geq x]$ can be represented as*

$$\mathbb{E}[Y_A \mid X \geq x] = p_A \, \mathbb{E}[\mathcal{E}_A] \, \mathbb{P}_\alpha[X \geq x]^{-1}$$
$$\times \left(w_{A0} \, \mathbb{P}_\alpha[X \geq x - \eta_A] + \sum_{j=1}^{N} w_{Aj} \, \mathbb{P}_{\alpha(j)}[X \geq x - \eta_A] \right). \quad (3.33)$$

Remark 3. The probabilities in the numerator of (3.33) have to be computed by means of convolution, i.e.

$$\mathbb{P}_{\alpha(j)}[X \geq x - \eta_A] = \sum_{i=0}^{\infty} \mathbb{P}_{\alpha(j)}[X = i] \, \mathbb{P}[\eta_A \geq x - i] \qquad (3.34a)$$

$$= \sum_{i=0}^{\infty} \mathbb{P}[\eta_A = i] \, \mathbb{P}_{\alpha(j)}[X \geq x - i]. \qquad (3.34b)$$

In practice, $\max \eta_A$ will be finite and small compared with x if for instance we have $x = q_\theta(X)$. As a consequence, (3.34b) will be more appropriate for the calculation of the probabilities $\mathbb{P}_{\alpha(j)}[X \geq x - \eta_A]$. □

Assume that the obligors A and B are identically modelled with respect to all CreditRisk$^+$ input parameters except the variances of the exposures $\mathcal{E}_A$ and $\mathcal{E}_B$ (the expectations are assumed to be equal). If, for instance, $\mathrm{var}[\mathcal{E}_A] > \mathrm{var}[\mathcal{E}_B]$ then one observes

$$\mathbb{E}[\eta_A] = \frac{\mathbb{E}[\mathcal{E}_A]^2 + \mathrm{var}[\mathcal{E}_A]}{\mathbb{E}[\mathcal{E}_A]} > \frac{\mathbb{E}[\mathcal{E}_B]^2 + \mathrm{var}[\mathcal{E}_B]}{\mathbb{E}[\mathcal{E}_B]} = \mathbb{E}[\eta_B]. \qquad (3.35)$$

η_A and η_B denote here the random variables defined by (3.25b) for use in (3.25c) and (3.33). Hence, we have got a dependence of the VaR and ES contributions upon the exposure volatilities. Since $\mathbb{E}[\eta_A] > \mathbb{E}[\eta_B]$ in a lot of cases entails that η_A is stochastically greater[9] than η_B, we see from (3.34a) that obligor A's ES contribution will be greater than obligor B's. This observation fits intuition very well. However, this monotonicity property need not hold for the VaR$_\theta$ contributions if the probabilities $\mathbb{P}[X = t]$ are not monotonous in a neighbourhood of $q_\theta(X)$ (cf. (3.25c)).

References

1. C. Acerbi. Spectral measures of risk: a coherent representation of subjective risk aversion. *Journal of Banking & Finance*, 26(7):1505–1518, 2002.
2. C. Acerbi and D. Tasche. On the coherence of expected shortfall. *Journal of Banking & Finance*, 26(7):1487–1503, 2002.
3. P. Artzner, F. Delbaen, J.-M. Eber, and D. Heath. Coherent measures of risk. *Mathematical Finance*, 9(3):203–228, 1999.
4. C. Bluhm, L. Overbeck, and C. Wagner. *An Introduction to Credit Risk Modeling*. CRC Press, Boca Raton, 2002.
5. F. Delbaen. Coherent Risk Measures. Lecture Notes, Scuola Normale Superiore di Pisa, 2001.
6. M. Denault. Coherent allocation of risk capital. *Journal of Risk*, 4(1):1–34, 2001.
7. M. Gordy. A risk-factor model foundation for ratings-based bank capital rules. *Journal of Financial Intermediation*, 12(3):199–232, 2003.
8. C. Gouriéroux, J. P. Laurent, and O. Scaillet. Sensitivity analysis of values at risk. *Journal of Empirical Finance*, 7:225–245, 2000.
9. H. Haaf and D. Tasche. Credit portfolio measurements. *GARP Risk Review*, (7):43–47, 2002.
10. M. Kalkbrener. An axiomatic approach to capital allocation. Technical document, Deutsche Bank AG, 2002.
11. A. Kurth and D. Tasche. Contributions to credit risk. *RISK*, 16(3):84–88, 2003.
12. G. Lemus. *Portfolio optimization with quantile-based risk measures*. PhD thesis, Sloan School of Management, MIT, 1999.
13. R. Litterman. Hot spots^{TM} and hedges. *The Journal of Portfolio Management*, 22:52–75, 1996.
14. R. Martin, K. Thompson, and C. Browne. VAR: Who contributes and how much? *RISK*, 14(8):99–102, 2001.
15. R. T. Rockafellar and S. Uryasev. Conditional value-at-risk for general loss distributions. *Journal of Banking & Finance*, 26(7):1443–1471, 2002.
16. T. Rolski, H. Schmidli, V. Schmidt, and J. Teugels. *Stochastic Processes for Insurance and Finance*. Wiley Series in Probability and Statistics. John Wiley & Sons, 1999.
17. D. Tasche. Risk contributions and performance measurement. Working paper, Technische Universität München, 1999.
http://citeseer.nj.nec.com/tasche99risk.html
18. D. Tasche. Expected shortfall and beyond. *Journal of Banking & Finance*, 26(7):1519–1533, 2002.

---

[9] That is, $\mathbb{P}[\eta_A \geq t] \geq \mathbb{P}[\eta_B \geq t]$ for all non-negative integers t.

4

Risk Factor Transformations Relating CreditRisk$^+$ and CreditMetrics

Christian Wieczerkowski*

Summary. CreditRisk$^+$ and CreditMetrics furnish special cases of general credit risk factor models. On a respective model space, there is a symmetry of factor transformations that relates identical though differently represented models. In the simplest case of homogeneous one-factor one-band-models, there is an approximate symmetry between consistently parametrized CreditRisk$^+$ and CreditMetrics. This can be viewed as evidence that there exists in general a consistent parametrization of both models that results in the same loss distribution.

Introduction

Since 1997, Credit Suisse's CreditRisk$^+$ and J. P. Morgan's CreditMetrics coexist as benchmarks for internal credit risk models. The two models, see [2] and [7], come in a rather different mathematical clothing. Furthermore, the interpretation of credit risk factors and thus the explanation of default correlations is surprisingly different in both models.

Not surprisingly, CreditRisk$^+$ and CreditMetrics have been compared by many authors. The findings of Gordy [6], translating each model to the other's language, Koyluoglu and Hickman [10, 11], comparing conditional default probabilities using a factor transformation, Finger [5], based on a comparison of factor models, Crouhy, Galei, and Mark [3], and followers, see for instance [1, 4, 9, 13, 14] and references therein can be summarized as follows:

(1) CreditRisk$^+$ and two-state CreditMetrics are both factor models with conditionally independent defaults, albeit with differing conditional default probabilities and risk factor distributions.

(2) CreditRisk$^+$ can be formulated analogously to two-state CreditMetrics and solved by Monte Carlo simulation. Two-state CreditMetrics can

---

* The views expressed herein are my own and do not necessarily reflect those of the HSH-Nordbank. I would like to thank Susanne Gögel, Jörg Lemm, and Klaus Pinn at the WGZ Bank for helpful discussions.

be formulated analogously to CreditRisk$^+$ in terms of a probability-generating function. Unfortunately, this has not led to a solution in closed form.

(3) CreditRisk$^+$ and two-state CreditMetrics are related through a factor transformation. In simple cases, the models can be effectively matched, but not using the standard parametrization.

Both CreditRisk$^+$ and CreditMetrics are based on a risk factor representation of the form

$$\mathbb{P}(\text{loss}) = \sum_{\text{risk conditions}} \mathbb{P}(\text{loss} \mid \text{risk condition}) \quad \mathbb{P}(\text{risk condition}),$$

with independent (factoring) conditional probabilities. Finding (3) means that each of these two probabilities by itself is a matter of choice. The risk factor distribution can be transformed into virtually anything else without changing the loss distribution provided that the conditional loss distribution is transformed conversely. It is the combination of conditional default probabilities and factor distribution that matters.

Both CreditRisk$^+$ and CreditMetrics use continuous risk factors so that the sum over risk conditions becomes an integral. This integral representation furnishes the statistical backbone in both models in the sense that the portfolio mean, standard deviation, value-at-risk, and other characteristics are derived from it. It is solved analytically in CreditRisk$^+$, whereas one resorts to Monte Carlo simulations in CreditMetrics. If both models are to represent the same portfolio distribution there must be – at least approximately – a transformation from one integral representation to the other.

Presently, the scarcity of available statistics about joint default events excludes ruling out either of the two models on the basis of their statistical assumptions. Therefore, neither of them can be considered to be right or wrong. This means that – within built-in error margins – there should exist a consistent parametrization of both models such that

$$\text{loss-distr.}_{\text{CreditRisk}^+}(\text{parameters}) \approx \text{loss-distr.}_{\text{CreditMetrics}}(\text{parameters}').$$

The shirt-sleeved approach is to match the mean, standard deviation, and value-at-risk by tuning the parameters in both models and then to hope that one comes as close as possible to the truth. The disadvantage of this approach is that it tests the parametric range of the two models rather than the validity of their architecture.

In this chapter, we propose to test the model matching on the intermediate level of the credit modelling stack

Fineness	Number of dimensions
Loss distribution	one
Risk factor distribution	intermediate
Obligor distribution	high

To do the matching, we choose a set-up where the conditional probabilities are defined with respect to the same credit events in both models. In particular, we require the number of risk factors to be the same. If the risk factors are to represent the same kind of uncertainty about the same kind of credit events, and if the models are indeed the same, then there should be a transformation of risk factors that transforms one model into the other. More general and loose model matching schemes are conceivable but will not be considered here.

Each of the two models comes together with an algorithm to assign values to its parameters:

(a) parameters of the conditional default probabilities: idiosyncratic weights, factor loadings;
(b) parameters of the factor distribution itself: means, standard deviations, correlations;
(c) parameters of the distribution of loss given default: recovery rates and standard deviations, spreads.

A statistical calibration of all parameters, for instance by means of a maximum likelihood priniciple, is presently out of reach.

Therefore, one resorts to additional, external data that invoke an identification of the risk factors, for example with a stock index or the oil price, that help to assign values to the distribution parameters, factor loadings, etc. This is interesting as the economical identification breaks the freedom of choice of factor distributions, which is only in an abstract risk factor space.

It remains to be proved that one's parametrization correctly accounts for default correlations, not at all an easy task. A minimal requirement is that the parametrization should be consistent. If CreditRisk$^+$ and CreditMetrics are both to be correct, the parameter values of the two models – applied to the same credit portfolio – should be related through a factor transformation. Below, we will compute such a symmetry between CreditRisk$^+$ and Credit-Metrics. There remains the question whether the standard parametrization schemes are consistent in this sense or not. It has been noticed that a default rate standard deviation of the order of the default rate itself (or even smaller) in CreditRisk$^+$ corresponds to unrealistically small asset correlations in CreditMetrics, see [10]. Further below, we will reproduce this result.

The method of choice of many rigorously tempered minds is to analyse and compare tail distributions. However, recall that a model does not have to be valid all the way to infinity (or even the maximal possible loss). To be useful, it should be valid beyond the quantile that defines the portfolio holder's solvency level. In this respect, we will be rather casual and settle for a few showcase VaR numbers. It remains to be seen whether factor transformations will generate any useful insights about the tails.

In these notes, we will review how the findings (1), (2), and (3) come about. En route, we show that both CreditRisk$^+$ and CreditMetrics (not just two-state CreditMetrics) furnish special cases of a general class of multi-state credit risk models. We will spend about half the chapter making precise how each of the models fits into the unified framework. The genealogy of credit risk models looks as follows:

General factor model with discretized losses

Multi-state CreditRisk$^+$ Multi-state CreditMetrics

Two-state CreditRisk$^+$ Rating-state CreditMetrics

Two-state CreditMetrics

Multiple states arise both in splitting up the pre-default state into rating classes, or finer sub-rating-states, and in splitting up the post-default state into severity classes. This may be of interest by itself to credit model builders. You can build exactly the same general multi-state model either the CreditRisk$^+$way or the CreditMetrics way. Formula-wise, we will remain as close as we can to CreditRisk$^+$, as we will use the probability-generating function in Poisson approximation. The other way, to solve CreditRisk$^+$ by means of Monte Carlo, is in fact both simpler and less restrictive but will not be considered here.

The second half of this chapter is devoted to factor transformations. We will transform the gamma-distributed risk factors into normally distributed risk factors in CreditRisk$^+$. Then we propose to use the goal

$$\mathbb{P}_{\text{CreditRisk}^+}\left(\text{loss}|\text{transformed factors}\right) \approx \mathbb{P}_{\text{CreditMetrics}}\left(\text{loss}|\text{factors}\right)$$

to match the parameters. We should add that in this formulation, where the risk factors are transformed in a universal form, all parameters reside in the conditional probabilities. The converse scheme, where the whole parameter dependence is shifted into the risk factor distribution, is also possible. On the intermediate credit modelling level, it seems more natural to put the parameters into the conditional probabilities and to think of the factor as a (universal) source of randomness.

The transformation part will remain technically rather incomplete. For instance, we will not give a precise definition for the distance of conditional default probabilities but restrict our attention to a few characteristic numbers. Having presented the concept, we will finish with a sample calculation for the simplest versions of CreditRisk$^+$ and CreditMetrics. In particular, we will show that one-band one-factor CreditRisk$^+$ and CreditMetrics are effectively equivalent.

4.1 Discretized Losses

As the point of departure, we choose a general class of credit risk models. To make the relationship between CreditRisk⁺ and CreditMetrics as clear as possible, it will be formulated as an increasing set of assumptions.[1]

4.1.1 Basic assumptions

Consider the following data comprising a one-period credit portfolio model with discrete loss functions:

(A) A finite set $\mathbb{A} = \{1, \ldots, K\}$ representing the obligors (counterparties) in a credit portfolio;

(B) a finite set $\mathbb{L} \subset \{L : \mathbb{A} \to \mathbb{Z}\}$ representing the possible losses (portfolio states) at the end of the time period (risk horizon), where each loss vector L assigns an aggregated loss L_A to every obligor A in the portfolio;

(C) a probability measure $\mu_{\mathbb{L}}$ on an admissible subset of set $\mathbb{L}$.

CreditRisk⁺ belongs to this class of models. CreditMetrics also belongs thereto, with the harmless modification to discretize[2] losses in terms of an external loss unit.

Ad (A): This is nomenclature and no restriction apart from the assumption that there is a discrete set of obligors that can separately default (or change their credit state). What is not well captured are directed relationships, for example a company and its subsidiaries.

Ad (B): To make contact with reality, one has to multiply these losses by a loss unit (e.g., $ 1000), see Chapter 2.

Negative values of L_A are understood as gains. We follow the bearish custom to speak of losses rather than gains in the context of credit risk. Gains are not included in standard CreditRisk⁺. Since they are included in CreditMetrics and since they are easily accomodated for, we have included them from the beginning.

Ad (C): Note that we have not yet introduced any credit event space apart from the portfolio loss itself. Credit portfolio models often start by writing the loss caused by each single obligor as a random variable that is given by the product of an exposure variable, the indicator of a credit event, and a severity variable, the first and the last factors being optionally stochastic.

We should concede that the loss is not always the outcome of an instantaneous default event but rather of a lengthy and painful post-default process. For simplicity, CreditRisk⁺ and CreditMetrics choose to work with a single time-scale (or risk horizon).

---

[1] Our assumptions will be such that they form a minimal common framework for CreditRisk⁺ and CreditMetrics. Only occasionally, we will take the bait and hint at even more general credit models.

[2] Both models have a discrete credit event space. In addition, CreditRisk⁺ discretizes the exposure and thus the loss at a credit event.

4.1.2 Probability-generating function (PGF)

The goal of any credit portfolio model is to compute expectation values of the portfolio loss variable $X = \sum_{A \in \mathbb{A}} \mathbf{L}_A$ and functions thereof. For the purpose of our model matching, we will restrict our attention to the expected loss (mean), standard deviation, and the VaR (quantile).

Expectation values of X are subsumed into the probability-generating function (PGF), which becomes a sum when losses are discrete, i.e.,

$$G_X(z) = \mathbb{E}\left[z^X\right] = \sum_{n \in \mathbb{Z}} \mu_X(n)\, z^n. \tag{4.1}$$

Here $\mu_X(P) = \sum_{L \in \mathbb{L}} \delta_{P, \sum_A L_A}\, \mu_{\mathbf{L}}(L)$ denotes the probability of a given portfolio loss as the sum of the probabilities of all (independent) portfolio states that constitute this loss (see also Chapter 2).

The present formulation slightly generalizes standard CreditRisk$^+$ and modifies standard CreditMetrics: in standard CreditRisk$^+$, there are no negative exponents and fewer states. With gains included as negative losses, the PGF becomes a Laurent polynomial.[3] Standard CreditMetrics is not formulated in terms of its PGF. However, as noticed in [6], it can easily be done. Furthermore, in standard CreditMetrics, losses are not discrete. The pricing of exposures at the risk horizon is a fully exchangeable module, which is separate from the migration dynamics. It is not forbidden to use losses that are coarsely discrete. They are simply not helpful in the standard Monte Carlo approach.

4.2 Conditional Independence

To arrive at computable numbers, it is indispensable to write down a probability measure that leads to manageable expressions. At this point, there is no way around simplifying assumptions.

4.2.1 Risk factor hypothesis

CreditRisk$^+$ and CreditMetrics both assume that there is a set of risk factors such that the probabilities of joint credit events factorize conditioned upon the risk factors. The fourth and fifth assumptions are that

(D) there exists a random variable S (the risk factors) such that

$$\mathrm{Prob}(\mathbf{L} = L | S) = \prod_{A \in \mathbb{A}} \mathbb{P}[\mathbf{L}_A = L_A | S], \tag{4.2}$$

---

[3] If one represents gains as differences from a maximal gain, the PGF becomes again an ordinary polynomial.

(E) that S takes values in $\mathbb{R}^N$ (or a subset thereof) and has a probability density $d\mu_S(s)$ so that

$$\mathbb{P}[S \in B] = \int_B d\mu_S(s). \qquad (4.3)$$

Ad (D): In a nutshell, it means that the correlations can be traced back to common risk factors. Risk factors are particularly appealing if they provide an economic explanation of credit events. Conversely to CreditRisk$^+$, CreditMetrics shows its colours claiming that one should use country-industry indices as risk factors. This is still debated. Also, there are practical difficulties, see [15].

Ad (E): the risk factors in CreditRisk$^+$ and CreditMetrics are real valued respectively. In CreditRisk$^+$ they are supported on the subspace $\mathbb{R}_+^D$. Discrete risk factors are likewise conceivable but will not be considered here. They arise for instance in models where one sums over scenarios.

4.2.2 Integral representation for the PGF

Conditional independence implies that the conditional PGF completely factorizes. Integrating out the risk factors, one is rewarded with an integral representation for the PGF. From this formula, the PGF is solved in closed form in CreditRisk$^+$.

As a warm-up, consider the mini-portfolio consisting of a single obligor. The PGF becomes

$$G_{\mathbf{L}_A}(z) = \sum_{L_A \in \mathbb{Z}} \mathbb{P}[\mathbf{L}_A = L_A]\, z^{L_A}.$$

Risk factors are introduced as conditions and (conversely) removed by integrating out according to

$$G_{\mathbf{L}_A}(z) = \int G_{\mathbf{L}_A|S}\left(z|s\right)\, d\mu_S(s)$$

where

$$G_{\mathbf{L}_A|S}\left(z|s\right) = \sum_{L_a \in \mathbb{Z}} z^{L_a}\, \mathbb{P}[\mathbf{L}_a = L_a|S = s].$$

The conditional independence (D) implies that the conditional portfolio PGF is given by a product over the portfolio (cf. Chapter 2)

$$G_{X|S}(z|s) = \prod_{A \in \mathbb{A}} G_{\mathbf{L}_A|S}(z|s).$$

Therefrom, the unconditional PGF is recovered by integration over the conditions, i.e.

$$G_X(z) = \int_{\mathbb{R}^N} \left[\prod_{A \in \mathcal{A}} G_{\mathbf{L}_A|S}(z|s) \right] d\mu_S(s). \qquad (4.4)$$

Credit portfolio modelling has thereby been reduced to a multiplication of polynomials and a subsequent risk factor integral.[4]

The integral (4.4) is explicitly solvable only in very special cases. CreditRisk$^+$ is one of them. Gordy was first to notice that also CreditMetrics can be formulated in terms of a PGF representation (4.4). Unfortunately, there the integral is not explicitly solvable and has to be solved by Monte Carlo simulation.

Please note that, at this stage, we have not yet specified the nature of the states that correspond to the losses. The set-up is sufficiently general to include anything from stochastic recovery rates to superfine sub-rating states. In particular, it enables us to include recovery rate correlations. Please note furthermore that we assume that there be sufficiently many factors to achieve complete conditional independence.

4.2.3 Binary state model

One of our basic assumptions is that only finitely many admissable values of $\mathbf{L}_A$ have a probability different from zero. The simplest case is a two-state model, for example the model defined by

1. $\mathbf{L}_A = 0$ in the case of non-default;
2. $\mathbf{L}_A = \nu_A$ in the case of default.

Here ν_A is a single number that is estimated in advance. In the present setting, the two-state model corresponds to a conditional probability of the special form[5]

$$G_{\mathbf{L}_A|S}^{(\nu_a)}(s) = \delta_{L_a,0}\mathbb{P}[\mathbf{L}_A = 0|S = s] + \delta_{L_a,\nu_a}\mathbb{P}[\mathbf{L}_A = \nu_A|S = s], \qquad (4.5)$$

where it is understood that the two probabilities sum up to one. Therefore, in the binary model

$$G_{\mathbf{L}_A|S}(z|s) = 1 + (z^{\nu_a} - 1) \, G_{\mathbf{L}_A|S}^{(\nu_a)}(s). \qquad (4.6)$$

Standard CreditRisk$^+$ is such a binary model. Standard CreditMetrics becomes a binary model in the special case of a rating system that distinguishes no further than between default and non-default.

---

[4] (4.4) is central in all credit risk factor models. The differences are mainly in how the product and the integral are actually computed.

[5] Here L_A is a variable, whereas ν_A is a constant. To save space, we will make use of the abbreviation $G_{\mathbf{L}_A|S}^{(L_A)}(s) = \mathbb{P}[\mathbf{L}_A = L_A|S = s]$.

4.2.4 Poisson approximation

The product in (4.4) is a computational hurdle. For instance, in the binary model the number of terms coming from (4.6) is two to the power of the number of obligors. There are different proposals how to deal with this product, one example being fast Fourier transformation. See also Hickman and Wollman [8] for an interesting alternative.

The Poisson approximation is an additional assumption that converts the product into a double sum by means of exponentiation. If for example the conditional default probability in (4.6) is small, then

$$G_{\mathbf{L}_A|S}\left(z|s\right) \approx \exp\left[(z^{\nu_a} - 1)\, G^{(\nu_a)}_{\mathbf{L}_A|S}(s)\right]$$

and the product in (4.4) becomes easy to compute. This trick applies whenever one state is much more likely than all others. To be precise,[6] we assume that

(F) for each obligor A and all values of the risk factors, the zero state[7] (neither loss nor gain) is more likely than all non-zero states (for typical values of s),

$$G^{(0)}_{\mathbf{L}_A|S}(s) \gg G^{(L_A)}_{\mathbf{L}_A|S}(s) \quad \Leftarrow \quad 0 \neq L_A. \qquad (4.7)$$

This assumption is stated explicitly in standard CreditRisk$^+$. Although it is not a necessary ingredient in standard CreditMetrics, in practice, the diagonal elements of the migration matrix are always much larger than the off-diagonal elements.

In the following, we count losses with respect to the most likely state, which we declare to be the zero state, and which most likely corresponds to the economical state where the obligor essentially remains unchanged. The probability of the zero state is

$$G^{(0)}_{\mathbf{L}_A|S}(s) = 1 - \sum_{L_A \in \mathbb{Z}\setminus\{0\}} G^{(L_A)}_{\mathbf{L}_A|S}(s)$$

so that

$$G_{\mathbf{L}_A|S}\left(z|s\right) = 1 + \sum_{L_A \in \mathbb{Z}\setminus\{0\}} (z^{L_A} - 1)\, G^{(L_A)}_{\mathbf{L}_A|S}(s) \qquad (4.8)$$

updates (4.6). As a consequence of assumption (F), we may substitute

$$G_{\mathbf{L}_A|S}\left(z|s\right) \approx \exp\left[\sum_{L_A \in \mathbb{Z}\setminus\{0\}} (z^{L_A} - 1)\, G^{(L_A)}_{\mathbf{L}_A|S}(s)\right]$$

in (4.4). The result is

---

[6] We leave aside what "typical value of x" and "$\gg$" precisely mean.

[7] The generalization to the situation when the most-likely state is non-zero is straightforward.

$$G_X(z) \approx \int_{\mathbb{R}^N} \exp\left[\sum_{A\in\mathcal{A}} \sum_{L_A\in\mathbb{Z}\backslash\{0\}} \left(z^{L_A}-1\right) G_{L_A|S}^{(L_A)}(s) \right] d\mu_S(s). \qquad (4.9)$$

For the remainder of these notes, we will restrict our attention to this Poisson approximation. (4.9) will serve as common denominator for the comparison of CreditRisk$^+$ and CreditMetrics. Consult Chapter 2 for further information about the Poisson approximation and how to interpret it.

4.3 Mounting the CreditRisk$^+$-Engine

Everything said so far is common to both CreditRisk$^+$ and CreditMetrics. At this point the two models branch into different assumptions.

To reassemble CreditRisk$^+$, we have to write down the model-specific expressions for the conditional probabilities and the risk factor distribution. CreditRisk$^+$ assumes that

(G) the risk factors are independently gamma-distributed, i.e.,

$$d\mu_S(s) = \prod_{n=1}^{N} \left\{ \frac{(s_n)^{\alpha_n-1}}{\Gamma(\alpha_n)\,\beta_n^{\alpha_n}} \exp\left(\frac{-s_n}{\beta_n}\right) ds_n \right\}$$

with mean $\mu_n = \alpha_n\beta_n$ and standard deviation $\sigma_n = \alpha_n\beta_n^2$;

(H) the conditional probabilities are approximately given by

$$G_{L_A|S}^{(L_A)}(s) \approx \theta_{A,0}^{(L_A)}\mu_0 + \sum_{n=1}^{D} \theta_{A,n}^{(L_A)}\, s_n \qquad (4.10)$$

for real non-negative numbers $\theta_{A,n}^{(L_A)}$, $n = 0,\dots,N$.

The constant term corresponds to a non-stochastic risk factor (singular limit of gamma distribution). It is customary to include it for the sake of generality. (4.10) can be viewed as a first-order Taylor formula for some exact non-linear conditional probabilities.

The extra index L_A in (4.10) is due to the fact that we do not restrict obligors to ending up in one of only two states[8] as is the case of standard CreditRisk$^+$. The multi-state model is easily reduced to the two-state model. Standard CreditRisk$^+$ assumes additionally that

(I) for each obligor A, the loss is either zero (in the case of non-default) or assumes a positive value ν_A (in the case of default):

$$\theta_{A,n}^{(L_A)} = w_{A,n}\, \delta_{L_A,\nu_A}$$

with factor loadings $w_{A,n}$.

---

[8] Recall that multiple states can be used to model both fluctuations of loss given default (post-default) and changes in value due to migration (pre-default).

As a consequence of (I), the L_A-sum in (4.9) shrinks to a single term for each address and we arrive at the model in Chapter 2. The original CreditRisk$^+$-model thus corresponds to one particular choice of $\theta_{A,n}^{(L_A)}$, $n = 0, \ldots, N$. At the expense of one more index, we will continue to use the more general model.

Solving the model has become a low-hanging fruit: insert (4.10) into (4.9). Then the exponent under the integral becomes

$$\sum_{a \in \mathcal{A}} \sum_{L_a \in \mathbb{Z} \backslash \{0\}} \left(z^{L_a} - 1\right) \left[\theta_{a,0}^{(L_a)} + \sum_{n=1}^{N} \theta_{a,n}^{(L_a)} s_n\right] = f_0(z) + \sum_{n=1}^{N} f_n(z) s_n,$$

and the s_n-integral becomes elementary. The result is the famous CreditRisk$^+$ PGF,

$$G_X^{\mathrm{CreditRisk}^+}(z) = \exp\left[f_0(z)\mu_0\right] \prod_{n=1}^{N} \left[1 - \beta_n f_n(z)\right]^{-\alpha_n}. \tag{4.11}$$

Interestingly, the final formula is again a product over the risk factors.[9] The moments of X are readily available as Taylor coefficients at $z = 1$. To illustrate what is different with multiple states and to prepare the ground for the comparison with CreditMetrics, the first moment is

$$\mathbb{E}[X] = \sum_{A \in \mathcal{A}} \sum_{L_A \in \mathbb{Z} \backslash \{0\}} L_A \sum_{n=0}^{N} \theta_{A,n}^{(L_A)} \mu_n,$$

while the second moment is

$$\mathbb{E}[X^2] = \sum_{A \in \mathcal{A}} \sum_{L_A \in \mathbb{Z} \backslash \{0\}} (L_A)^2 \sum_{n=0}^{N} \theta_{A,n}^{(L_A)} \mu_n + \sum_{n=1}^{D} \left(\sum_{A \in \mathcal{A}} \sum_{L_A = 1}^{\infty} L_A \theta_{A,n}^{(L_A)}\right)^2 \sigma_n^2.$$

4.4 Mounting the CreditMetrics Engine

To switch from CreditRisk$^+$ to discretized CreditMetrics, assumptions (G) and (H) of the previous section need to be exchanged. For CreditMetrics one assumes instead that

(G') the risk factors are multivariately normally distributed with mean zero and covariance C, i.e.

$$d\mu_S(s) = \det(2\pi C)^{-\frac{1}{2}} \exp\left\{-\frac{1}{2} \sum_{n,m=1}^{D} s_n (C^{-1})_{n,m} s_m\right\} \prod_{n=1}^{N} ds_n;$$

---

[9] One can interpret this replication as a superportfolio of pairwise uncorrelated subportfolios, where each replicate comes with a default probability that is reduced by the factor loadings and one has a one-factor-correlation model within each subportfolio.

(H′) the conditional probabilities are given by

$$G_{L_A|S}^{(L_A)}(s) = \phi\left[\left(Z_{A,L_A} - \psi_A\sum_{n=1}^{N}\theta_{A,n}s_n\right)\bigg/\kappa_A\right]$$

$$- \phi\left[\left(Z_{A,L_A+1} - \psi_A\sum_{n=1}^{D}\theta_{A,n}s_n\right)\bigg/\kappa_A\right],$$

where
(i) ϕ is the standard cumulative normal distribution function

$$\phi(z) = \frac{1}{\sqrt{2\pi}}\int_{-\infty}^{z}\exp\left(-\frac{r^2}{2}\right)dr;$$

(ii) $Z : \mathbb{A}\times\mathbb{Z}\to\mathbb{R}$ is a matrix of thresholds such that[10]

$$-\infty \leq \ldots \leq Z_{A,L_A+1} \leq Z_{A,L_A} \leq Z_{A,L_A-1} \leq \ldots \leq \infty; \qquad (4.12)$$

(iii) κ_A is a parameter between zero and one that is called the firm specific volatility;
(iv) $\theta_{A,n}$ is a matrix of factor loadings, normalized such that the sum over all risk factors is $\sum_{n=1}^{N}\theta_{A,n} = 1$;
(v) ψ_A is an abbreviation for the normalization

$$\psi_A = \sqrt{\frac{1-\kappa_A^2}{\sum_{n,m=1}^{D}\theta_{A,n}\,C_{n,m}\,\theta_{A,m}}}.$$

This is called a PROBIT representation for the loss distribution. Admittedly, it looks very different from the CreditRisk+model at first sight.

Ad (H′): A discontinuity of the threshold function $L \mapsto Z_{A,L}$ at $L = \lambda$ signals that obligor A has a potential state given by a loss λ that will occur with marginal (unconditional) probability $\phi(Z_{A,L}) - \phi(Z_{A,L+1})$. At this level of generality,

(i) obligors can have any number of different potential states with any probabilities;
(ii) different obligors can have different numbers of potential states with different sets of probabilities.

The restriction that there be only finitely many such states requires the threshold function to decrease from plus infinity for all losses below the lowest loss to minus infinity for all losses above the highest loss in a finite number of steps.

We should concede that the original CreditMetrics model is not formulated in terms of discretized losses, although, strictly speaking, machine precision sets a cutoff. Discretization is unnecessary if CreditMetrics is solved as in the standard version using Monte Carlo simulation instead of a PGF recursion.

---

[10] The larger losses correspond to thresholds that are further to the left.

4.4.1 Risk factor interpretation

Differing from CreditRisk$^+$, the CreditMetrics risk factors come with a specific interpretation as equity indices associated with country and industry sectors.

For each obligor, one assembles a normally distributed random variable with mean zero and variance one according to

$$\mathbf{R}_A = \psi_A \sum_{n=1}^{N} \theta_{A,n} S_n + \kappa_A \mathbf{Y}_A, \tag{4.13}$$

which represents the relative change of the value of all assets of a firm.[11] Here $\mathbf{Y}_A$ is an idiosyncratic normally distributed random variable with mean zero and variance one that is uncorrelated with all other random variables. It takes care of obligor-specific risk. The parameter κ_A denotes the obligor-specific volatility.

CreditMetrics is based on the picture that a firm defaults when the value of its assets falls below a default threshold. Suppose that the maximal loss caused by default is ν_A, then the threshold is set such that the unconditional probability of default is given by

$$\mathbb{P}[\mathbf{L}_A = \nu_A] = \mathbb{P}[\mathbf{R}_A \leq Z_{A,\nu_A}] = \phi(Z_{A,\nu_A}). \tag{4.14}$$

Since there can be no loss greater than the maximal loss, all previous thresholds Z_{A,L_A} with $L_A > \nu_A$ equal $-\infty$.

Iterating this scheme from loss to loss, one integrates the normal distribution from threshold to threshold and reserves an interval for the return variable with each loss such that the respective probability is the integral of the normal distribution over this interval.

The conditional probabilities follow from a change of variables according to (4.13). When the risk factors S are kept fixed, then we are left with the idiosyncratic random variables $\mathbf{Y}_A$. It is then clear that the conditional probabilities are given as in assumption (H').

As for CreditRisk$^+$, the simplest case is again a two-state model. Consider the general binary model, where obligor A causes loss ν_A in the case of default and zero in the case of no default. This corresponds to a threshold function

$$Z_{A,L} = \begin{cases} -\infty & L > \nu_A \\ \xi_A & \nu_A \geq L > 0 \\ +\infty & 0 \geq L \end{cases} . \tag{4.15}$$

The obligor-specific default thresholds ξ_A correspond to default probabilities $\phi(\xi_A)$.

---

[11] More generally, one interprets it as an aggregated fitness variable.

4.4.2 Reduction to rating states

In the rating universe, obligors within rating classes migrate, and in particular default with uniform probabilities. Compared with the PROBIT model with general threshold functions, the rating models form a small subspace. We will briefly describe this reduction, since

(i) the term CreditMetrics is commonly used in connection with ratings and migration matrices rather than general PROBIT models;

(ii) it is desirable to match CreditRisk$^+$ with a rating model rather than a general model with an abundance of additional parameters.

Therefore, let $\mathcal{R}_A^T \in \{0, 1, \ldots, r\}$ denote the rating of obligor A at time T so that 0 be default and S the highest ranking. Then

$$\mathbb{P}[\mathcal{R}_A^T = \rho_A | \mathcal{R}_A^0 = t_A] \ = \ M_{\rho_A | t_A}$$

with an obligor-independent migration matrix[12] $M_{\rho|t}$. The migration matrix has itself a PROBIT representation encoded in a (much smaller) threshold matrix $\zeta = (\zeta_{\rho|t})_{0 \leq \rho, t \leq S}$ which is defined by

$$\sum_{\rho \leq u} M_{\rho|t} \ = \ \phi\left(\zeta_{u|t}\right).$$

The reduction is based upon the picture that every obligor A

1. has a rating t_A today;
2. migrates to a future rating ρ_A with probability $M_{\rho_A | t_A}$;
3. thereby causes a loss l_{A, ρ_A}.

The loss given future rating depends on all kinds of data that we prefer to sweep under the carpet of a precomputation. We conclude that rating models correspond to particular threshold functions, namely

$$Z_{A,L} \ = \ \zeta_{\varphi_A(L) | \rho_A},$$

where the function

$$\varphi_A(L) \ = \ \sup\left\{s \in \{-1, 0, \ldots, r\} : l_{A,s} \geq L\right\}$$

is a kind of inverse of the loss given future rating. Here -1 is an auxiliary additional state corresponding to an infinite loss $l_{A,-1} = \infty$ that fortunately occurs with probability zero due to $\zeta_{-1|t} = -\infty$. It takes care of $Z_L = -\infty$ when $L > l_{A,0}$ exceeds the worst possible loss.

The two-state rating model has the same kind of threshold function as (4.15) but with uniform default thresholds ξ for all obligors. In practice, the two-state rating model is too crude. An intermediate approach consists of using a set of rating systems instead of a single one. The general model is recovered as the limit when every obligor comes with his own rating system.

---

[12] A discrete time migration matrix is a stochastic matrix with (i) $M_{s|t} \in [0, 1]$ (probability), (ii) $\sum_s M_{s|t} = 1$ (completeness), and (iii) $M_{s|0} = \delta_{s,0}$ (no return).

4.5 Transformations of Risk Factors

We have presented a general credit risk model that contains both CreditRisk$^+$ and CreditMetrics as special cases that appear rather different. The question remains whether it is possible to parametrize the two models such that they yield the same loss distribution.

As we mentioned in the introduction, we will not approach this problem directly, say, by matching the first few moments or some other characteristic numbers of the loss distribution. Instead, we will compare the models in terms of their factor distributions and conditional default probabilities. To do so, one first has to remove the ambiguity of the factor model representation.

We propose to gauge fix the factor distribution by means of the following scheme:

(i) Transform the risk factors into a normal form. Here we will choose independent normally distributed risk factors with mean zero and variance one as the normal form.[13]

(ii) Then compare the conditional probabilities.

The matching problem is now to find parametrizations such that the conditional probabilities of CreditRisk$^+$ and CreditMetrics in the normal form are as close as possible.

Our comparison of conditional probabilities will remain vague. To measure model distances, one should use a suitable metric on the space of conditional probabilities, perhaps with an emphasis on the tails.

Factor transformations have been considered previously in credit risk modelling by Koyluoglu and Hickman [10]; see also [8] by Hickman and Wollman, which attempts to build a better and more robust credit risk model using this methodology.

4.5.1 Probability-conserving transformations

For the sake of notational simplicity, we will restrict our attention to one-factor models. The generalization to multi-factor models is straightforward, at least concerning the basic notions. We will be rather casual regarding the mathematical assumptions underlying the existence of factor transformations. However, the idea should become quite clear.

Recall that the integral representation for the PGF of a general one-factor model is given by

$$G_X(z) = \int_{\mathbb{R}} \left[\prod_{A \in \mathbb{A}} G_{L_A|S}(z|s) \right] d\mu_S(s). \tag{4.16}$$

---

[13] This means that we will also have to transform CreditMetrics as we will have to rotate and scale the risk factors in order to get independent normally distributed variables.

This representation is non-unique. We can perform coordinate transformations of the following kind that leave the structure of (4.16) intact.

Let S and $\mathbf{Y}$ be two different real-valued random variables with continuous distributions and probability densities $d\mu_S(s)$ and $d\mu_{\mathbf{Y}}(y)$. Define a transformation $s = f(y)$ by

$$\int_{-\infty}^{s} d\mu_S(s') = \int_{-\infty}^{y} d\mu_{\mathbf{Y}}(y'). \tag{4.17}$$

This transformation makes S and $f(\mathbf{Y})$ agree on the σ-algebra of Borel sets in $\mathbb{R}$, i.e., equal in probability. Using (4.17), we can transform (4.16). The transformation formula for this change of variables is

$$G_X(z) = \int_{\mathbb{R}} \left[\prod_{A \in A} G_{\mathbf{L}_A | S}(z | f(y)) \right] d\mu_{\mathbf{Y}}(y). \tag{4.18}$$

Conditional independence is preserved. Indeed, (4.18) has exactly the same form as (4.16) with:

(i) the factor distribution $d\mu_S(s)$ replaced by $d\mu_{\mathbf{Y}}(y)$;
(ii) the conditional probabilities $G_{\mathbf{L}_A | S}(z | s)$ replaced by $G_{\mathbf{L}_A | S}(z | f(y))$.

This transformation not only leaves invariant the mean, standard deviation, and value-at-risk, but respects the whole loss distribution. Suppose that we have a second factor integral representation with respect to $\mathbf{Y}$. Then the crucial question is whether

$$G_{\mathbf{L}_A | S}(z | f(y)) = G_{\mathbf{L}_A | \mathbf{Y}}(z | y) \tag{4.19}$$

holds.

4.5.2 Application to the one-factor case

Factor transformations still apply when the random variables have different ranges. For instance, we can transform a normally distributed variable, whose range is $\mathbb{R}$, into a gamma-distributed variable, whose range is the half-line $\mathbb{R}^+$, as in the case for CreditRisk$^+$ and CreditMetrics. To fix the sides, we identify the bad loss ends of the distributions. Let us do this for the simplest case of a homogeneous portfolio with N obligors and:

1. one factor;
2. two states of rating type;
3. one band.

All obligors in this portfolio default with the same probability and cause a unit loss when they default. The general matching principle is to match the conditional probabilities in normal form individually for each obligor. In the homogeneous case, these are simply all the same.

Homogeneous CreditRisk$^+$

With all the above simplifications, (4.10) simply becomes s. Inserted into (4.5), the conditional probabilities in CreditRisk$^+$ read

$$(1 - \delta_{L_a,0}) \; G_{L_A|S}^{(L_a)}(s) \; = \; \delta_{L_a,1} \; s.$$

In the Poisson approximation, the conditional PGF (4.6) is therefore given by

$$G_{L_A|S}(z|s) \approx \exp\left[(z-1)\,s\right]. \tag{4.20}$$

Notice that it is independent of A in our simple case. Setting the factor distribution to

$$d\mu_S(s) = \frac{s^\alpha}{\Gamma(\alpha)\beta^\alpha} \, \exp\left(-\frac{s}{\beta}\right) \, ds$$

the probability-generating function becomes

$$G_X(z) \; = \; \frac{1}{\Gamma(\alpha)\beta^\alpha} \int_0^\infty s^\alpha \, \exp\left[N\,(z-1)\,s - \frac{s}{\beta}\right] \, ds.$$

This integral is trivial. It yields

$$G_X(z) \; = \; \left[1 - \beta\,N\,(z-1)\right]^{-\alpha}.$$

The factor mean is $\alpha\beta$, the factor standard deviation is $\alpha\beta^2$.

Homogeneous CreditMetrics

In the homogeneous two-state model, a firm defaults if the fitness variable

$$\mathbf{R} = \kappa\mathbf{Y} + \sqrt{1-\kappa^2}S$$

falls below a threshold Z such that $\phi(Z) = p$ is the average default probability. Therefore, we find

$$(1 - \delta_{L_a,0}) \; G_{L_A|S}^{(L_a)}(s) = \delta_{L_a,1} \; \phi\left(\frac{Z - \sqrt{1-\kappa^2}\,s}{\kappa}\right)$$

and in Poisson approximation

$$G_{L_A|S}(z|s) \approx \exp\left[(z-1)\,\phi\left(\frac{Z - \sqrt{1-\kappa^2}\,s}{\kappa}\right)\right]. \tag{4.21}$$

The conditional probability is again independent of A. The factor distribution is here

$$d\mu_S(s) = \frac{1}{\sqrt{2\pi}} \, \exp\left(-\frac{s^2}{2}\right) \, ds$$

wherefore

$$G_X(z) = \frac{1}{\sqrt{2\pi}} \int_{-\infty}^{\infty} \exp\left[N(z-1)\phi\left(\frac{Z - \sqrt{1-\kappa^2}s}{\kappa} \right) - \frac{s^2}{2} \right] ds.$$

For the same portfolio, we have two rather different-looking formulas. Let us recall that the distribution parameters are (α, β) in CreditRisk$^+$ and (Z, κ) in CreditMetrics. Therefore, we need two equations to relate the parameters to each other. As one equation, we choose to match the means:

$$\alpha\beta = \mu = \phi(Z).$$

One way to obtain another equation that relates σ and κ is to match standard deviations. We leave this to the reader. Instead, we will here match the conditional probabilities and verify that the standard deviations are approximately equal. This will turn out to be an excellent matching condition.

Transformation for the homogeneous models

We identify the limit $y \to \infty$ in CreditMetrics with the point $x = 0$ in CreditRisk$^+$. The transformation is thus defined by

$$\frac{1}{\Gamma(\alpha)\beta^\alpha} \int_0^x (s')^{\alpha-1} \exp\left(-\frac{s'}{\beta}\right) ds' = \frac{1}{\sqrt{2\pi}} \int_y^\infty \exp\left(-\frac{(y')^2}{2}\right) dy'. \qquad (4.22)$$

The transformation from a gamma-distributed into a Gaussian-distributed risk factor therefore reads

$$s = \psi_{\alpha,\beta}^{-1}(\phi(-y)) \qquad (4.23)$$

where $\psi_{\alpha,\beta}(s)$ denotes the probability function of the gamma distribution, i.e. the left-hand side of (4.22).

If we apply the matching principle (4.19) to the conditional probabilities (4.20) and (4.21) using the factor transformation (4.23), we find the goal

$$\psi_{\alpha,\beta}^{-1}(\phi(-s)) \approx \phi\left(\frac{Z - \sqrt{1-\kappa^2}\,s}{\kappa} \right).$$

The crucial question now is whether there exists a parameter mapping $(\alpha, \beta) \to (Z, \kappa)$ so that this goal is satisfied.

One-point match

Transform the gamma-distributed risk factor into a normally distributed risk factor. The conditional default probability of CreditRisk$^+$ becomes

$$G_{L_A|S}^{\text{CreditRisk}^+}\left(z\big|\psi_{\alpha,\beta}^{-1}(\phi(s))\right) \approx \exp\left[(z-1)\,\psi_{\alpha,\beta}^{-1}(\phi(-s))\right] \qquad (4.24)$$

which should be compared with

$$G_{L_A|S}^{\text{CreditMetrics}}(z|s) \approx \exp\left[(z-1)\,\phi\left(\frac{Z-\sqrt{1-\kappa^2}\,s}{\kappa}\right)\right]. \qquad (4.25)$$

There is no exact symmetry between the two models as the two conditional probabilities[14] (4.24) and (4.25) are different functions of s.

Is there at least an approximate symmetry? Recall that the risk factor S is normally distributed with mean zero and variance one. We match (4.24) and (4.25) at the point $s = 0$ (mean, median, and maximum of the factor distribution) requiring that[15]

$$\frac{1}{2} = \psi_{\alpha,\beta}\left[\phi\left(\frac{Z}{\kappa}\right)\right].$$

This equation can be solved numerically.[16] Then we test whether (4.24) and (4.25) become approximately equal and yield approximately equal portfolio distributions.

Numerical results

We have carried out the above for default rates ranging from one to four percent and default rate standard deviations ranging from one to four times the default rate itself. The computations were done using computer algebra and the numerical integration routines in Maple 7 [12].

Table 4.1 shows the numerical results for a homogeneous portfolio with 500 obligors. Apart from μ, σ, in both models we display the obligor-specific volatility κ, and

(a) the portfolio mean $M = \mathbb{E}[X]$;
(b) the portfolio standard deviation $\mathcal{S}$;
(c) the probability P of a loss larger than $M + 2\mathcal{S}$ (with the $s\mathcal{S}$-values taken from CreditRisk$^+$).

---

[14] A proper way to proceed is to minimize a suitable distance of conditional probabilities $\|G_{L|S}^{\text{CreditRisk}^+} - G_{L|S}^{\text{CreditMetrics}}\|$, optionally incorporating weight functions for the variables (or parameters) A, z, and s.

[15] In the homogeneous case, we have only one equation to solve. Otherwise, we have one equation for every obligor A. The simplest CreditRisk$^+$ variant with varying default probabilities is a two-factor model with one non-stochastic idiosyncratic factor. It can be matched with a general two-state CreditMetrics model. We do not write the matching equations here as they do not lead to any new insights.

[16] Here $\psi_{\alpha,\beta}(s) = 1 - \frac{\Gamma\left(\alpha,\frac{s}{\beta}\right)}{\Gamma(\alpha)}$ and $\phi(x) = \frac{1}{2}\left[1 + \operatorname{erf}\left(\frac{x}{\sqrt{2}}\right)\right]$.

Notice that the asset correlation is $\sqrt{1 - \kappa^2}$. For $\sigma = \mu$, the asset correlation is surprisingly low. It is much lower than typical asset correlations one would expect for instance from real-world historical data.

500 obligor one-factor models				
μ	σ	M_{CR}	S_{CR}	P_{CR}
κ	$[M + 2S]_{CR}$	M_{CM}	S_{CM}	P_{CM}
0.01000	0.01000	5.0	5.477	0.04507
0.9454	16.0	5.0	5.479	0.04227
0.01000	0.02000	5.0	10.25	0.04722
0.7965	25.0	5.0	12.91	0.04624
0.01000	0.03000	5.0	15.17	0.03885
0.6287	35.0	5.0	21.38	0.03583
0.01000	0.04000	5.0	20.12	0.03156
0.4972	45.0	5.0	27.93	0.02773
0.02000	0.02000	10.0	10.49	0.04736
0.9330	31.0	10.0	10.43	0.04471
0.02000	0.04000	10.0	20.25	0.04724
0.7614	50.0	10.0	23.62	0.04807
0.02000	0.06000	10.0	30.17	0.03897
0.5834	70.0	10.0	36.07	0.03968
0.02000	0.08000	10.0	40.12	0.03165
0.4529	90.0	10.0	44.62	0.03289
0.03000	0.03000	15.0	15.49	0.04816
0.9230	46.0	15.0	15.34	0.04578
0.03000	0.06000	15.0	30.25	0.04724
0.7348	75.0	15.0	33.38	0.04937
0.03000	0.09000	15.0	45.17	0.03902
0.5513	105.0	15.0	48.54	0.04246
0.03000	0.1200	15.0	60.12	0.03169
0.4228	135.0	15.0	58.23	0.04191
0.04000	0.04000	20.0	20.49	0.04856
0.9140	61.0	20.0	20.21	0.04648
0.04000	0.08000	20.0	40.25	0.04724
0.7123	100.0	20.0	42.49	0.05041

Table 4.1. CreditRisk$^+$ (CR) vs. CreditMetrics (CM)

Figures 4.1 and 4.2 show full loss distributions at a two percent average default probability. The thick line is from CreditRisk$^+$, the thin line from CreditMetrics. For $\sigma \geq \mu$, the CreditRisk$^+$-distribution becomes monotonically decreasing. Figure 4.3 compares the conditional default probabilities themselves. For numerical reasons, we show the transformed functions $\psi_{\alpha,\beta}\left[\frac{Z - \sqrt{1-\kappa^2}\,s}{\kappa}\right]$ and $\phi(-s)$.

All our results point to a very good match between the two models. We conclude that, apart from approximation errors, e.g., from using conditional

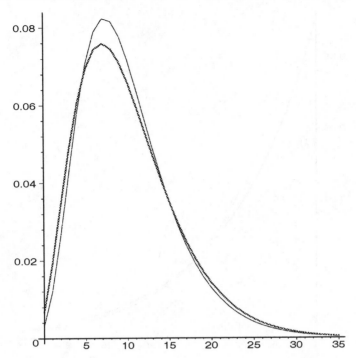

Fig. 4.1. Loss distributions for $\mu = 0.02$, $\sigma = 0.010$, $\kappa = 0.983192731117$

probabilities that may exceed one, the two models are identical in this simple case. A unified point of view of CreditRisk$^+$ and CreditMetrics is thus not only possible but very natural. The practical mapping of both models certainly requires more work than presented here, but the procedural path of the mapping should have become clearer.

4.6 Conclusions

CreditRisk$^+$ and CreditMetrics both belong to a general class of credit risk models that can be described as factor models with conditionally independent probabilities and discretized losses.

There is a large symmetry group of factor transformations acting on this model space. The same model, i.e. with the same loss distribution, may be represented by very different factor distributions and conditional probabilities. It is the product of both that matters.

Factor transformations can be used to match the parameters in a model that is differently represented. This opens a way to check whether one has parametrized the model consistently. This is important in particular since there is presently no way to parametrize CreditRisk$^+$ or CreditMetrics directly by means of, e.g., maximum likelihood methods.

66   Christian Wieczerkowski

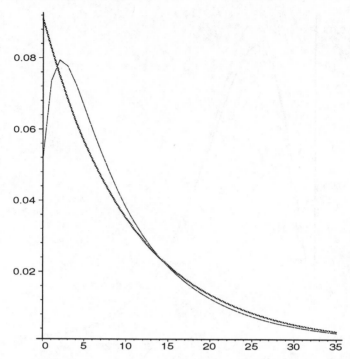

Fig. 4.2. Loss distributions for $\mu = 0.02$, $\sigma = 0.020$, $\kappa = 0.933037388209$

For a realistic credit portfolio, the factor transformation between CreditRisk[+] and CreditMetrics is rather complicated and unlikely to be solved analytically. Standard parameter values in CreditRisk[+] correspond to rather low asset correlations in CreditMetrics. From these results, the two models look very comparable. The uncertainty is in the parametrization rather than in the model assumptions.

Among the ends that are left open in this investigation, we mention

- the development of parametrization schemes for the general credit risk model that incorporate, for instance, severity variations;
- the matching and comparison of realistic portfolios in multi-factor multi-state CreditRisk[+] and CreditMetrics;
- the investigation of suitable metrics on the space of conditional event probabilities;
- the matching in the saddlepoint approximation, the relationship with expansions, and the asymptotic tail distribution.

It is often said that the main difference between CreditRisk[+] and CreditMetrics is that the former is a binary (default state) model, whereas the latter is a mark-to-market multi-state model. Although these attributes are certainly true for the standard implementations, this statement is in our opinion misleading as both models occur in both phenotypes.

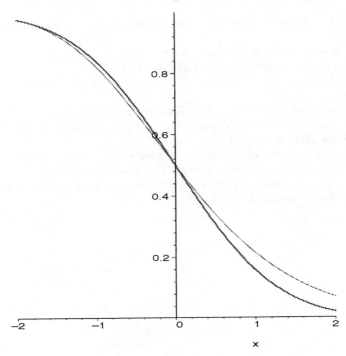

Fig. 4.3. Conditional default probabilities for $\mu = 0.02$, $\sigma = 0.020$, $\kappa = 0.933037388209$

References

1. F. Bröker. *Quantifizierung von Kreditportfoliorisiken.* Fritz Knapp Verlag, Frankfurt am Main, 2000.
2. Credit Suisse Financial Products. CreditRisk⁺: A credit risk management framework. London, 1997. Available at http://www.csfb.com/creditrisk.
3. M. Crouhy, D. Galei, and R. Mark. A comparative analysis of current credit risk models. *Journal of Banking and Finance,* 24:57–117, 2000.
4. D. Diaz-Ledezma and G. Genmill. A systematic comparison of two approaches to measuring credit risk: CreditMetrics versus CreditRisk+. *Trans. 27th ICA,* 2002.
5. C. Finger. Sticks and stones. Working Paper, RiskMetrics Group, 1999.
6. M. B. Gordy. A comparative anatomy of credit risk models. *Journal of Banking & Finance,* 24:119–149, 2000.
7. G. M. Gupton, C. C. Finger, and M. Bhatia. CreditMetrics – Technical Document. J. P. Morgan, New York, 1997.
   http://www.riskmetrics.com/creditdocs.html
8. A. Hickman and J. Wollman. An evolutionary leap in credit portfolio risk modelling. *ERisk Report,* 2002.
9. M. Kern and B. Rudolph. Comparative analysis of alternative credit risk models – an application on German middle market loan portfolios. CFS

Working Paper No. 2001/03, Center of Financial Studies, Frankfurt am Main, 2001.

10. U. Koyluoglu and A. Hickman. A generalized framework for credit risk portfolio models. Working Paper, OWC and CSFP, 1998.

11. H. Koyluoglu and A. Hickman. Reconcilable differences. *Risk* 11(10):56–62, 1998.

12. Maple 7 documentation. http://www.waterloomaple.com

13. M. Nyfeler. Modelling dependencies in credit risk management. Diploma Thesis, ETH Zürich, 2000.

14. M. Wahrenburg and S. Niethen. Vergleichende Analyse alternativer Kreditrisikomodelle. *Kredit und Kapital*, Heft 2/2000:235–257, 2000.

15. B. Zeng, J. Zhang. An empirical assessment of asset correlation models. Working Paper, KMV LLC, 2001.

Numerically Stable Computation of CreditRisk$^+$

Hermann Haaf, Oliver Reiß* and John Schoenmakers

Summary. We present an alternative numerical recursion scheme for CreditRisk$^+$, equivalent to an algorithm recently proposed by Giese, based on well-known expansions of the logarithm and the exponential of a power series. We show that it is advantageous to the Panjer recursion advocated in the original CreditRisk$^+$ document, in that it is numerically stable. The crucial stability arguments are explained in detail. Furthermore, the computational complexity of the resulting algorithm is stated.

5.1 Resume of the Classical CreditRisk$^+$ Model

We assume familiarity with the basic principles of CreditRisk$^+$, restrict ourselves to a concise resume and refer for a more detailed description to Chapter 2, from where we adapt the notation, or to [1, 4, 8]. We recall that we are considering a portfolio of K obligors that are denoted by A and that exhibit one-period probabilities of default p_A. These are affected by N risk factors S_k (corresponding to sectors $\mathcal{S}_k$). The impact of the risk factors is given by factor loadings $w_{Ak} \geq 0$ such that $\sum_{k=1}^{N} w_{Ak} \leq 1$, and $w_{A0} = 1 - \sum_{k=1}^{N} w_{Ak}$ describes the idiosyncratic risk affiliation of obligor A.

In addition to these notations we introduce the polynomial degree M at which we truncate the power series expansion of the probability-generating function (PGF) and call it the *degree of truncation*.

According to Chapter 2, the aggregate portfolio loss in CreditRisk$^+$ as a multiple of the basic loss unit L_0 is given by

$$X = \sum_A \nu_A D_A$$

* Supported by the DFG Research Center "Mathematics for key technologies" (FZT 86) in Berlin.

with ν_A denoting integer-valued multiplicities of L_0 corresponding to the A-th obligor and D_A being Poisson-distributed random variables with stochastic intensities

$$p_A^S = p_A \left(w_{A0} + \sum_{k=1}^{N} w_{Ak} S_k \right),$$

conditional on independent gamma-distributed random variables

$$S = (S_1, \ldots, S_N)$$

with parameters $\mathbb{E}[S_k] = 1$ and $\sigma_k^2 := \mathrm{var}(S_k)$, $(k = 1, \ldots, N)$. Note that

$$\mathbb{E}[D_A] = \mathbb{E}[p_A^S] = p_A \quad \text{for all } A.$$

Recall from (2.20) that the PGF of the CreditRisk$^+$ model $G_X(z) = \mathbb{E}[z^X]$ can be expressed in closed analytical form

$$G(z) = \exp\left(\sum_A w_{A0}\, p_A \left(z^{\nu_A} - 1 \right) \right.$$

$$\left. - \sum_{k=1}^{N} \frac{1}{\sigma_k^2} \ln\left[1 - \sigma_k^2 \sum_A w_{Ak}\, p_A \left(z^{\nu_A} - 1 \right) \right] \right), \qquad (5.1)$$

with $G := G_X$ and z being a formal variable. On the other hand, from the definition of the PGF of a discrete, integer-valued random variable, we know that G may also be represented as

$$G(z) = \sum_{j=0}^{\infty} P[X = j]\, z^j. \qquad (5.2)$$

The central problem in this chapter is the efficient and numerically stable computation of the probabilities $P[X = j]$ in (5.2) from (5.1).

5.2 Panjer Recursion

It is known that the algorithm advocated in the original CreditRisk$^+$ document in order to obtain the probabilities $p_n := P[X = n]$, the Panjer recursion scheme, is numerically unstable. The Panjer recursion is derived by requiring that the log-derivative of G is a rational function of the form $\frac{A(z)}{B(z)}$, with polynomials A and B. Its numerical instability arises from an accumulation of numerical roundoff errors, which is nicely explained in Gordy [6] and is connected with the summation of numbers of similar magnitude but opposite sign, as both the polynomials A and B contain coefficients of both signs. The reasons for the numerical instability are unravelled in the next section.

Several remedies have been offered in order to avoid the instability of the Panjer recursion. Amongst others we mention the saddlepoint approximations to the tail of the loss distribution proposed by Gordy [6], see Chapter 7, and Martin, Thompson and Browne [9], constituting an asymptotic result specific to the chosen quantile.

5.3 Propagation of Numerical Roundoff Errors

We explain the numerical stability and instability of an algorithm in terms of the propagation of numerical roundoff errors of the elementary (arithmetic) operations.

Recall that the relative error ε_{x+y} of the addition operation is given by

$$\varepsilon_{x+y} = \frac{x}{x+y}\varepsilon_x + \frac{y}{x+y}\varepsilon_y \quad \text{if} \quad x+y \neq 0,$$

in terms of the relative errors ε_x and ε_y of its arguments x and y, respectively.

If the summands x and y are of the same sign, we have that $|\varepsilon_{x+y}| \leq \max\{|\varepsilon_x|, |\varepsilon_y|\}$. On the other hand, if the arguments of the addition are of opposite sign, at least one of the terms $|\frac{x}{x+y}|$, $|\frac{y}{x+y}|$ is greater than 1 and hence at least one of the coefficients of the relative errors ε_x or ε_y gets amplified. This amplification becomes particularly big if $|x| \approx |y|$, and so a cancellation in the denominator term $x+y$ occurs, leading to an explosion of the relative error ε_{x+y}.

From the above it is clear that the error propagation of the addition of two numbers of equal sign can be considered as harmless, leading even under repeated application to no amplification of the original error terms.

On the other hand, if under repeated summation (e.g., in a recursive algorithm) there is a single constellation of summands that are of similar magnitude, but opposite sign, cancellation effects will occur leading at least to spurious results, if not to a complete termination of the algorithm.

Furthermore, the relative error of a multiplication $x \cdot y$ is approximately given by

$$\varepsilon_{x \cdot y} \approx \varepsilon_x + \varepsilon_y,$$

i.e. the relative errors of the arguments simply add up. Therefore, we conclude that a recursive algorithm relying exclusively on the summation and multiplication of non-negative numbers can be considered as numerically stable.

We refer to standard textbooks on numerical analysis, e.g. Stoer and Bulirsch [11], for more details on the subject.

5.4 Numerically Stable Expansion of the PGF

We introduce the portfolio polynomial of the k-th sector to be

$$Q_k(z) := \sum_A w_{Ak}\, p_A\, z^{\nu_A}, \quad k \in \{0, \ldots, N\}.$$

For further analysis, it is important to note that the coefficients of Q_k are all non-negative. In terms of Q_k, G can be re-expressed as

$$G(z) = \exp\left[-Q_0(1) + Q_0(z) - \sum_{k=1}^{N} \frac{1}{\sigma_k^2} \ln\left(1 + \sigma_k^2 Q_k(1) - \sigma_k^2 Q_k(z)\right)\right]. \quad (5.3)$$

Observe that (5.2) can be interpreted as the power series representation of the analytical representation of G around $z = 0$, having a radius of convergence R strictly greater than 1, see Haaf and Tasche [7]. Therefore, it is natural to calculate the coefficients, i.e. the probabilities p_n, directly, by applying standard algorithms for the logarithm and exponential of power series, which can be found in the analysis and mathematical physics literature, see e.g. Brent and Kung [3] and the references therein. We systematically derive a method for calculating the coefficients of the power series expansion of (5.3) and present a two-step recursive scheme, where the sign structures of the coefficients involved are such that numerical stability of the two steps is ensured by two lemmas. For the convenience of the reader we provide detailed proofs of both lemmas. In fact, a basically equivalent recursion algorithm in this spirit was previously suggested by Giese [5], see also Chapter 6. However, in [5] the numerical stability is not analysed.

Thus, we firstly look at the power series expansion of the logarithm of a power series.[1] Secondly, having gained information about the sign structure of the coefficients of the resulting series, we investigate in a further step the power series expansion of its exponential.

We will show that the coefficients of the power series of $G(z)$ can be computed in a numerically stable way by this method. In particular, using Lemma 1 and Lemma 2, it will be shown that the stability follows from the particular sign structure of the polynomials under consideration. In fact, in the crucial operations of the recursion scheme, only non-negative terms are summed.

Lemma 1 (Expansion of the logarithm). *Consider a sequence $(a_k)_{k\geq 0}$ with $a_0 > 0$, $a_k \geq 0$ for all $k \geq 1$ and the function $g(z) := -\ln(a_0 - f(z))$, where $f(z) := \sum_{k=1}^{\infty} a_k z^k$. Let us assume that f has a positive convergence radius, so that g is analytic in a disc $\{z : |z| < R\}$ for some $R > 0$ and thus can be expanded as $g(z) =: \sum_{k=0}^{\infty} b_k z^k$ on this disc. Then, for the coefficients of*

---

[1] We present this in a slightly more general context; for a mere application to the right-hand side of (5.3) it would have been sufficient to consider the logarithm of a polynomial rather than of an (infinite) power series. However, if stochastic severities in the sense of Chapter 3 are introduced, arbitrary high exposure might be realized, leading naturally to the infinite power series formulation.

g we have $b_k \geq 0$ for $k \geq 1$ and their computation by means of the following recursively defined sequence[2]

$$b_0 = \ln(a_0),$$

$$b_k = \frac{1}{a_0} \left[a_k + \frac{1}{k} \sum_{q=1}^{k-1} q \, b_q \, a_{k-q} \right] \qquad for \quad k \geq 1 \qquad (5.4)$$

is numerically stable.

Proof. Note that $g'(z) = f'(z)/(a_0 - f(z))$, hence

$$\left(a_0 - \sum_{k=1}^{\infty} a_k \, z^k \right) \sum_{k=0}^{\infty} (k+1) b_{k+1} \, z^k = \sum_{k=0}^{\infty} (k+1) \, a_{k+1} \, z^k.$$

Performing the Cauchy product of the power series on the left-hand side of the preceding equation and comparing coefficients, it follows that $(b_k)_{k \geq 0}$ is given by (5.4) for $k \geq 1$. Substituting $z = 0$ gives $g(0) = \ln(a_0)$.

From the assumptions on the sequence (a_k) it follows by (5.4) that $b_k \geq 0$ for $k \geq 1$. So the recursive computation of $(b_k)_{k \geq 0}$ by (5.4) is numerically stable, as exclusively sums of non-negative terms are involved.   □

Lemma 2 (The exponential of a power series). *Let* $f(z) = \sum_{k=0}^{\infty} a_k z^k$ *and*

$g(z) := \exp(f(z)) = \sum_{n=0}^{\infty} b_n z^n$ *in a disc* $\{z : |z| < R\}$ *for some* $R > 0$. *Then*

$$b_0 = \exp(a_0),$$

$$b_n = \sum_{k=1}^{n} \frac{k}{n} b_{n-k} a_k \qquad for \quad n \geq 1. \qquad (5.5)$$

Moreover, the recursion (5.5) is numerically stable, if the coefficients of f *satisfy* $a_k \geq 0$ *for* $k \geq 1$.

Proof. The relation $b_0 = \exp(a_0)$ follows by substituting $z = 0$. For the j-th derivative we have

$$f^{(j)}(0) = j! \, a_j \qquad and \qquad g^{(j)}(0) = j! \, b_j. \qquad (5.6)$$

On the other hand, for $n \geq 1$ one obtains

$$g^{(n)}(z) = \frac{d^n}{dz^n} \exp(f(z)) = \left(\frac{d}{dz} \right)^{n-1} [g(z) \cdot f'(z)].$$

Hence by Leibniz's rule for the higher derivative of a product

---

[2] As usual, an empty sum, if $k = 1$, is defined to be zero.

$$g^{(n)}(z) = \sum_{k=0}^{n-1} \binom{n-1}{k} f^{(k+1)}(z) g^{(n-(k+1))}(z) \qquad (5.7)$$

holds. Then (5.5) follows straightforwardly by substituting $z = 0$ in (5.7) and using (5.6). Finally, the stability assertion is clear, since from $a_k \geq 0$ for $k \geq 1$ and $b_0 > 0$, it follows that $b_n \geq 0$ and so in (5.5) only positive terms are involved. $\qquad\square$

Remark 1. In fact, the results in Lemma 1 and Lemma 2 may be derived from one another. However, in order to clearly reveal the sign structures of the power series involved and their impact on numerical stability, we have chosen to treat them separately.

The Algorithm

Let M be a pre-specified order. Define for $k = 1, \ldots, N$,

$$a_0^{(k)} := 1 + \sigma_k^2 Q_k(1),$$
$$a_j^{(k)} := \sigma_k^2 \sum_A w_{Ak} p_A \mathbf{1}_{\{\nu_A = j\}}, \quad (j = 1, \ldots, M),$$

hence $1 + \sigma_k^2 Q_k(1) - \sigma_k^2 Q_k(z) = a_0^{(k)} - \sum_{j=1}^M a_j^{(k)} z^j + \mathcal{O}(z^{M+1})$. We then compute for each $k = 1, \ldots, N$ the truncated series expansion

$$-\ln\left(1 + \sigma_k^2 Q_k(1) - \sigma_k^2 Q_k(z)\right) = \sum_{j=0}^M b_j^{(k)} z^j + \mathcal{O}(z^{M+1}),$$

by means of the procedure defined in Lemma 1

$$b_0^{(k)} = \ln(a_0^{(k)}) \quad \text{and} \quad b_j^{(k)} = \frac{1}{a_0^{(k)}}\left[a_j^{(k)} + \frac{1}{j}\sum_{l=1}^{j-1} l b_l^{(k)} a_{j-l}^{(k)}\right],$$

for $j = 1, \ldots, M$. Hence,

$$\ln(G(z)) = -Q_0(1) + Q_0(z) + \sum_{k=1}^N \frac{1}{\sigma_k^2}\sum_{j=0}^M b_j^{(k)} z^j + \mathcal{O}(z^{M+1}) =: \sum_{j=0}^M \alpha_j z^j + \mathcal{O}(z^{M+1})$$

with

$$\alpha_0 = -Q_0(1) + \sum_{k=1}^N \frac{1}{\sigma_k^2} b_0^{(k)}, \quad \alpha_j = \sum_A w_{A,0} p_A \mathbf{1}_{\{\nu_A = j\}} + \sum_{k=1}^N \frac{1}{\sigma_k^2} b_j^{(k)}.$$

Note that Lemma 1 guarantees that $\alpha_j \geq 0$ for $j \geq 1$. In the last step we apply Lemma 2. We recursively compute by Lemma 2 the coefficients β_j of the expansion

$$G(z) = \sum_{j=0}^{M} \beta_j z^j + \mathcal{O}(z^{M+1})$$

by

$$\beta_0 = \exp(\alpha_0) \quad \text{and} \quad \beta_j = \sum_{l=1}^{j} \frac{l}{j} \alpha_l \beta_{j-l}.$$

The numerical stability of the algorithm follows from Lemma 1 and Lemma 2, due to the sign structure of the coefficients $a_j^{(k)}$ and α_j, respectively. Note that the coefficients $\beta_j = P[X = j]$ correspond to loss probabilities and are exact up to $j = M$.

Remark 2. A conservative upper bound for M, in the absence of multiple defaults, constitutes $\sum_A \nu_A$, corresponding to the case that each loan in the entire portfolio defaults. In practice, however, one is usually interested in the loss distribution up to some pre-specified upper tail quantile with (small) probability α. For this, one may run the above algorithm for $M = 1, 2, \ldots$, and stop as soon as $\sum_{j=0}^{M} P[X = j] > 1 - \alpha$. More generally, when for varying portfolio sizes K the corresponding credit portfolios have a homogeneous structure in the sense that $1 \leq \nu_A \leq \nu_{\max}$ and $0 < p_{\min} \leq p_A \leq p_{\max}$ for all A with $p_{\min}, p_{\max}$ and $\nu_{\max}$ being independent of the portfolio size K, one can show that for $K \to \infty$ the respective portfolios are almost gamma distributed (Bürgisser, Kurth and Wagner [2]) and, as a consequence, that $M = \mathcal{O}(K)$, where M is the required order for reaching a fixed tail quantile with probability α.

5.5 Conclusion

We conclude that the calculation of the coefficients of the power series representation of G gives rise to a numerically stable algorithm. The computational complexity is obtained as follows: each series expansion due to Lemma 1 or Lemma 2 requires a computation time $\frac{1}{2} M^2 (\text{op}_+ + 2\text{op}_\times)$, where op_+ denotes the time needed for an addition and $\text{op}_\times$ the time for a multiplication.[3] These expansions have to be performed $N + 1$ times, so that the total time required for the algorithm is given by

$$\frac{1}{2}(N + 1)M^2(\text{op}_+ + 2\text{op}_\times) + \mathcal{O}(KN + NM)\max(\text{op}_+, \text{op}_\times).$$

As a consequence, the loss distribution of CreditRisk$^+$ in the standard setting can be determined fast and reliably. Therefore, the method presented

---

[3] Note that on a modern PC the time for a multiplication is roughly comparable to the time for an addition.

for accurately determining the CreditRisk$^+$ loss distribution or a pre-assigned quantile of it, is hard to beat using any other technique.[4]

For generalizations of CreditRisk$^+$-type models we refer to the work of Reiß [10], which can be found in Chapters 8 and 13 and in which Fourier inversion techniques are consequently applied, allowing more freedom in the modelling. In addition, there is no need to introduce a basic loss unit L_0 anymore. In fact, for practical purposes we essentially obtain using fast Fourier transformation (FFT) techniques the loss distribution on a continuous scale.

Of course, the Fourier inversion algorithm can also be applied to the standard CreditRisk$^+$ model. The computational effort of the Fourier inversion algorithm based on n sample points (corresponding to a pre-assigned numerical accuracy) is given by

$$\mathcal{O}(KNn) + \mathcal{O}(n \log n).$$

Hence for fixed numbers of sample points n and sectors N the Fourier method is of order $\mathcal{O}(K)$. On the other hand, the computational effort of the algorithm presented in this chapter is of order $\mathcal{O}(K^2)$ for homogeneously structured portfolios, since for such portfolios M ought to be chosen of order $\mathcal{O}(K)$, see Remark 2. Hence the Fourier method is faster for very large and homogeneous credit portfolios. Conversely, the series expansion of the PGF presented is usually faster for smaller portfolios or portfolios with a relatively small amount of large exposures.

Acknowledgements. This chapter was developed in co-operation with the bmbf project "Effiziente Methoden zur Bestimmung von Risikomaßen", which is supported by Bankgesellschaft Berlin AG. The authors thank D. Tasche for helpful suggestions.

References

1. C. Bluhm, L. Overbeck, and C. Wagner. *An Introduction to Credit Risk Modeling*, CRC Press, Boca Raton, 2002.
2. P. Bürgisser, A. Kurth, and A. Wagner. Incorporating severity variations into credit risk. *Journal of Risk*, 3(4):5–31, 2001.
3. R.P. Brent and H.T. Kung. Fast algorithms for manipulating formal power series. *J. ACM*, 25:581–595, 1978.
4. Credit Suisse Financial Products. CreditRisk$^+$: A credit risk management framework. London, 1997. Available at http://www.csfb.com/creditrisk.

---

[4] Of course, the saddlepoint approximation [6, 9] still keeps its importance with a view towards modifications of CreditRisk$^+$, particularly with regard to the original setting, where the default indicators are binomially distributed.

5. G. Giese. Enhancing CreditRisk$^+$. *Risk*, 16(4):73–77, 2003.

6. M. B. Gordy. Saddlepoint approximation of CreditRisk$^+$. *Journal of Banking & Finance*, 26:1335–1353, 2002.

7. H. Haaf and D. Tasche. Credit portfolio measurements. *GARP Risk Review*, (7):43–47, 2002.

8. F. B. Lehrbass, I. Boland, and R. Thierbach. Versicherungsmathematische Risikomessung für ein Kreditportfolio. *Blätter der Deutschen Gesellschaft für Versicherungsmathematik*, XXV:285–308, 2002.

9. R. J. Martin, K. E. Thompson, and C. J. Browne. Taking to the saddle. *Risk*, 14(6):91–94, 2001.

10. O. Reiß. *Fourier inversion algorithms for generalized CreditRisk$^+$ models and an extension to incorporate market risk*. WIAS-Preprint No. 817, 2003.

11. J. Stoer and R. Bulirsch. *Introduction to Numerical Analysis*. Springer-Verlag, Berlin, Heidelberg, New York, 2002.

6

Enhanced CreditRisk$^+$

Götz Giese

Summary. In this chapter we discuss the link between the CreditRisk$^+$ loss distribution and the moment-generating function (MGF) of the risk factors. We show that the probability-generating function (PGF) of the loss variable is the MGF of the factors, evaluated at a particular "point". This approach has two major advantages: it leads to a new recursion formula for the portfolio loss distribution that is faster and more accurate than the standard approach. It also allows us to extend the modelling framework to a wider class of factor distributions incorporating sector correlations. At the end of this chapter we show that risk contributions are related to the partial derivatives of the MGF. We derive the formula for exact risk contributions in this generalized modelling framework and highlight the differences from the corresponding result obtained in the saddlepoint approximation.

Introduction

One of the major advantages of the CreditRisk$^+$ model is that the portfolio loss distribution can be computed analytically so that Monte Carlo simulation can be avoided. However, as is well known ([3, 8]), the standard recursion relation for the loss distribution in CreditRisk$^+$, which goes back to Panjer [7], tends to be numerically unstable for large portfolios and many risk factors, see also Chapter 5. Instead of using a saddlepoint approximation (as proposed in [3], see Chapter 7), we present a new way to calculate the true loss distribution that is numerically stable, fast and accurate, even under rather extreme conditions. The approach centres on what we call *nested evaluation* of the moment-generating function (MGF) of the risk factor distribution and its partial derivatives. An essential advantage of the new method is that the assumption of independent factors in standard CreditRisk$^+$ can be relaxed without losing analytical tractability.

In this chapter, we discuss the generalized formulae for the loss distribution and risk contributions in a setting that makes no explicit assumptions on the factor distribution. The new recursion algorithm will be illustrated with the example of standard CreditRisk$^+$, i.e. by using the MGF of independent gamma-distributed variables. In Chapter 10 we will introduce other (dependent) factor distributions, which are also amenable to this approach.

6.1 An MGF Point of View on CreditRisk$^+$

Let us begin by briefly summarizing the definitions and properties of the MGF and the cumulant-generating function (CGF) of a real N-dimensional random variable $\mathbf{Y}$. The MGF of $\mathbf{Y}$, $M_\mathbf{Y}(\mathbf{t})$, is a function of $\mathbf{t} \in \mathcal{D} \subseteq \mathbb{R}^N$ defined as

$$M_\mathbf{Y}(\mathbf{t}) = \mathbb{E}\left[\exp\left(\sum_{k=1}^N t_k Y_k\right)\right] = \mathbb{E}\left[e^{\mathbf{t}\cdot\mathbf{Y}}\right],$$

where $\mathcal{D}$ is the (open) definition range, for which the expectation exists. Here we have introduced the "·" to denote the inner product of vectors. As the name suggests, the MGF is directly linked to the moments of the joint distribution by its partial derivatives at $\mathbf{t} = 0 \in \mathcal{D}$. For the n-th-order mixed moment one finds

$$\mathbb{E}[Y_1^{i_1} Y_2^{i_2} \cdot \ldots \cdot Y_N^{i_N}] = \frac{\partial^n}{\partial t_1^{i_1} \partial t_2^{i_2} \cdots \partial t_N^{i_N}} M_\mathbf{Y}(\mathbf{t} = 0)\,, \text{ where } \sum_{k=1}^N i_k = n.$$

The CGF of $\mathbf{Y}$, $K_\mathbf{Y}(\mathbf{t})$, is defined as the logarithm of the MGF,

$$K_\mathbf{Y}(\mathbf{t}) = \ln M_\mathbf{Y}(\mathbf{t}) = \ln \mathbb{E}[e^{\mathbf{t}\cdot\mathbf{Y}}].$$

The partial derivatives of the CGF at $\mathbf{t} = 0$ are called cumulants. The first-order cumulants correspond to the first moments of the distribution,

$$\frac{\partial}{\partial t_k} K_\mathbf{Y}(\mathbf{t} = 0) = \mathbb{E}[Y_k].$$

The second-order cumulants are the elements of the covariance matrix of the distribution

$$\frac{\partial^2}{\partial t_k \partial t_l} K_\mathbf{Y}(\mathbf{t} = 0) = \text{cov}[\mathbf{Y}]_{kl}.$$

It is interesting to note that for a multivariate Gaussian distribution all cumulants of order $n > 2$ vanish, as the CGF of an N-dimensional Gaussian distribution with mean $\boldsymbol{\mu}$ and covariance matrix $\text{cov}[\mathbf{Y}]_{kl}$ is simply $K_\mathbf{Y}^{Gauss} = \mathbf{t} \cdot \boldsymbol{\mu} + \frac{1}{2}\sum_{k,l} t_k t_l \text{cov}[\mathbf{Y}]_{kl}$. For a general multivariate distribution, the higher-order cumulants can therefore be interpreted as a measure of the deviation from the Gaussian case. Other important properties that follow directly from the definitions are:

- Both MGF and CGF are convex functions of $\mathbf{t}$.

- If the Y_k are independent the MGF factorizes and becomes the product of the MGFs of the marginal distributions,

$$M_{\mathbf{Y}}(\mathbf{t}) = \prod_{k=1}^{N} M_{Y_k}(t_k),$$

so that the CGF is the sum of the marginal CGFs,

$$K_{\mathbf{Y}}(\mathbf{t}) = \sum_{k=1}^{N} K_{Y_k}(t_k).$$

- For the trivial case that $\mathbf{Y}$ is deterministic with $\mathbf{Y} \equiv \mathbf{c} \in \mathbb{R}^N$ we obtain

$$M_{\mathbf{Y}}(\mathbf{t}) = e^{\mathbf{t} \cdot \mathbf{c}}, \qquad K_{\mathbf{Y}}(\mathbf{t}) = \mathbf{t} \cdot \mathbf{c}. \tag{6.1}$$

Finally, if Y is a one-dimensional random variable that can take on only natural numbers (such as the portfolio loss X in CreditRisk$^+$), the following relationship between CGF and PGF holds:

$$K_Y(t) = \log G_Y(e^t). \tag{6.2}$$

After these preliminaries we can now discuss the CreditRisk$^+$ model from the viewpoint of the MGF of the risk factors. In Chapter 2, the key assumptions of what we will refer to as the "CreditRisk$^+$ framework" have been discussed:

- Poisson approximation for the default variables;
- conditional independence of these variables;
- linear relationship between the systematic risk factors S_k and conditional default rates, p_A^S (cf. (2.9)).

Under these assumptions the conditional PGF of the loss distribution has been derived as

$$G_X(z|S) = \exp\left(\sum_{k=0}^{N} S_k \mu_k (\mathcal{P}_k(z) - 1)\right). \tag{6.3}$$

Here $\mu_k = \sum_A w_{Ak} p_A$ is the expected number of defaults in sector S_k and

$$\mathcal{P}_k(z) = \frac{1}{\mu_k} \sum_A w_{Ak} p_A z^{\nu_A}.$$

(6.3) also includes the specific sector, which is modelled by the deterministic variable $S_0 \equiv 1$. For reasons that will shortly become clear, we slightly extend the notation and introduce the $(N+1)$-dimensional vector $\mathbf{P}(z)$ of sector polynomials $P_k(z)$ by setting

$$P_k(z) = \mu_k \left(\mathcal{P}_k(z) - 1\right) = \sum_A w_{Ak} p_A (z^{\nu_A} - 1). \tag{6.4}$$

The conditional PGF can now be simply written as

$$G_X(z|S) = e^{\mathbf{S} \cdot \mathbf{P}(z)}. \tag{6.5}$$

Let us for the moment ignore the systematic risk component and consider all default rates as constant. This defines the deterministic limit where all $S_k \equiv 1$ and obligor defaults are independent. In this case

$$G_X^{Ind}(z) = \exp\left(\sum_{k=0}^{N} P_k(z)\right),$$

which reveals that the sector polynomials – evaluated at $z = e^t$ – are simply the sector CGFs of the independent loss distribution,

$$K_X^{Ind}(t) = \sum_{k=0}^{N} P_k(e^t) \equiv \sum_{k=0}^{N} K_{Xk}^{Ind}(t)$$

(cf. (6.2)). Each $P_k(z)$ therefore provides a condensed representation of the obligor-specific risk in sector S_k. As pointed out in Chapter 3 by Tasche, the functional form of (6.5) and its interpretation remain valid even if independent, obligor-specific severity variations are included in the model.

On the other hand, if we introduce systematic risk by considering an arbitrary (potentially dependent) factor distribution subject to the restrictions

$$\mathbb{E}[S_k] = 1, \qquad S_k \geqslant 0, \tag{6.6}$$

we obtain the unconditional PGF by taking the expectation

$$G_X(z) = \mathbb{E}\left[e^{\mathbf{S} \cdot \mathbf{P}(z)}\right] = M_{\mathbf{S}}\left(\mathbf{P}(z)\right). \tag{6.7}$$

In other words, the PGF of the loss distribution is the MGF of the risk factor distribution, evaluated at the particular "point" $\mathbf{t} = \mathbf{P}(z)$. This is how specific risk components couple to systematic factors S_k in the CreditRisk$^+$ framework, independent of the particular form of the factor distribution. This relationship becomes even more transparent if we formulate the above equation in terms of the respective CGFs:

$$K_X(t) = K_{\mathbf{S}}\left(\mathbf{K}_X^{Ind}(t)\right),$$

where $\mathbf{K}_X^{Ind}(t)$ contains the sector CGFs $K_{Xk}^{Ind}(t) = P_k(e^t)$ as components. (6.7) is the cornerstone of the new recursion algorithm, which will be introduced in the next section and which allows us to calculate the PGF also for dependent factor distributions – as long as their MGFs have a simple algebraic form.

For illustration purposes let us return to the standard CreditRisk$^+$ model and consider the case that for $k \geq 1$ the S_k are independent and gamma distributed. The MGF of the univariate gamma distribution with shape parameter α_k and scale parameter β_k is

$$M_{S_k}^{\Gamma}(t_k) = (1 - \beta_k t_k)^{-\alpha_k} . \tag{6.8}$$

In our setting we have $\alpha_k = \frac{1}{\beta_k} = \frac{1}{\sigma_k^2}$, ensuring that the distribution has mean 1 and variance σ_k^2. Including the 0-th (deterministic) factor and using stochastic independence the MGF for the standard model reads

$$M_{\mathbf{S}}^{CR+}(\mathbf{t}) = e^{t_0} \prod_{k=1}^{N} M_{S_k}^{\Gamma}(t_k).$$

Slightly rewriting this expression and applying (6.7) we arrive at the analogue of (2.19), in our notation

$$G_X^{CR+}(z) = \exp\left(P_0(z) - \sum_{k=1}^{N} \frac{1}{\sigma_k^2} \ln\left(1 - \sigma_k^2 P_k(z)\right) \right) \tag{6.9}$$

for the PGF of the loss distribution in standard CreditRisk$^+$.

6.2 Nested Evaluation

The simple analytic form of (6.9) lays the ground for a new and very convenient numerical calculation of the PGF. The key idea is that the equation represents the PGF as a series of nested operations on the polynomials $P_k(z)$. Once we have "taught" the computer how to perform elementary arithmetic operations (addition, multiplication) on polynomials and how to take the exponential and logarithm of a polynomial, the equation can be calculated term by term in much the same way as if the $P_k(z)$ were ordinary real numbers.

In more technical terms, we represent $P_k(z)$ and $G_X^{CR+}(z)$ (and all other power series that occur as intermediate results) as polynomials of fixed degree M_{max} and store the coefficients in arrays of length $M_{max} + 1$ (including the 0-th coefficient). Addition, multiplication and even division of polynomials now translate into simple operations on arrays. Likewise, both the exp and the ln function are mappings from one array to another and can be defined by recursion relations as we will see in more detail in the following.

For notational simplicity, let us first define the formal coefficient operator $\mathbf{C}^{[n]}$, which simply maps a power series $P(z) = \sum_{n=0}^{\infty} p_n z^n$ to its n-th coefficient, p_n:

$$\mathbf{C}^{[n]} P(z) = \frac{1}{n!} \frac{d^n}{dz^n} P(z = 0) = p_n. \tag{6.10}$$

Let $Q(z) = \sum_{n=0}^{\infty} q_n z^n$ be another power series and $r \in \mathbb{R}$. In coefficient operator notation the rules for scalar multiplication, polynomial addition and multiplication read

$$\mathbf{C}^{[n]}\left(r\,Q(z)\right) = r\,q_n$$
$$\mathbf{C}^{[n]}\left(P(z)+Q(z)\right) = p_n + q_n \qquad (6.11)$$
$$\mathbf{C}^{[n]}\left(P(z)\cdot Q(z)\right) = \sum_{j=0}^{n} p_{n-j}\,q_j.$$

The last relation corresponds to the convolution of distributions if $P(z)$ and $Q(z)$ are PGFs. Inverting the equation we obtain the coefficients for the quotient of power series (provided $q_0 \neq 0$):

$$\mathbf{C}^{[n]}\left(\frac{P(z)}{Q(z)}\right) = \frac{1}{q_0}\left(p_n - \sum_{j=1}^{n} q_j\,\mathbf{C}^{[n-j]}\left(\frac{P(z)}{Q(z)}\right)\right). \qquad (6.12)$$

Here we have adopted the convention that empty sums vanish (i.e. sums where the lower bound is larger than the upper bound). Note that in contrast to the first three relations this equation defines a recursion rule, since the result of the operation appears on the left- *and* on the right-hand side of the equation.

In general, there is no straightforward way to derive the coefficients of an analytic function of a power series (see [1] for a general discussion). For the exp function, however, we can exploit the fact that $\exp(y)$ is invariant under differentiation. Employing Leibniz's rule we obtain (cf. [2, p. 38])

$$\mathbf{C}^{[0]}\left(e^{P(z)}\right) = e^{p_0}$$
$$\mathbf{C}^{[n]}\left(e^{P(z)}\right) = \frac{1}{n!}\frac{d^{n-1}}{dz^{n-1}}\left[e^{P(z)}\frac{d}{dz}P(z)\right]_{z=0} \qquad (6.13)$$
$$= \frac{1}{n!}\sum_{j=1}^{n}\binom{n-1}{j-1}\frac{d^{n-j}}{dz^{n-j}}\left[e^{P(z)}\right]_{z=0}\underbrace{\frac{d^j}{dz^j}\left[P(z)\right]_{z=0}}_{=j!\,p_j}$$
$$= \sum_{j=1}^{n}\frac{j}{n}\,p_j\,\mathbf{C}^{[n-j]}\left(e^{P(z)}\right) \qquad \text{for } n > 0.$$

We can again invert these equations to obtain the recursion relations for the ln function:

$$\mathbf{C}^{[0]}\left(\ln P(z)\right) = \ln(p_0), \qquad (6.14)$$
$$\mathbf{C}^{[n]}\left(\ln P(z)\right) = \frac{1}{p_0}\left(p_n - \sum_{j=1}^{n-1}\frac{j}{n}\,p_{n-j}\,\mathbf{C}^{[j]}\left(\ln P(z)\right)\right) \qquad \text{for } n > 0,$$

which of course requires $p_0 > 0$.

Having defined all necessary elementary operations, we can now evaluate the right-hand side of (6.9) term by term as a sequence of nested operations on polynomials (arrays). Object-oriented programming languages like C++, which allow overloading of operators and functions, provide a very elegant framework for this approach. Once the necessary class operations have been defined, expressions like (6.9) can be more or less directly cast into programming code.

Replacing power series by polynomials of fixed length M_{max} is of course an approximation. The recursion scheme for division as well as those for the exp and ln function do not break off at M_{max}, even if the arguments are polynomials. The choice of this parameter therefore determines the truncation error of the calculation. The important point is, however, that in each operation the n-th coefficient is a function only of the first n coefficients in the input. This means that cutting off higher coefficients does not introduce errors in the lower coefficients. The first M_{max} coefficients of the PGF will always be calculated exactly and the truncation error only consists of losing the far tail of the distribution, which can be monitored rather easily.[1]

Apart from its simplicity, the benefits of the "nested evaluation" approach lie in its speed and stability. Note that polynomial addition and scalar multiplication (cf. (6.11)) involve only $O(M_{max})$ computation steps, whereas the recursion relations are computationally much more expensive ($O(M_{max}^2)$ steps). Nested evaluation of (6.9) obviously requires $(N + 1)$ function calls, resulting in $O((N + 1)M_{max}^2)$ computation steps. This has to be contrasted with the standard Panjer recursion discussed in Section 2.6, which consists of representing $\frac{d}{dz} \ln G_X^{CR+}(z)$ as the rational polynomial

$$\frac{A(z)}{B(z)} = \frac{d}{dz} P_0(z) + \sum_{k=1}^{N} \frac{\frac{d}{dz} P_k(z)}{1 - \sigma_{kk} P_k(z)}. \tag{6.15}$$

Solving for $A(z)$ and $B(z)$ involves at the minimum $3N - 2$ multiplications of polynomials, each requiring $O(M_{max}^2)$ computation steps. The final recursion (2.25) that derives the coefficients of $G_X^{CR+}(z)$ from those of $A(z)$ and $B(z)$ is also of order $O(M_{max}^2)$. In summary, nested evaluation is approximately $\frac{3N-1}{N+1}$ times faster[2] than the Panjer approach.

More importantly, it is well known ([3, 8]) that the repeated multiplication of polynomials in (6.15) involves the addition of many small numbers with alternating sign, which is a numerically dangerous operation and can lead to significant round-off errors, particularly if N, M_{max}, and the number of obligors are large. Nested evaluation, on the other hand, is insensitive to this problem. As explained in detail in Chapter 5, the new recursion involves only

---

[1] An upper bound for M_{max} can be obtained from Chebyshev's inequality (cf. [6]). We monitor the tail truncation error by comparing the first moments of the loss distribution with their theoretical values, which – as shown in Chapter 2 – can be derived analytically from the input parameters. In practical applications, $M_{max} = 5 \cdot \nu_{max}$ (ν_{max} being the maximal exposure size) is usually a good starting point for diversified portfolios.

[2] As the formula indicates, in the case $N = 1$ (i.e. one specific and one systematic sector) both algorithms are actually equally fast.

sums of strictly positive numbers, which is a numerically harmless operation. We routinely run our algorithm on portfolios consisting of more than 1 million obligors with $N = 65$ factors and $M_{max} \geq 10,000$ and have never observed any numerical instabilities. Figure 6.1 provides a practical example and illustrates that in "real world" applications the Panjer recursion may sometimes not even be able to correctly determine the 95% loss quantile.

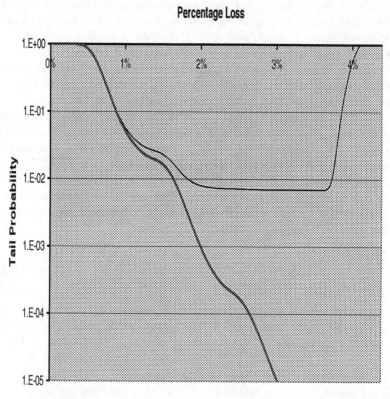

Fig. 6.1. Tail probabilities $\mathbb{P}[X \geq x]$ for a very large portfolio (1.4 million obligors distributed over 65 sectors) in the standard CreditRisk$^+$ framework. The x-axis represents the portfolio loss x as a percentage of the maximum loss $\sum_A \nu_A$. The graph demonstrates the breakdown of the Panjer recursion (thin line) due to numerical instability. Numerical errors cumulate long before the 1% tail has been reached, producing even negative probabilities in the far tail (area where the tail probability seems to increase). The thick line represents the exact distribution generated by nested evaluation of polynomials of degree $M_{max} = 20,000$.

The most important advantage of nested evaluation, however, is that it allows us to extend the model to other more realistic distributions, which include factor dependence. Any multivariate factor distribution, whose MGF is a combination of elementary arithmetic operations and exp and ln functions, can be plugged into (6.7), and the PGF can be calculated as described above. Two examples will be discussed in Chapter 10.

6.3 Risk Contributions in the Generalized Framework

For a portfolio manager, perhaps the most important feature of a portfolio model is its ability to break down portfolio risk to obligor level in a risk-adjusted way. Being additive, risk contributions per obligor can be used to individually allocate economic capital. They can be aggregated across arbitrary dimensions in order to "slice and dice" through portfolio risk.

In this section, we extend the results provided in Chapter 3 (see also [4]) on exact risk contributions to the case of the generalized CreditRisk$^+$ model defined by (6.7). We limit ourselves to the case that exposures are deterministic (incorporating independent severity variations as discussed in Chapter 3 is also possible).

For a fixed confidence level $1 - \epsilon$ we denote the corresponding quantile by $x_\epsilon = \bar{\rho}_\epsilon(X) \in \mathbb{N}$ (cf. the definitions in Chapter 2). As discussed by Tasche in Chapter 3, the contribution of obligor A to the unexpected loss of the portfolio, $r_A(UL)$, can be defined as the conditional expectation

$$r_A(UL) = \nu_A \, \mathbb{E}[D_A \, | X = x_\epsilon] - EL_A = \nu_A \frac{\mathbb{E}\left[D_A \mathbf{1}_{\{X=x_\epsilon\}}\right]}{\mathbb{P}[X = x_\epsilon]} - p_A \nu_A, \quad (6.16)$$

where D_A is the default indicator variable in the Poisson approximation and $\mathbf{1}_B$ is the default indicator, which is equal to 1 on the event B and zero otherwise. The denominator is obviously given by the x_ϵ-th coefficient of the PGF in (6.7). Using the coefficient operator (6.10) we can write

$$\mathbb{P}[X = x_\epsilon] = \mathbf{C}^{[x_\epsilon]} G_X(z). \quad (6.17)$$

The difficult task is to evaluate the expectation in the numerator of (6.16). From Chapter 3 we also know that for Poisson variables, which are conditionally independent (on $\mathbf{S}$),

$$\mathbb{E}\left[D_A \, \mathbf{1}_{\{X=x_\epsilon\}}\right] = \mathbb{E}\left[p_A^S \, \mathbb{P}[X = x_\epsilon - \nu_A \, |S]\right] = \mathbb{E}\left[p_A^S \, \mathbf{C}^{[x_\epsilon - \nu_A]} G_X(z|S)\right].$$

Since the coefficient operator is defined by differentiation we can interchange it with the expectation operator. Using the linear relationship (2.9) between default rate and risk factors we obtain

$$\mathbb{E}\left[D_A \, \mathbf{1}_{\{X=x_\epsilon\}}\right] = p_A \sum_{k=0}^{N} w_{Ak} \, \mathbf{C}^{[x_\epsilon - \nu_A]} \mathbb{E}[S_k \cdot \exp\left(\mathbf{S} \cdot \mathbf{P}(z)\right)]$$

$$= p_A \sum_{k=0}^{N} w_{Ak} \, \mathbf{C}^{[x_\epsilon - \nu_A]} \frac{\partial}{\partial t_k} M_{\mathbf{S}} \left(\mathbf{t} = \mathbf{P}(z)\right),$$

88   Götz Giese

which reveals that risk contributions are related to the partial derivatives of the MGF (provided they exist). Note that

$$G_{Xk}(z) := \frac{\partial}{\partial t_k} M_S \left(\mathbf{t} = \mathbf{P}(z) \right)$$

is also a PGF because $\frac{\partial}{\partial t_k} M_S$ can be read as the MGF of the multivariate density $\hat{f}_S(s) = s_k f_S(s)$, which in turn is a member of the allowed class of distributions for the generalized model (up to a trivial normalization, cf. (6.6)). In particular, we have from (6.1)

$$G_{X0}(z) = G_X(z)$$

for the deterministic sector.

Computing risk contributions for the portfolio therefore requires calculating N additional power series. The final formula for the UL contribution per obligor reads

$$r_A(UL) = p_A \nu_A \left[\frac{\sum\limits_{k=0}^{N} w_{Ak}\, \mathbf{C}^{[x_\epsilon - \nu_A]} G_{Xk}(z)}{\mathbf{C}^{[x_\epsilon]} G_X(z)} - 1 \right] \qquad (6.18)$$

$$\equiv EL_A \sum\limits_{k=0}^{N} w_{Ak}\, F_k(\nu_A),$$

where we have represented the UL contribution as the product of the expected loss contribution EL_A and a weighted sum of sector-specific penalty factors

$$F_k(\nu_A) = \frac{\mathbf{C}^{[x_\epsilon - \nu_A]} G_{Xk}(z)}{\mathbf{C}^{[x_\epsilon]} G_X(z)} - 1. \qquad (6.19)$$

Penalty factors contain in a very condensed form the complete information about the risk profile of the portfolio and can also be used to approximate the marginal increase of portfolio risk due to (small) changes in the underlying positions. In practical applications (particularly when x_ϵ corresponds to a large quantile far in the tail of the loss distribution), penalty factors are typically increasing functions of exposure size, reflecting the fact that large exposures contribute overproportionately to tail risk. The mathematical reason for this size penalty is that with increasing ν_A we effectively run uphill the tail of the distribution that corresponds to the respective $G_{Xk}(z)$. The obligor's default rate has direct impact on the risk contribution only via the expected loss.[3] As portfolio-specific quantities, penalty factors do, of course, depend

---

[3] The particular functional form of (6.18) is a consequence of the Poisson approximation. For Bernoulli variables one would expect that individual UL contributions go to zero for $p_A \to 1$. In extreme cases (large exposure and default rate) the Poisson approximation may even lead to $EL_A + r_A(UL) > \nu_A$.

indirectly on all input parameters, including the default rates. If, however, we want to approximate the impact of a small position change ($\Delta \nu_A \ll \nu_A$ or $\Delta p_A \ll p_A$) on the UL of a large portfolio we can with high accuracy consider the functional form of the factors as constant.

It is interesting to compare the exact result with a formula for UL contributions in the saddlepoint approximation, which has been proposed by Martin, Thompson and Browne [5, Eq. (8)]. Using the CGF of the loss distribution, $K_X(t) = \ln G_X(e^t)$, the saddlepoint approximation reads

$$r_A^{SP}(UL) = \frac{\nu_A}{\hat{t}} \frac{\partial K_X(\hat{t})}{\partial \nu_A} - EL_A.$$

Here $\hat{t} < \ln(R)$ is the unique solution of the saddlepoint equation $\frac{d}{dt} K_X(t) = x_\epsilon$ ($R \geq 1$ is the radius of convergence of $G_X(z)$). Noting that

$$\frac{\partial P_k(e^{\hat{t}})}{\partial \nu_A} = \hat{t} \, w_{Ak} \, p_A \, e^{\hat{t} \nu_A}$$

we see that the saddlepoint result has the same structure as (6.18) with the penalty factors reading

$$F_k^{SP}(\nu_A) = e^{\hat{t} \nu_A} \frac{G_{Xk}(e^{\hat{t}})}{G_X(e^{\hat{t}})} - 1. \tag{6.20}$$

For a diversified portfolio, the saddlepoint method accurately determines the tail probability $P(X \geq x_\epsilon)$ [3]. Applied to risk contributions, however, the method produces only a uniform exponential penalty factor (with growth constant $\hat{t}$) for all sectors and all obligors. (The expressions on the fraction in the above equation are constants.) Comparing with the exact result (6.19) it becomes clear that all distributions generated by the $G_{Xk}(z)$ must have the *same* decay behaviour should the saddlepoint result be accurate.

As a final remark, we mention that using UL contributions can lead to counterintuitive results for portfolios with a very heterogeneous exposure distribution, since the analysis concentrates on a single-point event ($X = x_\epsilon$) of the distribution. One can construct exposure compositions where simply for combinatorial reasons some obligors may never contribute to this particular loss event, even if their default rate is rather high. These problems can be circumvented by choosing a risk measure that reflects the entire tail of the distribution. An example is what we define as unexpected shortfall,[4]

$$US_\epsilon[X] = \mathbb{E}[X|X \geq x_\epsilon] - EL. \tag{6.21}$$

---

[4] The more common definition is that of expected shortfall, $ES_\epsilon[X] = \mathbb{E}[X|X \geq x_\epsilon]$. In order to gain a risk measure that is comparable to the loss variance and unexpected loss we have, however, to separate off the expected loss component from this definition.

90   Götz Giese

Contributions to unexpected shortfall can be directly derived from (6.18) by summing over all relevant loss events (see again Chapter 3). Expressed in terms of penalty factors, the formula for exact US contributions can be written as

$$F_k^{US}(\nu_A) = \frac{1}{\delta}\left(1 - \delta - \sum_{n=0}^{x_\epsilon - \nu_A - 1} \mathbf{C}^{[n]} G_{Xk}(z)\right), \qquad (6.22)$$

where we have used that PGFs are normalized (i.e. $\sum_n \mathbf{C}^{[n]} G_{Xk}(z) = 1$) and denoted the tail probability by $\delta = \mathbb{P}[X \geq x_\epsilon] > \epsilon$.

In this chapter we have derived a toolbox for efficiently calculating the loss distribution and risk contributions in the generalized CreditRisk+ framework. In Chapter 10 we will apply the concept and consider factor distributions that allow modelling factor dependence. For the enhanced model, we will also provide numerical examples for the different contribution methods discussed in this section.

References

1. R. P. Brent and H. T. Kung. Fast algorithms for manipulating power series. *Journal of the Association of Computing Machinery*, 25(4):581–595, 1978.
2. Credit Suisse Financial Products. CreditRisk+: A credit risk management framework. London, 1997. Available at http://www.csfb.com/creditrisk.
3. M. B. Gordy. Saddlepoint approximation of CreditRisk+. *Journal of Banking & Finance*, 26:1335–1353, 2002.
4. H. Haaf and D. Tasche. Credit portfolio measurements. *GARP Risk Review*, (7):43–47, 2002.
5. R. Martin, K. Thompson, and C. Browne. VaR: Who contributes and how much? *Risk* 14(8):99–102, 2001.
6. S. Merino and M. Nyfeler. Calculating portfolio loss. *Risk* 15(8):82–86, 2002.
7. H.H. Panjer and G.E. Willmot. *Insurance Risk Models*. Society of Actuaries, Schaumberg, IL, 1992.
8. T. Wilde. CreditRisk+. In S. Das, editor, *Credit derivatives and credit linked notes, second edition*, New York, 2000. John Wiley & Sons.

7

Saddlepoint Approximation

Michael B. Gordy*

Summary. Saddlepoint approximation offers a robust and extremely fast alternative to Panjer recursion for the solution of the CreditRisk$^+$ loss distribution. This chapter shows how saddlepoint approximation can be applied to an extended version of CreditRisk$^+$ that incorporates idiosyncratic severity risk. Regardless of the number of sectors and without any need for discretizing loss exposures, both value-at-risk and expected shortfall are easily calculated.

Introduction

Analytical tractability is perhaps the most compelling advantage of CreditRisk$^+$ over competing models of portfolio credit risk. The recurrence algorithm introduced in the CreditRisk$^+$ manual [6, §A.10] eliminates the need for Monte Carlo simulation. Computation time is reduced dramatically, and simulation error avoided entirely.

In practical application, however, reliance on the Panjer recursion algorithm comes with significant costs. First, as has been recognized from early on, the need to round loss exposures to integer multiples of the "loss unit" L_0 may introduce a trade-off between speed and accuracy. The distribution of exposure sizes within a credit portfolio is typically quite skew.[1] When this is the case, choosing a high value for L_0 implies crude measurement of loss exposure for the bulk of the obligors. Choosing a low value for L_0 comes at significant computational cost. If L_0 is cut in half, the recurrence algorithm polynomials $A(z)$ and $B(z)$ roughly double in length and, since the number of loss units in VaR doubles, the recurrence equation must be executed twice

---

* This chapter draws heavily on the article [14]. The views expressed herein are my own and do not necessarily reflect those of the Board of Governors or its staff. I thank Dirk Tasche for helpful comments.

[1] Carey [5] finds that the largest 10% of exposures account for roughly 40% of total exposure in bank portfolios.

as many times. Therefore, a halving of L_0 roughly quadruples the execution time.

A second problem, also due to discretizing losses, is that marginal contributions to VaR can be poorly behaved. If VaR is defined in the usual manner for a discrete model, that is,

$$\mathrm{VaR}_q[X] \equiv \inf\{n \in \mathbb{N} : \mathbb{P}[X \leq n] \geq q\}$$

for target solvency probability q, then in general $\mathbb{P}[X \leq \mathrm{VaR}_q[X]] > q$. When a new loan is added to a portfolio, the "steps" in the loss distribution are realigned.[2] If adding the new loan brings $\mathbb{P}[X \leq \mathrm{VaR}_q[X]]$ closer to q, then it is possible to have a negative marginal contribution to VaR. Similarly, one can have a disproportionately large marginal contribution. A closely related problem is that we can have $\mathbb{P}[X = n] = 0$ for certain loss levels n. This leads to capital charge of zero for some loans in the allocation technique of [16].[3]

Third, it is increasingly recognized that the Panjer algorithm is numerically fragile. Numerical error may be introduced in the convolution steps needed to produce the polynomials $A(z)$ and $B(z)$ and in summing terms in the recurrence equation, see Chapter 5. The latter may be executed thousands of times before the target solvency probability is reached, so the cumulative effect of these errors may be quite large. The scope for error increases with the maximum exposure size (which thus limits how small L_0 can be set) and with the number of sectors N.

A fourth drawback is the difficulty of introducing severity risk (also known as recovery risk). At the level of the individual loan or bond, the uncertainty in the recovery process is typically quite large, see, e.g. [22]. Even if this risk is purely idiosyncratic to the instrument, defaults are sufficiently rare in portfolios of modest size or high quality that it cannot be assumed to be diversified away. The competing leading models of credit risk all incorporate idiosyncratic severity risk. The only barrier to introducing severity risk in CreditRisk$^+$ is the requirement that losses be integer multiples of the loss unit. Unless a loan's expected loss exposure is many multiples of L_0, variation of even one or two units of L_0 may be unrealistically large. This limits the usefulness of severity risk extensions such as that of [4].

Value-at-risk is not the only useful metric of portfolio risk. Indeed, as has been emphasized in recent literature on risk measures, VaR has significant theoretical and practical shortcomings. In particular, VaR is not sub-additive. That is, if X_A and X_B are losses on bank portfolios A and B, then it is possible to have $\mathrm{VaR}_q[X_A + X_B] > \mathrm{VaR}_q[X_A] + \mathrm{VaR}_q[X_B]$, which would imply

[2] This measurement issue is most serious (a) the coarser the discretization of exposure sizes, (b) the smaller the variances of the systemic risk factors, and (c) the higher the credit quality of the marginal asset. One perhaps can ameliorate this problem by interpolating VaR as suggested by Gundlach in Chapter 2.

[3] The focus of this chapter is on computation of portfolio-level risk measures. Martin et al. [19] show how to allocate capital to individual positions via saddlepoint approximation.

that a merger of bank A and bank B could *increase* VaR; see [11, §2.3] for an example based on credit risk measurement. Sub-additivity is one of the four requirements for a coherent risk measure, as defined by [2]. Among the coherent measures, expected shortfall is a leading alternative to VaR, and is increasingly used in practice for portfolio risk measurement and optimization. With minimal modification, the standard Panjer recursion algorithm can calculate expected shortfall and VaR at the same time, but then all of the drawbacks associated with Panjer recursion solution of VaR would apply here as well.

This chapter shows how all of the shortcomings of Panjer recursion are avoided when the tail of the loss distribution is instead solved by saddlepoint approximation. Saddlepoint techniques rely on having the cumulant-generating function (CGF) in tractable form much the way Panjer recursion relies on the tractability of the probability-generating function. I show that the CreditRisk$^+$ CGF can be calculated quite easily and poses no numerical challenges. Increasing the number of sectors does not degrade the accuracy of saddlepoint approximation, and leads to only a linear increase in solution time. Perhaps most importantly, the need to discretize losses is eliminated entirely. Idiosyncratic severity risk can be introduced to the model at negligible cost in complexity and speed.

In Section 7.1, I derive the CreditRisk$^+$ cumulant-generating function and show how to obtain from it the mean, variance, and indexes of skewness and kurtosis. Saddlepoint approximation of value-at-risk is described and tested in Section 7.2. I extend the methodology to incorporate idiosyncratic severity risk in Section 7.3. A simple saddlepoint approximation for expected shortfall is provided in Section 7.4.

7.1 The CreditRisk$^+$ Cumulant-Generating Function

The cumulant-generating function (CGF) of a distribution is the log of its moment-generating function. The cumulants are obtained from the CGF in the same manner as moments are obtained from the moment-generating function, i.e., the j-th cumulant of Y is given by the j-th derivative of the CGF of Y evaluated at $z = 0$; see also Chapter 8 by Reiß. For the first four cumulants of a distribution, the relationship to the central moments is straightforward. Let m be the mean of the distribution for some random variable Y, and let m_j be the j-th centred moment of Y, i.e., $m_j = \mathbb{E}[(Y - m)^j]$. The first four cumulants are given by $\kappa_1 = m$, $\kappa_2 = m_2$, $\kappa_3 = m_3$, and $\kappa_4 = m_4 - 3m_2^2$. Thus, the first cumulant is the mean and the second is the variance. The index of skewness and index of kurtosis can be written

$$\text{Skewness}(Y) \equiv \frac{m_3}{m_2^{3/2}} = \frac{\kappa_3}{\kappa_2^{3/2}}, \qquad \text{Kurtosis}(Y) \equiv \frac{m_4}{m_2^2} = \frac{\kappa_4}{\kappa_2^2} + 3.$$

After the first four terms, the relationship becomes increasingly complex; e.g., $\kappa_5 = m_5 - 10m_2 m_3$. As an example, the normal distribution with mean μ and variance σ^2 has CGF $\mu z + \sigma^2 z^2/2$, so $\kappa_1 = \mu$, $\kappa_2 = \sigma^2$, and $\kappa_j = 0$ for all $j > 2$.

The CGF can be expressed in terms of the probability-generating function (PGF). If $G_Y(z)$ is the PGF of Y, then the CGF $\psi_Y(z)$ is given by

$$\psi_y(z) = \log(G_Y(\exp(z))).$$

Using the CreditRisk$^+$ PGF (2.19), we find

$$\psi_X(z) = \log(G_X(\exp(z)))$$

$$= \mu_0(\mathcal{P}_0(\exp(z)) - 1) + \sum_{k=1}^{N} \alpha_k \log\left(\frac{1 - \delta_k}{1 - \delta_k \mathcal{P}_k(\exp(z))}\right)$$

where $\mu_k \equiv \sum_A w_{Ak} p_A$ and $\delta_k \equiv \mu_k/(\alpha_k + \mu_k)$. Define the function

$$\mathcal{Q}_k(z) \equiv \mu_k \mathcal{P}_k(\exp(z)) = \sum_A w_{Ak} p_A \exp(\nu_A z) \qquad (7.1)$$

and substitute into the CGF to arrive at

$$\psi_X(z) = (\mathcal{Q}_0(z) - \mu_0) + \sum_{k=1}^{N} \alpha_k \log\left(\frac{\mu_k(1 - \delta_k)}{\mu_k - \delta_k \mathcal{Q}_k(z)}\right) \equiv \psi_0(z) + \sum_{k=1}^{N} \psi_k(z). \quad (7.2)$$

The sector CGF components $\psi_k(z)$ are introduced for notational convenience.

Let D denote the differential operator (i.e., $D^j f(x)$ is the j-th derivative of f with respect to x). For the specific risk sector, the j-th derivative of $\psi_0(z)$ is simply the j-th derivative of $\mathcal{Q}_0(z)$, which can be evaluated using the general equation

$$D^j \mathcal{Q}_k(z) = \sum_A w_{Ak} p_A \nu_A^j \exp(\nu_A z) \quad \forall j \geq 0, k \in \{0, 1, \dots, N\}. \qquad (7.3)$$

For the systematic risk sectors, the first derivative of ψ_k is

$$\psi_k'(z) = \alpha_k \left(\frac{\delta_k D\mathcal{Q}_k(z)}{\mu_k - \delta_k \mathcal{Q}_k(z)}\right).$$

It is useful to generalize the expression in parentheses as

$$V_{j,k}(z) \equiv \frac{\delta_k D^j \mathcal{Q}_k(z)}{\mu_k - \delta_k \mathcal{Q}_k(z)}$$

and to observe that derivatives of V have the simple recurrence relation

$$DV_{j,k}(z) = V_{j+1,k}(z) + V_{j,k}(z) V_{1,k}(z). \qquad (7.4)$$

Therefore, using

$$\psi_k'(z) = \alpha_k V_{1,k}(z) \qquad (7.5a)$$

and (7.4), higher derivatives of ψ_k can be generated mechanically. The second, third and fourth are given by

$$\psi_k''(z) = \alpha_k(V_{2,k}(z) + V_{1,k}(z)^2) \tag{7.5b}$$

$$\psi_k'''(z) = \alpha_k(V_{3,k}(z) + 3V_{2,k}(z)V_{1,k}(z) + 2V_{1,k}(z)^3) \tag{7.5c}$$

$$\psi_k''''(z) = \alpha_k(V_{4,k}(z) + 4V_{3,k}(z)V_{1,k}(z) + 3V_{2,k}(z)^2$$
$$+ 12V_{2,k}(z)V_{1,k}(z)^2 + 6V_{1,k}(z)^4). \tag{7.5d}$$

Higher-order derivatives can obtained in an automated fashion using *Mathematica* [14, Appendix A]. The $V_{j,k}(z)$ are easily evaluated using (7.1) and (7.3).

So long as all the w_{Ak}, p_A and ν_A are non-negative and there exist loans for which the product $w_{Ak}p_A\nu_A$ is strictly positive, then $\mathcal{Q}_k$ and its derivatives are positive, continuous, increasing, and convex functions of z. As $z \to -\infty$, $\mathcal{Q}_k(z)$ and all its derivatives go to zero, so $V_{j,k}(z)$ goes to zero as well for all $j \geq 1$. Let z_k^* be the unique solution to $\mathcal{Q}_k(z) = \mu_k + \alpha_k$. For all $j \geq 1$, the denominator in $V_{j,k}(z)$ is positive and decreasing for all $z < z_k^*$ and equal to zero at z_k^*. The numerator in $V_{j,k}(z)$ is positive and increasing, so $V_{j,k}(z)$ is positive and increasing for all $z < z_k^*$. Solving for z_k^* is not difficult. I show in Appendix 1 that for each systematic sector k,

$$0 < z_k^* \leq -\log(\delta_k) \Big/ \sum_A \frac{w_{Ak}p_A}{\mu_k}\nu_A. \tag{7.6}$$

As $\mathcal{Q}_k(z)$ is strictly increasing in z, finding z_k^* in this bounded interval is computationally trivial.

These results for the $V_{j,k}(z)$ functions imply that, regardless of portfolio composition, the CGF always satisfies certain properties. Let $z^* = \min\{z_1^*, \ldots, z_N^*\}$ be the supremum of the valid domain of ψ_x, and observe that $z^* > 0$. We have:

- $\psi_x'(z)$ and higher derivatives are positive, continous, increasing and convex for all $-\infty < z < z^*$;
- the valid domain of ψ_x contains the origin;
- as $z \to -\infty$, $\psi_x'(z)$ and all higher derivatives tend to zero;
- as $z \to z^*$, $\psi_x(z)$ and all its derivatives tend to infinity.

To obtain the cumulants, we evaluate the derivatives of ψ_x at $z = 0$. Using the equality $\mathcal{Q}_k(0) = \mu_k$ and the definition of δ_k, we get

$$V_{j,k}(0) = (1/\alpha_k)D^j\mathcal{Q}_k(0),$$

and substitute into (7.5a)–(7.5d).

Calculation of the CGF and the cumulants is extremely fast and accurate. Though it may appear somewhat tedious, the expressions for $V_{j,k}(z)$ are trivially simple to program. Calculating the first four cumulants for a portfolio of thousands of obligors takes only a small fraction of a second. Observe that numerical issues do not arise in direct calculation of cumulants, because all the summations used in calculating the cumulants contain only non-negative terms of roughly similar magnitude. These calculations are fast and accurate, regardless of model parameters and portfolio composition.

7.2 Saddlepoint Approximation of Value-at-Risk

Edgeworth expansions are frequently used to approximate distributions that lack a convenient closed-form solution but for which cumulants are available. The density is expanded as an infinite series in terms of the normal density and derivatives of the normal density (i.e., the Hermite polynomials), where the coefficient on each term depends on the cumulants. Taking the first few terms of the series gives the Edgeworth approximation. It is widely recognized that this classic technique works well in the centre of a distribution, but can perform very badly in the tails. Indeed, it often produces negative values for densities in the tails.

Saddlepoint expansion can be understood as a refinement of Edgeworth expansion. While not as widely applied, it offers vastly improved performance in every respect. In saddlepoint approximation, the density of a distribution $f(y)$ at each point y is obtained by "tilting" the distribution around y to get a new distribution $\tilde{f}(z; y)$ such that y is in the centre of the new distribution. The tilted distribution is approximated via Edgeworth expansion, and the mapping is inverted to obtain the approximation to $f(y)$. Because Edgeworth expansion is used only at the centre of the tilted distributions, the approximation is in general uniformly good, in the sense of having a small relative error even in the tails. A consequence is that saddlepoint approximation of the cumulative distribution function is often quite good throughout the support. Indeed, for some distributions, including the gamma, the approximation is in fact exact, cf. [8]. Barndorff-Nielsen and Cox [3] discuss the relationship between the Edgeworth and saddlepoint approximations, and Jensen [17] provides an excellent textbook treatment of saddlepoint approximation more broadly.

Although saddlepoint approximation appeared first in the 1930s and was fully developed by the mid-1980s, it is not widely used because it requires that the CGF have tractable form. It is not often the case that a distribution of interest with intractable cumulative distribution function has a known and tractable CGF. Nonetheless, there have been efforts to apply saddlepoint approximation to problems in risk management. Feuerverger and Wong [10] derive a CGF of the distribution of a delta-gamma approximation to loss in a VaR model of market risk, and apply saddlepoint approximation using this approximated CGF. Martin et al. [18] provide a simple model of credit risk that is suitable for saddlepoint approximation. However, their model is designed around the need to maintain tractability rather than ease and reliability of calibration. In this section, I show that saddlepoint approximation applies quite well to CreditRisk$^+$. Accurate approximations to the tail percentiles of the loss distribution can be obtained with trivial computation time.

Different expansions lead to different forms for the saddlepoint approximation. An especially parsimonious and easily implemented variant is the Lugannani–Rice approximation [17, §3.3]. Let Y be a random variable with CGF $\psi_Y(z)$, and let $\hat{z}$ denote the real root of the equation $y = \psi_Y'(z)$. The value $\hat{z}$ is known as the saddlepoint, and its existence and uniqueness is demonstrated under very general conditions by [7, Theorem 6.2]. The Lugannani–Rice formula for tail probabilities when Y is continuous is

$$\mathbb{P}[Y \geq y] \approx \mathrm{LR}_Y(y; w, u) \equiv 1 - \Phi(w) + \phi(w)\left(\frac{1}{u} - \frac{1}{w}\right)$$

where

$$w = \mathrm{sign}(\hat{z})\sqrt{2(\hat{z}y - \psi_Y(\hat{z}))} \quad \text{and} \quad u = \hat{z}\sqrt{\psi_Y''(\hat{z})}$$

and where Φ and ϕ denote the cumulative distribution function and probability density function of the standard normal distribution.[4] For discrete Y taking on integer values, Daniels [9] shows that the Lugannani–Rice formula can still be used if we replace u with $\tilde{u} = (1 - \exp(-\hat{z}))\sqrt{\psi_Y''(\hat{z})}$. That is, we approximate $\mathbb{P}[Y \geq n] \approx \mathrm{LR}_Y(n; w, \tilde{u})$.

When y is in the neighbourhood of $\mathbb{E}[Y]$, we have $\hat{z} \approx 0$. Both w and u go to zero as well, which causes numerical instability in the term $1/u - 1/w$. In Appendix 2, I derive an approximation

$$\frac{1}{u} - \frac{1}{w} \approx \frac{-\kappa_3}{6\kappa_2^{3/2}} + \left(\frac{5\kappa_3^2 - 3\kappa_2\kappa_4}{24\kappa_2^{5/2}}\right)\hat{z} \tag{7.7}$$

for use when $\hat{z} \approx 0$. The same issue arises when $\tilde{u}$ is used in place of u. For $\hat{z} \approx 0$, we have

$$\frac{1}{\tilde{u}} \approx \frac{1}{u} + \frac{1}{2\sqrt{\kappa_2}} + \left(\frac{1}{12\sqrt{\kappa_2}} - \frac{\kappa_3}{4\kappa_2^{3/2}}\right)\hat{z}.$$

Substitution into (7.7) gives a local approximation for $1/\tilde{u} - 1/w$.

Application of saddlepoint approximation to CreditRisk$^+$ is straightforward. Assuming for the moment that X is continuous in the neighbourhood of $\mathrm{VaR}_q[X]$, then we find the saddlepoint $\hat{z}$ that solves $\mathrm{LR}_X(\psi_X'(\hat{z}); w, u) = 1 - q$ and set $\mathrm{VaR}_q[X] = \psi_X'(\hat{z})$. So long as q is a tail probability (i.e., $q > \mathbb{P}[X \leq \mathbb{E}[X]]$), we know that $\hat{z} \in (0, z^*)$, where z^* is the supremum of the valid domain of ψ_X. As both $\psi_X'(z)$ and the Lugannani–Rice formula are continuously

---

[4] The normal distribution arises in the approximation because of the normal base in the Edgeworth expansion of the tilted distributions. Changing the base distribution in the expansion of the tilted distributions leads to alternative forms for the Lugannani–Rice formula [23].

differentiable and strictly monotonic functions, solving for $\hat{z}$ is straightforward and fast.[5] When X is restricted to integer loss units, we choose $\hat{z}$ to solve $\mathrm{LR}_X(\psi'_X(\hat{z}); w, \tilde{u}) = 1 - q$, and then set $\mathrm{VaR}_q[X]$ to the smallest integer greater than or equal to $\psi'_X(\hat{z})$.

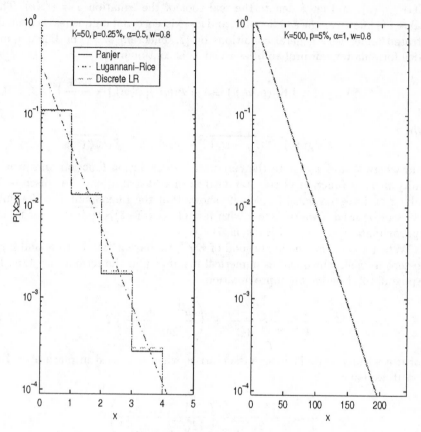

Fig. 7.1. Saddlepoint approximation of the loss distribution

Exact and approximated distributions for two stylized portfolios are shown in Figure 7.1. The left-hand panel represents the case of a small high-quality portfolio. There are $K = 50$ equal-sized loans in the portfolio, each with default probability $p = 0.25\%$. There is a single systematic risk factor S with

---

[5] The saddlepoint approximation to VaR always exists. As $\hat{z} \to z^*$, w, u and x tend to infinity, so $\lim_{\hat{z} \to z^*} \mathrm{LR}_X(\psi'_X(\hat{z}); w, u) = 0$. Therefore, for any $q \in (\mathbb{P}[X \le \mathbb{E}[X]], 1)$, there is always a unique solution to $\hat{z}$ in $(0, z^*)$. In [14], I suggest an alternative solution strategy based on interpolation that eliminates the need to find $\hat{z}$. This alternative is especially efficient when one wants to calculate VaR for several target solvency probabilities.

$\alpha = 1/2$, and each borrower has factor loading $w_{A1} = w = 0.8$ on S. The right-hand panel represents the case of a medium-sized low-quality portfolio. There are $K = 500$ equal-sized loans in the portfolio, each with default probability $p = 5\%$. There is a single systematic risk factor S with $\alpha = 1$, and each borrower has factor loading $w = 0.8$ on S. Loss units are plotted on the x-axis, and $\mathbb{P}[X \geq x]$ is plotted on a log-scale on the y-axis.

Both the continuous and discrete saddlepoint approximations are plotted. In the left-hand panel, the discrete Lugannani–Rice formula provides a very good fit throughout the distribution. The continuous Lugannani–Rice formula smooths through the steps in the discrete distribution function. Under a literal interpretation of the model, the resulting discrepancy implies a poor fit. However, if we wish to relax the assumption of fixed loss exposures (as in Section 7.3 below), then this smoothing may be desirable. As the number of loans in the portfolio grows larger, the discrete distribution can be more and more closely approximated as continuous, so the difference between the two approximations vanishes. Even for relatively modest values of K, as in the right-hand panel, both the discrete and continuous Lugannani–Rice formulae provide nearly perfect fit.

Probability mass at zero is an unavoidable feature of an actuarial model such as CreditRisk$^+$, as it is always possible that no defaults will occur. If one is interested only in tail probabilities, then no difficulties arise. However, the continuous Lugannani–Rice formula "blows up" in the neighbourhood of $x = 0$.[6] An expedient way to handle this problem is to find the point x^* that maximizes $\mathrm{LR}_X(x; w, u)$, and then set $\mathbb{P}[X \geq x] = \mathrm{LR}_X(x^*; w, u)$ for $x \leq x^*$. This correction has been applied in plotting Figure 7.1.

In [14], I assessed the accuracy and robustness of saddlepoint approximation of VaR on 15,000 simulated heterogeneous portfolios. The portfolios varied in number of borrowers (K as low as 200 and as high as 5000), in the skewness of the exposure size distribution, and in the average credit quality of borrowers.[7] I also varied the number of systematic sectors ($N \in \{1, 3, 7\}$) and the variances ($1/\alpha$) of the systematic risk factors, and assigned random vectors of factor loadings to the obligors. For each portfolio, I solved for VaR at the 99.5% percentile via the standard Panjer recursion algorithm. Whenever the distribution produced by that algorithm was inconsistent with the first two cumulants (as obtained directly from the CGF), I concluded that the Panjer algorithm had failed, and estimated VaR via Monte Carlo simulation with

---

[6] As $x \to 0$, the saddlepoint $\hat{z} \to -\infty$. It is easily shown that u then converges to zero from below, but that w converges to $-\sqrt{-2 \cdot \log(\mathbb{P}[X = 0])}$, and thus is bounded away from zero. This implies $\lim_{x \to 0} \mathrm{LR}_X(x; w, u) = -\infty$.

[7] The average default probability $\bar{p}$ was chosen from $\{0.001, 0.005, 0.01, 0.03, 0.05\}$, and then the individual p_A were drawn as iid exponential random variables with mean $\bar{p}$.

50,000 trials.[8] Lastly, I estimated VaR via the continuous Lugannani–Rice formula.

Several conclusions can be drawn from this exercise. First, the reliability of Panjer recursion depends strongly on the nature of the portfolio and the complexity of model specification. For simple risk factor specifications and small to medium-sized portfolios, the Panjer algorithm rarely failed. However, when $N = 7$ and $K = 5000$, it failed in 222 out of 1000 cases. As a practical matter, the consequences for VaR estimation were usually drastic. Compared against the Monte Carlo estimate of VaR, reported VaR from the Panjer algorithm was off by 30% or more in 200 of the 222 detected failures and by 60% or more in 141 of the failures. This result is not unexpected. The greater the number of sectors and the larger the portfolio, the longer the polynomials $A(z)$ and $B(z)$ in the recurrence equation, so the greater the opportunity for round-off error to accumulate.[9]

Second, the saddlepoint approximation was typically quite accurate. For the study as a whole, median absolute relative error was 0.67% of VaR, and absolute relative error was under 4.2% of VaR in 95% of the cases. When discrepancies are this small, it is not clear which result should be viewed as more accurate, because the saddlepoint approximation allows us to use exact values for the loss exposures, rather than discretized values. When Monte Carlo simulation was substituted for Panjer recursion, discretization could be avoided, but simulation noise was introduced.

Finally, outliers in absolute relative error were much more common for the smaller portfolios ($K \in \{200, 300\}$) than in medium-sized and larger portfolios. The examples in Figure 7.1 suggest an explanation: Even when the continuous Lugannani–Rice approximation cuts through the discrete distribution at the mid-point of each step, estimated VaR can differ by as much as half a unit of loss. When the VaR is a small number of loss units, such differences are large in relative terms. This discrepancy could have been avoided by interpolating VaR from the discrete distribution as suggested in Chapter 2, or by using the discrete Lugannani–Rice formula.

There are alternatives to the Lugannani–Rice formula. The Barndorff-Nielsen formula, see [10], appears to be somewhat less robust for CreditRisk+. Other saddlepoint approximations make use of higher-order derivatives of the CGF, e.g. [17, §6.1] or non-Gaussian base distributions, e.g. [23], but do not improve performance in our application.

---

[8] Details of the moment-based diagnostic test are given in [14] and summarized in Chapter 2.

[9] The number of obligors is not the problem *per se*, but rather the maximum loss exposure among the ν_A. In my test portfolios, higher K were associated with a larger ratio between largest and smallest loans. If I maintained a reasonably fine discretization of the loss exposures, the maximum ν_A became quite large.

7.3 Incorporating Severity Risk

Severity risk can be introduced to the CreditRisk$^+$ framework without sacrificing any of the computational facility of the saddlepoint approximation. In the event of default, assume that loss exposure for obligor A is $\nu_A \xi_A$, where the severity multiplier ξ_A is gamma distributed with mean one and variance $1/\theta_A$. I assume that the ξ_A are independent across obligors and independent of the default events $\mathbf{1}_{A'}$ for all obligors A'. This is a variant on the treatment of specific severity risk in [4].[10]

A minor disadvantage of this specification is that it allows loss to exceed the outstanding exposure $\mathcal{E}_A$. In practice, this is not a problem. For loans with high expected loss given default (LGD) ($\nu_A \approx \mathcal{E}_A$), the variance of LGD should be small, so one expects to see a high value of θ_A assigned to this loan. More generally, so long as the θ_A is not too small, aggregate losses in the portfolio will be well-behaved, so the problem can be ignored.

Conditional on the systematic factors S and the severity multiplier ξ_A, the moment-generating function for loss on obligor A is

$$M_A(z|S, \xi_A) = \mathbb{E}[\exp(z\nu_A \xi_A \mathbf{1}_A)|S, \xi_A] = 1 - p_A^S + p_A^S \exp(z\nu_A \xi_A).$$

We integrate out the ξ_A to get the moment-generating function conditional only on S:

$$M_A(z|S) = \mathbb{E}[M_A(z|S, \xi_A)|S] = 1 - p_A^S + p_A^S \mathbb{E}[\exp(z\nu_A \xi_A)].$$

Observe that $\mathbb{E}[\exp(z\nu_A \xi_A)]$ is the moment-generating function for ξ_A evaluated at $z\nu_A$. Using a standard result for the gamma distribution, we have

$$\mathbb{E}[\exp(z\nu_A \xi_A)] = (1 - z\nu_A/\theta_A)^{-\theta_A}.$$

Conditional independence across the obligors is preserved, and terms of the form $(1 - z\nu/\theta)^{-\theta}$ do not depend on S. Therefore, the CGF takes on exactly the same form as (7.2), where the $\mathcal{Q}_k(z)$ functions are redefined as

$$\mathcal{Q}_k(z) = \sum_A w_{Ak} p_A (1 - z\nu_A/\theta_A)^{-\theta_A}.$$

This model embeds the standard model as a limiting case. It is easily verified that

$$\lim_{\theta \to \infty} (1 - z\nu/\theta)^{-\theta} = \exp(z\nu).$$

Therefore, the generalized $\mathcal{Q}_k(z)$ functions converge to the expression in (7.1) when the variance $1/\theta_A$ goes to zero for all obligors.

---

[10] There may also be systematic risk in loss severity, as has been argued by Frye [12, 13]. Systematic risk in severities could be accommodated by introducing a factor model for the ξ_A. One simple way to do this is demonstrated by Bürgisser et al. [4], see Chapter 9.

To simplify notation and to allow explicitly for some (or all) of the θ_A to be infinite, define the function

$$\Lambda(t;\theta) \equiv \begin{cases} (1-t/\theta)^{-\theta} & \text{if } \theta \in \mathbb{R}_+, \\ \exp(t) & \text{if } \theta = \infty. \end{cases}$$

The generalized $\mathcal{Q}_k(z)$ functions can then be compactly written as

$$\mathcal{Q}_k(z) = \sum_A w_{Ak} p_A \Lambda(z\nu_A;\theta_A).$$

The derivatives of the $\mathcal{Q}_k(z)$ are given by

$$D^j \mathcal{Q}_k(z) = \sum_A w_{Ak} p_A \nu_A^j \Lambda_j(z\nu_A;\theta_A),$$

where the Λ_j functions are defined by

$$\Lambda_j(t;\theta) \equiv \begin{cases} \frac{\Gamma(\theta+j)}{\theta^j \Gamma(\theta)}(1-t/\theta)^{-(\theta+j)} & \text{if } \theta \in \mathbb{R}_+, \\ \exp(t) & \text{if } \theta = \infty. \end{cases}$$

We can also write the definition of Λ_j recursively as

$$\Lambda_{j+1}(t;\theta) = \frac{\theta+j}{\theta}(1-t/\theta)^{-1}\Lambda_j(t;\theta)$$

where $\Lambda_0(t;\theta) \equiv \Lambda(t;\theta)$. Otherwise, all calculations are as in Section 7.1. In particular, the definition of the $V_{j,k}(z)$ functions in terms of the $\mathcal{Q}_k(z)$ and the expressions for the derivatives of $\psi_x(z)$ in terms of the $V_{j,k}(z)$ are all unchanged.

The properties of the CGF remain as described in Section 7.1, except that the supremum of the valid domain of ψ_x may change. The bounds provided by (7.6) are valid in the generalized model, but there are poles as well at $z = \theta_A/\nu_A$ for each obligor. As shown in Appendix 1, these additional constraints are easily incorporated in the procedure for finding z^*. The loss distribution is now continuous for $x > 0$ (though there remains mass at $x = 0$), so the continuous Lugannani–Rice formula is used for saddlepoint approximation.

Introducing idiosyncratic severity risk increases value-at-risk. The smaller is θ, the greater the amount of risk added to the portfolio. The larger is K, the more the additional risk is diversified away, so the smaller the impact on VaR. Figure 7.2 demonstrates these effects on homogeneous portfolios that differ only in K and θ. There is a single systematic factor ($N = 1$) with $\alpha = 1/2$, and I set $p_A = p = 0.01$, $w_{A1} = w = 0.9$ and $\nu_A = 1$ for all A. The target solvency probability is $q = 99.5\%$. Loss is expressed as a percentage of total expected loss exposure, i.e., define

$$X_K = \frac{1}{K}\sum_A \mathbf{1}_A \xi_A.$$

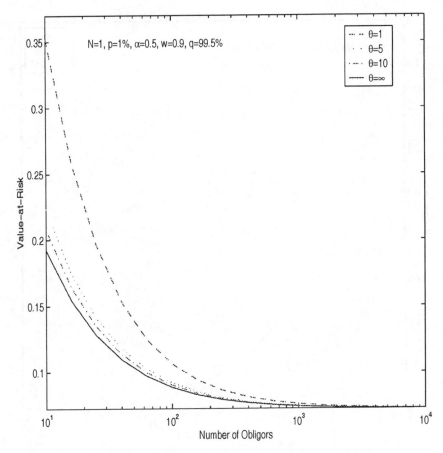

Fig. 7.2. Effect of severity risk on value-at-risk

The x-axis (log scale) shows the number of obligors K. The y-axis shows $\text{VaR}_q[X_K]$.

Observe that VaR converges to the same asymptotic value regardless of θ. In this simple one-factor model, the asymptotic VaR is given by $\text{VaR}_q[X_\infty] = p(1 + w(S_q - 1))$, where S_q is the q-th percentile of the distribution of S. Gordy [15] shows how generalized versions of this result follow from asymptotic properties of single-factor models of credit risk.

In Figure 7.3, the results for the extended model are plotted in terms of the "premium" over the model with no severity risk. Let $\text{VaR}_q[X_K; \theta]$ denote VaR for X_K with severity risk parameter θ. The premium plotted on the y-axis is then $(\text{VaR}_q[X_K; \theta] - \text{VaR}_q[X_K; \infty])/\text{VaR}_q[X_K; \infty]$. In the case of $p = 0.01$ and $\theta = 5$, for example, VaR is increased by 16.3% over the same model with no severity when $K = 10$. When $K = 100$, the premium falls to under 4%.

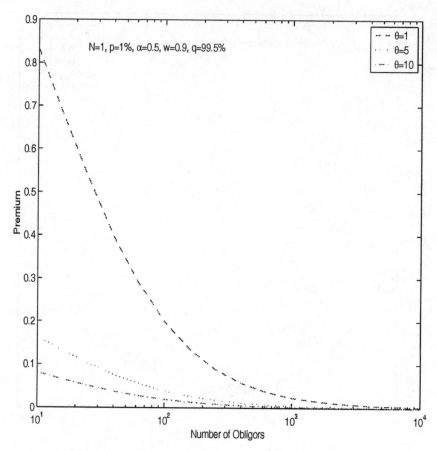

Fig. 7.3. Premium associated with severity risk

The relative impact of introducing severity risk depends on the credit quality of the obligors. When p is high, there are many defaults in the event that $X = \text{VaR}_q[X]$, so any idiosyncratic severity risk will tend to be diversified away. When p is low, then the default rate is relatively low, even in the stress event represented by $X = \text{VaR}_q[X]$. Idiosyncratic severity risk is less effectively diversified when defaults are few, so remains a significant contributor to total portfolio risk. Figure 7.4 is similar to Figure 7.3, except that p is varied across the plotted curves and $\theta = 5$. When $p = 0.2\%$, the premium is 11.6% for a portfolio of $K = 100$. If the portfolio of 100 obligors has $p = 5\%$, then the premium is under 1%. Thus, whether idiosyncratic severity risk can be ignored depends not only on the size of the portfolio, but also on its quality. In any case, the computational cost to including severity risk is always quite small.

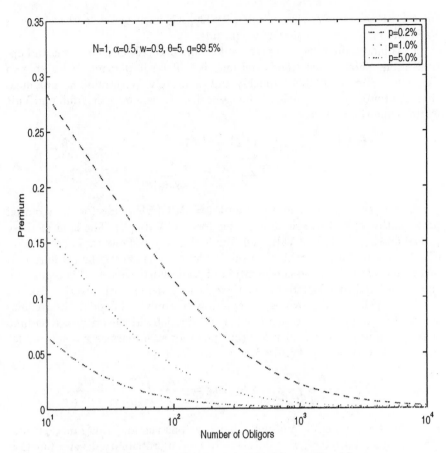

Fig. 7.4. Premium associated with severity risk

7.4 Saddlepoint Approximation of Expected Shortfall

As an alternative to VaR, Acerbi and Tasche [1] propose that risk be measured using generalized expected shortfall (ES), defined by

$$\mathrm{ES}_q[X] = \frac{\mathbb{E}[X\mathbf{1}_{X \geq \mathrm{VaR}_q[X]}] + \mathrm{VaR}_q[X](q - \mathbb{P}[X < \mathrm{VaR}_q[X]])}{1 - q}. \qquad (7.8)$$

The first term is often used as the definition of expected shortfall for continuous variables. It is also known as "tail conditional expectations" and is equivalent to

$$\mathrm{ES}_q[X] = \mathbb{E}[X | X \geq \mathrm{VaR}_q[X]].$$

The second term is a correction for mass at the q-th quantile of Y. Under this definition, Acerbi and Tasche [1] show that ES is coherent and equivalent to the condtional value-at-risk (CVaR), see [20]. ES is always at least as large as VaR, and typically is of similar magnitude.

ES is increasingly used in practice for portfolio risk-measurement and optimization. Should this trend continue, it will be important to be able to calculate ES in CreditRisk$^+$ quickly and accurately. With minimal modification, the Panjer recursion algorithm can calculate expected shortfall and VaR at the same time. Observe that

$$\mathbb{E}[X\mathbf{1}_{X\geq \mathrm{VaR}_q[X]}] = \mathbb{E}[X] - \mathbb{E}[X\mathbf{1}_{X<\mathrm{VaR}_q[X]}]$$
$$= \sum_A p_A \nu_A - \sum_{n<\mathrm{VaR}_q[X]} n\mathbb{P}[X=n].$$

As the Panjer recursion continues until $n = \mathrm{VaR}_q[X]$, values for the required probabilities $\mathbb{P}[X=n]$ are all available for $n < \mathrm{VaR}_q[X]$. The sum of these probabilities gives $\mathbb{P}[X < \mathrm{VaR}_q[X]]$. While this result allows rapid calculation of ES, the numerical instability of the Panjer scheme affects the computation of ES just as it does the computation of VaR. Furthermore, this algorithm cannot be used to calculate ES for the extended model with severity risk.

Studer [21, Ch. 2] provides a simple way to apply saddlepoint approximation to ES. As in the extended CreditRisk$^+$ model with severity risk, assume that X is continuous for all $x > 0$ (but possibly with mass as $x = 0$), and let $h(x)$ be the probability density function for X. Then

$$\mathrm{ES}_q[X] = \frac{1}{1-q}\int_{\mathrm{VaR}_q[X]}^\infty x\,h(x)\,dx = \frac{\mathbb{E}[X]}{1-q}\int_{\mathrm{VaR}_q[X]}^\infty \frac{x\,h(x)}{\mathbb{E}[X]}dx.$$

Let $\tilde{h}(x) \equiv x\,h(x)/\mathbb{E}[X]$, and note that (by construction) this function integrates to one. Therefore, we can take $\tilde{h}$ to be the probability density function for a new random variable $\tilde{X}$, and so can write

$$\mathrm{ES}_q[X] = \frac{\mathbb{E}[X]}{1-q}\int_{\mathrm{VaR}_q[X]}^\infty \tilde{h}(x)\,dx = \frac{\mathbb{E}[X]}{1-q}\mathbb{P}[\tilde{X} \geq \mathrm{VaR}_q[X]].$$

Saddlepoint approximation of $\mathbb{P}[\tilde{X} \geq \mathrm{VaR}_q[X]]$ requires the CGF of $\tilde{X}$. The moment-generating function for $\tilde{X}$ is

$$M_{\tilde{X}}(z) = \int_0^\infty \exp(zx)\tilde{h}(x)dx = \frac{1}{\mathbb{E}[X]}\int_0^\infty x\exp(zx)h(x)dx = \frac{M_X'(z)}{M_X'(0)}.$$

From the definition of the CGF, we have $\psi_X'(z) = M_X'(z)/M_X(z)$, which implies

$$\log(M_X'(z)) = \log(M_X(z)) + \log(\psi_X'(z)) = \psi_X(z) + \log(\psi_X'(z)).$$

Therefore, the CGF of $\tilde{X}$ is

$$\psi_{\tilde{X}}(z) = \psi_X(z) + \log(\psi_X'(z)) - \log(\psi_X'(0)). \tag{7.9}$$

The valid domain of $\psi_{\tilde{x}}(z)$ is the same as that of $\psi_x(z)$, and the derivatives of $\psi_{\tilde{x}}(z)$ are easily expressed in terms of the derivatives of $\psi_x(z)$.

Regardless of the number of sectors, saddlepoint approximation of ES requires only two calls to the saddlepoint algorithm. The first approximates $\text{VaR}_q[X]$ using the CGF ψ_x, and the second approximates $\mathbb{P}[\tilde{X} \geq \text{VaR}_q[X]]$ using the CGF $\psi_{\tilde{x}}$. Numerical exercises along the lines of the previous section are straightforward. Not surprisingly, one finds that the effect of severity risk on ES is very similar to the effect on VaR.

The algorithm for the discrete case of the standard CreditRisk$^+$ model is similar. We find

$$\mathbb{E}[X\mathbf{1}_{X \geq \text{VaR}_q[X]}] = \mathbb{E}[X] \cdot \mathbb{P}[\tilde{X} \geq \text{VaR}_q[X]]$$

where $\tilde{X}$ is a discrete random variable with $\mathbb{P}[\tilde{X} = n] = n \cdot \mathbb{P}[X = n]/\mathbb{E}[X]$. As in the continuous case, the CGF of $\tilde{X}$ is given by (7.9). The first-stage saddlepoint approximation provides

$$\mathbb{P}[X < \text{VaR}_q[X]] \approx 1 - \text{LR}_x(\text{VaR}_q[X]; w, \tilde{u})$$

as well as $\text{VaR}_q[X]$. Therefore, all terms in (7.8) can be obtained in our two-step algorithm.

Appendix 1: Bounds on z_k^*

Define z_k^* as the value of z such that the denominator term in the $V_{j,k}$ equals zero, that is, $Q_k(z_k^*) = \mu_k/\delta_k = \mu_k + \alpha_k$. We will establish bounds on z_k^* for the generalized $Q_k(z)$ functions introduced in Section 7.3.

The $\Lambda(t; \theta)$ function is increasing and convex in t and is decreasing in θ. Since $\Lambda(0; \theta) = 1$, we have $Q_k(0) = \sum_A w_{Ak}p_A = \mu_k$, which implies that $0 < z_k^*$. To obtain an upper bound, observe that

$$Q_k(z)/\mu_k = \sum_A \frac{w_{Ak}p_A}{\mu_k}\Lambda(\nu_A z; \theta_A) \geq \sum_A \frac{w_{Ak}p_A}{\mu_k}\exp(\nu_A z)$$

$$\geq \exp\left(z\sum_A \frac{w_{Ak}p_A}{\mu_k}\nu_A\right).$$

Substitute $Q_k(z_k^*)/\mu_k = 1/\delta_k$ and take logs to get

$$z_k^* \leq -\log(\delta_k)\bigg/\frac{1}{\mu_k}\sum_A w_{Ak}p_A\nu_A.$$

These lower and upper bounds on z_k^* apply as well to the "standard" CreditRisk$^+$ CGF of Section 7.1.

Another upper bound on z_k^* is implied by the pole in $\Lambda(z\nu; \theta)$ at $z = \theta/\nu$. Let ζ be the minimum such pole for the obligors in the portfolio, that is,

$$\zeta \equiv \min\{\theta_A/\nu_A\}.$$

As z^* can be no greater than ζ, the search for z_k^* can be limited to the range

$$0 < z_k^* \le \min\left\{\zeta, -\log(\delta_k)\Big/\frac{1}{\mu_k}\sum_A w_{Ak}p_A\nu_A\right\}.$$

When the obligor for which $\zeta = \theta_A/\nu_A$ has zero weight on sector k, it is possible to have $\mathcal{Q}_k(\zeta) < \mu_k + \alpha_k$. In this case, it is harmless to set $z_k^* = \zeta$ because this z_k^* will not be the binding one for z^*. In the case of no severity risk, we simply set $\zeta = \infty$.

Now let k denote the sector for which $z^* = z_k^*$. For this sector, we obtain $\lim_{z \to z^*} V_{1,k}(z) = \infty$. As $\psi'(z) \ge \psi_k'(z) = \alpha_k V_{1,k}(z)$, we have $\lim_{z \to z^*} \psi'(z) = \infty$. We also know that $\psi'(0)$ equals the mean of the loss distribution. Therefore, for any $x > \mathbb{E}[X]$, there must exist a $\hat{z} \in (0, z^*)$ such that $\psi_x'(\hat{z}) = x$.

Appendix 2: Lugannani–Rice Formula Near the Mean of a Distribution

In this appendix, we will examine the behaviour of $1/u - 1/w$ and $1/\tilde{u} - 1/w$ in the Lugannani–Rice formulae when $\hat{z} \approx 0$. By substituting $y = \psi'(\hat{z})$ into the expression for w, we obtain

$$w = \text{sign}(\hat{z})\sqrt{2(\hat{z}\psi'(\hat{z}) - \psi(\hat{z}))}.$$

If the CGF and all its cumulants exist in some open set that includes the origin, we can expand $\psi(z)$ as

$$\psi(z) = \kappa_1 z + \frac{1}{2}\kappa_2 z^2 + \frac{1}{3!}\kappa_3 z^3 + \frac{1}{4!}\kappa_4 z^4 + \mathcal{O}(z^5),$$

which implies

$$\psi'(z) = \kappa_1 + \kappa_2 z + \frac{1}{2}\kappa_3 z^2 + \frac{1}{3!}\kappa_4 z^3 + \mathcal{O}(z^4)$$

$$\psi''(z) = \kappa_2 + \kappa_3 z + \frac{1}{2}\kappa_4 z^2 + \mathcal{O}(z^3).$$

Thus, for $\hat{z} \approx 0$,

$$w = \text{sign}(\hat{z})\sqrt{\kappa_2\hat{z}^2 + \frac{2}{3}\kappa_3\hat{z}^3 + \frac{1}{4}\kappa_4\hat{z}^4 + \mathcal{O}(\hat{z}^5)}$$

$$= \hat{z}\sqrt{\kappa_2}\sqrt{1 + \frac{2}{3}\frac{\kappa_3}{\kappa_2}\hat{z} + \frac{1}{4}\frac{\kappa_4}{\kappa_2}\hat{z}^2 + \mathcal{O}(\hat{z}^3)}$$

and similarly

$$u = \hat{z}\sqrt{\kappa_2}\sqrt{1 + \frac{\kappa_3}{\kappa_2}\hat{z} + \frac{1}{2}\frac{\kappa_4}{\kappa_2}\hat{z}^2 + \mathcal{O}(\hat{z}^3)}.$$

Both $1/w$ and $1/u$ contain terms of the form $(1 + \eta_1 z + \eta_2 z^2 + \mathcal{O}(z^3))^{-1/2}$, which has Taylor expansion

$$(1 + \eta_1 z + \eta_2 z^2 + \mathcal{O}(z^3))^{-1/2} = 1 - \frac{\eta_1}{2}z + \left(\frac{3}{8}\eta_1^2 - \frac{1}{2}\eta_2\right)z^2 + \mathcal{O}(z^3).$$

Therefore, $1/u - 1/w$ can now be expanded as

$$\begin{aligned}
\frac{1}{u} - \frac{1}{w} &= \frac{1}{\hat{z}\sqrt{\kappa_2}}\left(\left(1 - \frac{1}{2}\frac{\kappa_3}{\kappa_2}\hat{z} + \left(\frac{3}{8}\frac{\kappa_3^2}{\kappa_2^2} - \frac{1}{4}\frac{\kappa_4}{\kappa_2}\right)\hat{z}^2 + \mathcal{O}(\hat{z}^3)\right)\right. \\
&\qquad\qquad \left. - \left(1 - \frac{1}{3}\frac{\kappa_3}{\kappa_2}\hat{z} + \left(\frac{1}{6}\frac{\kappa_3^2}{\kappa_2^2} - \frac{1}{8}\frac{\kappa_4}{\kappa_2}\right)\hat{z}^2 + \mathcal{O}(\hat{z}^3)\right)\right) \\
&= \frac{1}{\hat{z}\sqrt{\kappa_2}}\left(\frac{-\kappa_3}{6\kappa_2}\hat{z} + \left(\frac{5}{24}\frac{\kappa_3^2}{\kappa_2^2} - \frac{1}{8}\frac{\kappa_4}{\kappa_2}\right)\hat{z}^2 + \mathcal{O}(\hat{z}^3)\right) \\
&= \frac{-\kappa_3}{6\kappa_2^{3/2}} + \left(\frac{5\kappa_3^2 - 3\kappa_2\kappa_4}{24\kappa_2^{5/2}}\right)\hat{z} + \mathcal{O}(\hat{z}^2).
\end{aligned}$$

If one is using the Lugannani–Rice formula for discrete distributions, then u is replaced by $\tilde{u}$, which can be written as

$$\tilde{u} = \frac{1 - \exp(-\hat{z})}{\hat{z}}u = \left(1 - \frac{1}{2}\hat{z} + \frac{1}{6}\hat{z}^2 + \mathcal{O}(\hat{z}^3)\right)u,$$

which implies that

$$\begin{aligned}
\frac{1}{\tilde{u}} &= \left(1 + \frac{1}{2}\hat{z} + \frac{1}{12}\hat{z}^2 + \mathcal{O}(\hat{z}^3)\right)\frac{1}{u} \\
&= \frac{1}{u} + \frac{1}{2\sqrt{\kappa_2}} + \left(\frac{1}{12\sqrt{\kappa_2}} - \frac{\kappa_3}{4\kappa_2^{3/2}}\right)\hat{z} + \mathcal{O}(\hat{z}^2).
\end{aligned}$$

When applied to CreditRisk$^+$, I find these first-order approximations for $1/u - 1/w$ and for $1/\tilde{u} - 1/w$ highly accurate well beyond the range of $\hat{z}$ for which numerical instability is an issue.

References

1. C. Acerbi and D. Tasche. On the coherence of expected shortfall. *Journal of Banking & Finance*, 26(7):1487–1503, 2002.

2. P. Artzner, F. Delbaen, J.-M. Eber, and D. Heath. Coherent measures of risk. *Mathematical Finance*, 9(3):203–228, 1999.

3. O. Barndorff-Nielsen and D. R. Cox. Edgeworth and saddle-point approximations with statistical applications. *Journal of the Royal Statistical Society Series B*, 41(3):279–312, 1979.

4. P. Bürgisser, A. Kurth, and A. Wagner. Incorporating severity variations into credit risk. *Journal of Risk*, 3(4):5–31, 2001.

5. M. Carey. Dimensions of credit risk and their relationship to economic capital requirements. In F. S. Mishkin, editor, *Prudential Supervision: What Works and What Doesn't*. University of Chicago Press, 2001.

6. Credit Suisse Financial Products. CreditRisk$^+$: A credit risk management framework. London, 1997. Available at http://www.csfb.com/creditrisk.

7. H. Daniels. Saddlepoint approximations in statistics. *Annals of Mathematical Statistics*, 25(4):631–650, 1954.

8. H. Daniels. Exact saddlepoint approximations. *Biometrika*, 67(1):59–63, 1980.

9. H. Daniels. Tail probability approximations. *International Statistical Review*, 55(1):37–48, 1987.

10. A. Feuerverger and A. C. M. Wong. Computation of value-at-risk for nonlinear portfolios. *Journal of Risk*, 3(1):37–55, 2000.

11. R. Frey and A. J. McNeil. VaR and expected shortfall in portfolios of dependent credit risks: Conceptual and practical insights. *Journal of Banking and Finance*, 26(7):1317–1334, 2002.

12. J. Frye. Collateral damage: A source of systematic credit risk. *Risk*, 13(4):91–94, 2000.

13. J. Frye. Depressing recoveries. *Risk*, 13(11):108–111, 2000.

14. M. B. Gordy. Saddlepoint approximation of CreditRisk$^+$. *Journal of Banking and Finance*, 26(7):1335–1353, 2002.

15. M. B. Gordy. A risk-factor model foundation for ratings-based bank capital rules. *Journal of Financial Intermediation*, 12(3):199–232, 2003.

16. H. Haaf and D. Tasche. Credit portfolio measurements. *GARP Risk Review*, (7):43–47, 2002.

17. J. L. Jensen. *Saddlepoint Approximations*. Clarendon Press, Oxford, 1995.

18. R. Martin, K. Thompson, and C. Browne. Taking to the saddle. *Risk*, 14(6):91–94, 2001.

19. R. Martin, K. Thompson, and C. Browne. VaR: Who contributes and how much? *Risk*, 14(8):99–102, 2001.

20. R. T. Rockafellar and S. Uryasev. Conditional value-at-risk for general loss distributions. *Journal of Banking and Finance*, 26(7):1443–1471, 2002.

21. M. Studer. *Stochastic Taylor expansions and saddlepoint approximations for risk management*. PhD thesis, ETH Zürich, 2001.

22. K. van de Castle and D. Keisman. Recovering your money: Insights into losses from defaults. Special Report, Standard & Poor's, June 16, 1999.

23. A. T. A. Wood, J. G. Booth, and R. W. Butler. Saddlepoint approximations to the CDF of some statistics with nonnormal limit distributions. *Journal of the American Statistical Association*, 88(422), 680-686, 1993.

Fourier Inversion Techniques for CreditRisk$^+$

Oliver Reiß*

Summary. The CreditRisk$^+$ model is described in terms of characteristic functions, and two methods to determine the distribution of the credit loss based on Fourier inversion are presented. For the convenience of the reader, a short introduction to the theory of characteristic functions and the Fourier transformation is given. Then two general results are stated how to obtain the distribution of a random variable from its characteristic function. These general techniques, which are based on Fourier inversion, will be applied to the CreditRisk$^+$ model and yield efficient and numerically stable algorithms, which provide the loss distribution in the CreditRisk$^+$ framework. Advantages of this approach are that the algorithms are easy to implement and that a basic loss unit is not required.

Introduction

In the original description [3] of CSFP it was assumed that each exposure is an integer multiple of a so-called basic loss unit, and the mathematical description of the model is based on probability-generating functions. The first numerical method to handle this model was based on Panjer recursions, but these may be unstable. Other techniques for the CreditRisk$^+$ model have been developed, e.g. [5, 6]. In Chapter 17, Merino and Nyfeler present an approach based on the discrete Fourier transform, for which they work with the basic loss unit. In this chapter, the Fourier inversion for CreditRisk$^+$ is studied more generally, for instance without any loss unit, but in this setting one has to care about the existence of appearing integrals.

The CreditRisk$^+$ model will be described in terms of characteristic functions instead of probability-generating functions. Since the proper choice of the basic loss unit may be crucial, it is an advantage of this alternative presentation that no basic loss unit has to be introduced. The result of this analysis is the characteristic function of the loss of a loan portfolio.

* Supported by the DFG Research Center "Mathematics for key technologies" (FZT 86) in Berlin.

In order to obtain the distribution of the credit loss, a Fourier inversion of the characteristic function has to be performed. Unfortunately, the characteristic function in the CreditRisk$^+$ model is not integrable and the standard Fourier inversion integral does not exist. But even in this case there are two Fourier inversion methods that can be used to obtain the distribution of the credit loss. Hence an alternative algorithm is established that can be used for efficient computations in the framework of CreditRisk$^+$. In addition, this algorithm is easy to implement and is numerically stable, so it may also be applied for large loan portfolios.

Since the result of this algorithm is the loss distribution, one can easily compute some functionals of the loss. For risk management purposes, the main functionals are some risk measures like the value-at-risk or the expected shortfall. But of course, almost any other functional can be determined if the distribution of the loan portfolio is known.

The Fourier inversion techniques presented in this chapter do not rely on specific properties of the CreditRisk$^+$ model and hence the Fourier inversion methods can be used in a more general context. A possible application is the Fourier inversion of generalized CreditRisk$^+$ models, if these models are described in terms of characteristic functions [8].

The outline of this chapter is as follows. In Section 8.1 the basics of characteristic functions and the Fourier transform are recalled and the relationship between the characteristic function and some corresponding generating functions are given. In Section 8.2 two methods are presented that can be used to obtain the distribution of a random variable from its characteristic function even if the characteristic function is not integrable. In Section 8.3 the CreditRisk$^+$ model is briefly reviewed in terms of characteristic functions. The result is the characteristic function of the loss of a loan portfolio in this model. In Section 8.4 it is illustrated how the methods from Section 8.2 can be applied to the characteristic function in the CreditRisk$^+$ model. In addition, some remarks on numerical stability and the computation time are given. The final conclusions summarize the main results.

8.1 Introduction to the Theory of Characteristic Functions

Some basics about characteristic functions of real-valued distributions are recalled. The proofs of the statements are omitted, since these results can be found in several textbooks, e.g. [4].

Definition 1. *Let X be a real-valued random variable. Then the characteristic function of X is defined by*

$$\Phi_X(z) := \mathbb{E}[e^{izX}]$$

where i denotes the imaginary unit.

Note that the characteristic function $\Phi_X(z)$ exists for any real-valued random variable X and for all $z \in \mathbb{R}$. This is one reason why this function is a powerful tool in stochastics. The practical use of this function is based on relations presented below. There is a close connection between characteristic functions and the Fourier transform. Let us recall this notion, referring to functions $f : \mathbb{R} \to \mathbb{R}$ satisfying $\int\limits_{-\infty}^{\infty} |f(x)|\, dx < \infty$ as $\mathbf{L}^1$.

Definition 2. *Let f be $\mathbf{L}^1$. Then the Fourier transform of f exists and is defined by*

$$\mathcal{F}f(z) := \int\limits_{-\infty}^{\infty} e^{izx} f(x)\, dx.$$

Remark 1. If f is the probability density of a random variable X, then the characteristic function of X is given by the Fourier transform of f,

$$\Phi_X(z) = \mathcal{F}f(z).$$

Theorem 1. *Assume that f is $\mathbf{L}^1$ and also $\mathcal{F}f(z)$ is $\mathbf{L}^1$. Then*

$$f(x) = \frac{1}{2\pi} \int\limits_{-\infty}^{\infty} e^{-izx} (\mathcal{F}f)(z)\, ds.$$

This Fourier inversion theorem is very important. Assume that a characteristic function $\Phi_X(z)$ of a random variable X is given and $\Phi_X(z) \in \mathbf{L}^1$. Then this theorem not only tells us that X has a continuous density $f(x)$, but it also gives a formula to compute the density. Unfortunately, there is the precondition that the characteristic function is $\mathbf{L}^1$. In Section 8.2 some methods are presented that can be used in a more general context. Here we will focus on some properties of characteristic functions.

Theorem 2. *Let X, Y be two independent, real-valued random variables and Φ_X, Φ_Y their characteristic functions. Then for all $a, z \in \mathbb{R}$,[1]*

$$|\Phi_X(z)| \leq 1,$$
$$\Phi_X(-z) = \overline{\Phi_X(z)},$$
$$\Phi_{X+Y}(z) = \Phi_X(z) \cdot \Phi_Y(z),$$
$$\Phi_{X+a}(z) = e^{iza} \cdot \Phi_X(z),$$
$$\Phi_{a \cdot X}(z) = \Phi_X(a \cdot z).$$

The smoothness of the probability density is related to the decay behaviour of the characteristic function:

---

[1] The bar denotes the conjugate complex, i.e. for all $x, y, \in \mathbb{R}$, $\overline{x + iy} = x - iy$ holds.

Theorem 3. *Let X be a real-valued random variable with probability density f. If f has an n-th derivative $f^{(n)}$ and if $f^{(n)}$ is $\mathbf{L}^1$, then*

$$\lim_{z \to \pm\infty} |z|^n \Phi(z) = 0.$$

If a distribution is studied, one is mostly interested in the moments and the absolute moments of the random variable X, and if they exist they will be denoted by

$$m_n = \mathbb{E}[X^n], \qquad M_n = \mathbb{E}[\,|X|^n\,].$$

The moments of X and its characteristic function are closely related.

Theorem 4. *Let X be a real-valued random variable. If the n-th absolute moment M_n exists, then the characteristic function is n times continuous differentiable and is given by*

$$\Phi_X^{(n)}(z) = i^n \mathbb{E}[e^{izX} X^n].$$

In particular, $\Phi_X^{(n)}(0) = i^n m_n$.

An example of a distribution where even the first moment does not exist and where hence the characteristic function is not differentiable at 0 is given by the Cauchy distribution, which has the density $f(x) = \frac{1}{\pi}\frac{1}{1+x^2}$. The corresponding characteristic function is $\Phi(z) = e^{-|z|}$, which is obviously not differentiable at $z = 0$.

If all moments of a random variable X exist, then one may write down the following power series for $\Phi_X(z)$ in z:

$$\Phi_X(z) = \sum_{k=0}^{\infty} \frac{m_k}{k!} (iz)^k.$$

Define

$$\lambda := \limsup_{k \to \infty} \sqrt[k]{\frac{M_k}{k!}} = e \limsup_{k \to \infty} \frac{1}{k} \sqrt[k]{M_k},$$

then the Taylor series for the characteristic function converges for $\lambda|z| < 1$. Hence, if $\lambda < \infty$ the moments of X define the distribution of X uniquely. This is a sufficient condition; sufficient and necessary that the moments define the distribution uniquely is Carleman's condition:

$$\sum_{k=0}^{\infty} \frac{1}{\sqrt[2k]{M_{2k}}} = \infty.$$

An example of a distribution that is not uniquely defined by its moments is the lognormal distribution. Its characteristic function is infinitely often continuously differentiable, but not analytical in 0.

8.1.1 The moment-generating function

If the moment-generating function (MGF) of a random variable X exists, it is defined by

$$M_X(t) := \mathbb{E}[e^{tX}].$$

In general, $M_X(t)$ is only defined for t in a certain domain and not for all $t \in \mathbb{R}$. The relation to the characteristic function is given by

$$M_X(t) = \Phi_X(-it).$$

The name of this function is based on the relation

$$\mathbb{E}[X^n] = \frac{d^n M_X}{dt^n}(0).$$

Note that the characteristic function exists for real arguments in general, but only if it can be continued on the imaginary axis, the MGF exists. The lognormal distribution is one example where the MGF exists only for $t \leq 0$, and the Cauchy distribution has no MGF at all.

8.1.2 The cumulant-generating function

Let X be a real-valued random variable with existing moment-generating function $M_X(t)$. The cumulant-generating function (CGF), see also Chapter 7, is defined on the same domain as the MGF and given by

$$C_X(t) := \ln M_X(t) = \ln \Phi_X(-it).$$

If existing, the nth cumulant is defined by

$$\kappa_n := \frac{d^n C_X}{dt^n}(0) \tag{8.1}$$

and the following relation between the moments and the cumulants holds for all t:

$$\exp\left(\sum_{n=1}^{\infty} \kappa_n \frac{t^n}{n!}\right) = 1 + \sum_{n=1}^{\infty} m_n \frac{t^n}{n!}.$$

The name "cumulant" is based on the following property. Let $X_1, \ldots X_N$ be independent random variables, then

$$\kappa_n\left(\sum_{j=1}^{N} X_j\right) = \sum_{j=1}^{N} \kappa_n(X_j).$$

The first cumulant is the expectation, the second cumulant is the variance. The higher cumulants do not have such a clear interpretation, cf. Chapter 7.

8.1.3 The probability-generating function

Let X be a discrete random variable with values in $\mathbb{N}$. Recall from Chapter 2 that the probability-generating function (PGF) is given by

$$G_X(z) := \mathbb{E}[z^X],$$

and that this function exists at least for $|z| \leq 1$. The name and the practical relevance of this function is based on the property

$$\mathbb{P}[X = n] = \frac{1}{n!} \frac{d^n G_X}{dz^n}(0).$$

From $G_X(z)$ one easily obtains the corresponding characteristic function

$$\Phi_X(z) = G_X(e^{iz}). \tag{8.2}$$

8.2 Obtaining the Distribution from a Characteristic Function

For each real-valued distribution there is a characteristic function and the characteristic function determines the distribution uniquely. So it is quite natural to analyse the next question: How can one obtain the distribution function of a distribution which is given by its characteristic function?

In the special situation that the characteristic function is integrable, the answer to this question is simply given by Theorem 1 and one will obtain a continuous density of the distribution. But in general, and especially in application with CreditRisk$^+$, the characteristic function is not integrable. Also in this case there are possibilities to obtain the distribution function. The first method aims to find an approximative density using the special structure of the fast Fourier transformation (FFT) algorithm. Another method is based on the additional knowledge of the first two moments and only the difference between two distributions has to be determined by Fourier inversion.

8.2.1 The FFT-based Fourier inversion

Let $\Phi_X(z)$ be the known characteristic function of a a real-valued random variable X. Perhaps this random variable has no density, but let us assume that the distribution is "almost" continuous, that is, that X may be approximated by a random variable that has a density f. Further it is assumed that f vanishes outside a given interval $[a, b]$.

To describe the approximation one may require that the following condition holds for each g in a suitable class of functions $\mathcal{G}$:

$$\int f(x)g(x)dx = \mathbb{E}[g(X)] \tag{8.3}$$

and a proper choice of $\mathcal{G}$ is given by

$$\mathcal{G} := \{g(x) = e^{izx}|z \in \mathcal{Z}\}$$

where $\mathcal{Z}$ is a finite set of real numbers.

Note that the functions g are linearly independent, which is necessary to expect a suitable approximation. The expectation in (8.3) for these functions g is the characteristic function evaluated at some points $z \in \mathcal{Z}$. Since the characteristic function is known, the density f is the only unknown in (8.3) and hence one can use this equation to determine f. It is natural to expect that f will be a good approximation.

Now we are at the point where the structure of the FFT algorithm will be used, which is described e.g. in [7]. The unknown density f vanishes outside a given interval $[a, b]$ and so its Fourier transform can be numerically computed by the following algorithm. Let Z be the number of sample points (usually a power of 2) and define $\Delta x := \frac{b-a}{Z-1}$, $\Delta z := \frac{2\pi}{Z\Delta x}$ as well as three Z-dimensional vectors and a $Z \times Z$ matrix M for $j, k = 0, \ldots Z - 1$ by

$$x_k := a + k\Delta x,$$
$$f_k := f(x_k),$$
$$z_k := \begin{cases} k\Delta z & \text{if } k < \frac{Z}{2} \\ (k - Z)\Delta z & \text{else} \end{cases},$$
$$M_{jk} := \exp\left(2\pi i \frac{jk}{Z}\right).$$

Define the set $\mathcal{Z} := \{z_k | k = 0, \ldots Z - 1\}$. The Fourier transform of f at the points z_k will be denoted by $\Phi_k := \Phi_X(z_k)$. Then the Fourier integral can be computed by the approximation

$$\int\limits_{-\infty}^{\infty} e^{izx} f(x) dx \approx \Delta x e^{iza} \sum_{k=0}^{Z-1} e^{izk\Delta x} f(a + k\Delta x).$$

Using the vector and matrix notations, this equation yields[2]

$$\Phi_k = \Delta x e^{iaz_k} \sum_{j=0}^{Z-1} M_{kj} f_j.$$

Having chosen a suitable interval $[a, b]$ and the number of Fourier steps Z, the density f_k is the only unknown in (8.4). To determine f_k one has to solve the system of linear equations. This can be done easily, since the inverse of M_{jk} is given by

---

[2] Due to the special form of M_{jk} one can evaluate the matrix vector product by $\mathcal{O}(Z \ln Z)$ instead of the usual $\mathcal{O}(Z^2)$ operations, which is the reason why this kind of Fourier transformation is called "fast".

$$M_{jk}^{-1} = \frac{1}{Z} \exp\left(-2\pi i \frac{jk}{Z}\right).$$

A short calculation shows that this is in fact the inverse:

$$\sum_{k=0}^{Z-1} M_{jk} M_{kl}^{-1} = \frac{1}{Z} \sum_{k=0}^{Z-1} \left(e^{2\pi i \frac{i-l}{Z}}\right)^k = \begin{cases} 1 & j = l \\ \frac{1}{Z}\frac{1-e^{2\pi i(j-l)}}{1-e^{2\pi i(j-l)/Z}} = 0 & j \neq l \end{cases}.$$

Hence the inversion formula based on the linear equations is given by

$$f_j = \frac{1}{Z \Delta x} \sum_{k=0}^{Z-1} \exp\left(-2\pi i \frac{jk}{Z}\right) e^{-iaz_k} \Phi_k.$$

Since M^{-1} has a quite similar structure to M, one can compute the matrix vector multiplication by the FFT algorithm to save computation time. So even if the characteristic function is not integrable, the Fourier inversion by FFT works, since the algorithm does not compute an integral, but it solves a set of linear equations that describe the Fourier transformation of the unknown density f.

Smoothing the FFT result

In most cases one expects the density f to be a smooth function, but the result of the FFT algorithm may have a sawtooth pattern. This pattern is in fact a perturbation with the so-called Nyquist critical frequency $\frac{Z}{2}\Delta z = \frac{\pi}{\Delta x}$, which corresponds to the cut-off of the domain of the characteristic function. To eliminate this perturbation, one can apply the following algorithm, which performs a midpoint interpolation and so the number of sample points will be reduced by 1.

1. Smoothing algorithm: Elimination of a periodic perturbation
Let f be a density sampled on equidistant points x_k, $k = 0, \ldots, Z - 1$. Define

$$\tilde{f}_k := \frac{1}{2}(f(x_k) + f(x_{k-1})) \qquad k = 1, \ldots, Z - 1.$$

Then $\tilde{f}$ is a smooth version of the density f sampled on the points $x_k - \frac{1}{2}\Delta x$, $k = 1, \ldots, Z - 1$.

Another smoothing technique is needed at some positions where the density tends to infinity, which can happen in two situations. Either the density may not exist at a certain point x because $P[X = x] > 0$, hence the density is a Dirac δ function at x, or the density tends to infinity at x even though $P[X = x] = 0$. An example of such a density is given by the χ^2 distribution with one degree of freedom. Then the FFT result may be very oscillatory near x and for some k one may obtain $f_k < 0$ even though a density has to be non-negative! In such situations the following trick is helpful.

2. Smoothing algorithm: Smoothing near singularities

Let f be a density sampled on equidistant points x_k, $k = 0, \ldots, Z - 1$. If there are inner points such that the density is negative, hence $f_k < 0$ for a k in $1, \ldots, Z - 2$, then perform the following algorithm for each such k:[3]

$$\zeta := (f_{k-1})^+ + (f_{k+1})^+,$$

$$f_{k-1} := f_{k-1} + \frac{(f_{k-1})^+}{\zeta} f_k,$$

$$f_{k+1} := f_{k+1} + \frac{(f_{k+1})^+}{\zeta} f_k,$$

$$f_k := 0.$$

For applications in practice, these smoothing methods, which both have the computational effort of $\mathcal{O}(Z)$, are rather fast. Hence it is recommended to smooth a density by three steps:[4]

1. Apply the first smoothing algorithm.
2. Perform the second smoothing algorithm on the result.
3. Finally apply the first smoothing algorithm again.

As a result, the application of the FFT algorithm on the characteristic function $\Phi_X(z)$ and the smoothing of the FFT result yields a tabulated probability density that describes the law of the random variable X.

8.2.2 Fourier inversion using mean and variance

The idea of this method is not to obtain the density of X, but the integral of the distribution function. Even if the density is only defined in terms of Dirac δ functions, this function will be continuous and hence it is a good candidate for numerical computations. In order to use this method, one needs to know the first two moments of the unknown distribution, because the idea is to approximate the unknown distribution by a distribution with the same mean and variance. Then the Fourier inversion can be done using the following result.

Theorem 5. *Let F, G be the distribution functions of two random variables with the same mean and existing third absolute moment. Define*

$$\hat{F}(x) = \int_{-\infty}^{x} F(y)\,dy, \quad \hat{G}(x) = \int_{-\infty}^{x} G(y)\,dy.$$

---

[3] $(x)^+$ denotes the positive part, hence $(x)^+ = \max(0, x)$.
[4] Since the application of the first smoothing algorithm shifts the x-domain by $\frac{1}{2}\Delta x$, this three-step algorithm effects a shift of the domain by Δx.

Then the following inversion formula holds:

$$\hat{F}(x) = \hat{G}(x) - \frac{1}{2\pi} \int\limits_{-\infty}^{\infty} e^{-izx} \frac{\Phi_f(z) - \Phi_g(z)}{z^2} \, dz.$$

To understand the practical use of the theorem, consider a random variable X with known characteristic function Φ_X and suppose that the mean and the variance of X are also known. The task is now to determine the distribution function F of the random variable X.

First, one chooses a random variable Y with a well-known distribution. Typical candidates are the normal distribution or the gamma distribution, but other distributions will also work. The parameters of the distribution of Y are fitted is such a way that $\mathbb{E}[X] = \mathbb{E}[Y]$ and $\mathrm{var}[X] = \mathrm{var}[Y]$ hold. Under this condition the fraction under the integral is well defined and bounded for all z, especially if $z \to 0$. Since the distribution of Y is well known, the integral $\hat{G}$ of the distribution function can also be computed and hence the function $\hat{F}$ can be determined by Theorem 5. The distribution function and the density of X can be determined by differentiating $\hat{F}$.

The advantage of this approach is the independence on a special algorithm. Thus one can use more sophisticated integration methods than FFT for the valuation of the integral.

For the proof of Theorem 5 the following results are needed.

Lemma 1. *Let two random variables X, Y have an existing third absolute moment and let them have the same first moment. Then*

$$\lim_{z \to 0} \frac{\Phi_X(z) - \Phi_Y(z)}{z^2} = \frac{1}{2}(\mathbb{E}[Y^2] - \mathbb{E}[X^2]).$$

Proof. Note that due to the existence of the third moment the result can be obtained from the expansion of the characteristic functions near 0.   □

Lemma 2. *Let F and G be two distribution functions of two random variables with the same mean μ. Let $\hat{F}$ and $\hat{G}$ be their integrals. Then*

$$\lim_{x \to \pm\infty} (\hat{F} - \hat{G})(x) = 0.$$

Proof. From the existence of the expectation one can conclude

$$\int\limits_{x}^{\infty} y \, dF(y) = o(1) \qquad \text{as } x \to \infty.$$

Since $o(1) = \int_x^\infty y \, dF(y) \geq x \int_x^\infty dF(y) = x(1 - F(x))$, it follows that

$$F(x) = 1 - o\left(\frac{1}{x}\right) \qquad \text{as } x \to \infty.$$

The same asymptotic behaviour also holds for G. Since both distributions have the same mean, we get

$$o(1) = \lim_{x \to \infty} \int_{-\infty}^{x} y \, d(F - G)(y)$$

$$= \lim_{x \to \infty} [y(F(y) - G(y))]_{-\infty}^{x} - \int_{-\infty}^{x} (F(y) - G(y)) \, dy = \lim_{x \to \infty} \hat{G}(x) - \hat{F}(x).$$

The case $x \to -\infty$ follows from $\lim_{x \to -\infty} F(x) = \lim_{x \to -\infty} G(x) = 0$. $\qquad \square$

Proof (Proof of Theorem 5). Let F, G fulfil the conditions of the theorem. By definition of the characteristic function

$$\Phi_f(z) - \Phi_g(z) = \int_{-\infty}^{\infty} e^{izx} d(F - G)(x)$$

$$= [(F(x) - G(x))e^{izx}]_{-\infty}^{\infty} - iz \int_{-\infty}^{\infty} (F(x) - G(x))e^{izx} dx.$$

Since $\lim_{x \to \pm\infty} F(x) - G(x) = 0$ and e^{izx} is bounded, one obtains

$$\Phi_f(z) - \Phi_g(z) = -iz \int_{-\infty}^{\infty} (F(x) - G(x))e^{izx} dx.$$

Lemma 2 and another integration by parts yield

$$\Phi_f(z) - \Phi_g(z) = -iz \left[\left(\hat{F}(x) - \hat{G}(x) \right) e^{izx} \right]_{-\infty}^{\infty} + (iz)^2 \int_{-\infty}^{\infty} \left(\hat{F}(x) - \hat{G}(x) \right) e^{izx} dx$$

$$\frac{\Phi_f(z) - \Phi_g(z)}{(iz)^2} = \int_{-\infty}^{\infty} \left(\hat{F}(x) - \hat{G}(x) \right) e^{izx} dx. \qquad (8.4)$$

So $\frac{\Phi_f(z) - \Phi_g(z)}{(iz)^2}$ is the Fourier transform of $\hat{F}(x) - \hat{G}(x)$. The left-hand side of (8.4) is well defined (see Lemma 1) and integrable, since it decays with $\frac{1}{z^2}$, because Φ_f and Φ_g are bounded by 1. Therefore one may apply the Fourier inversion formula (8.1) to obtain

$$\hat{F}(x) - \hat{G}(x) = \frac{1}{2\pi} \int_{-\infty}^{\infty} \frac{\Phi_f(z) - \Phi_g(z)}{(iz)^2} e^{-izx} dz.$$

$\qquad \square$

8.3 CreditRisk$^+$ in Terms of Characteristic Functions

In the previous sections, characteristic functions have been introduced as a powerful tool to describe the distribution of a random variable. Now the CreditRisk$^+$ model is presented in terms of characteristic functions instead of PGFs. In the original description of CreditRisk$^+$, there is a basic loss unit L_0 such that the exposure is assumed to be an integer multiple of L_0. The use of this loss unit is essential for working with PGFs.

For the description of the CreditRisk$^+$ model in terms of characteristic functions, one does not need such a basic loss unit. Hence, the exposure of each obligor may have any value, which will not be rounded to an integer multiple of a basic loss unit. One also avoids the problem of determining a suitable value of the basic loss unit. The proper choice may be rather difficult if the exposures range from small credit card transactions to large corporate loans. If the basic loss unit is chosen rather small, the error due to rounding the exposures is also small, but the computational effort is rather large. If the basic loss unit is chosen large, the computational effort is small but there is a large impact of rounding the exposures.

Hence, in the following short description of the CreditRisk$^+$ model, one abstains from the use of a basic loss unit. The computations are kept brief, since they are quite similar to the computations in the original description.

8.3.1 Recalling the model

Let us consider a loan portfolio with K obligors as in Chapter 2. If obligor A defaults, the corresponding loss is given by $\nu_A > 0$. For the dependencies between the obligors, N sectors are used and the affiliation of each obligor to these sectors is described by the factor loadings w_{Ak}. The share of specific or idiosyncratic risk is denoted by w_{A0}. The N sectors are modelled by independent gamma-distributed random variables S_k that have the expectation 1 and the variance σ_k^2. Recall from (2.9) that the default intensity of the obligor A is given by

$$p_A^S = p_A \left(w_{A0} + \sum_{k=1}^{N} w_{Ak} S_k \right), \tag{8.5}$$

where p_A is the probability of default within one year. For a given state of the sector variables S_k the default events of the obligors are independent of each other. Recall that the loss of the loan portfolio is denoted by X and given by

$$X = \sum_A \mathbf{1}_A \nu_A.$$

8.3.2 The characteristic function of X

The aim is to obtain the distribution of X. Since the characteristic function contains all the information about the distribution, one method to determine the distribution is to compute Φ_X.

In [8] a derivation of Φ_X is given without using the PGF. Here the characteristic function is determined from the probability-generating function G_X, which is known from (2.20):

$$G_X(z) = e^{\sum_A p_A w_{A0}(z^{\nu_A} - 1)} \prod_{k=1}^{N} \left(\frac{1}{1 + \sigma_k^2 T \sum_A p_A w_{Ak}(1 - z^{\nu_A})} \right)^{\frac{1}{\sigma_k^2}}.$$

There is the general relationship between the PGF and the corresponding characteristic function, see (8.2). By this relation one obtains the characteristic function, which remains valid if no basic loss unit has been introduced:

Theorem 6. *The characteristic function of loss in standard CreditRisk$^+$ is given by*

$$\Phi_X(z) = e^{\sum_A p_A w_{A0}(e^{i\nu_A z} - 1)} \prod_{k=1}^{N} \left(\frac{1}{1 + \sigma_k^2 T \sum_A p_A w_{Ak}(1 - e^{i\nu_A z})} \right)^{\frac{1}{\sigma_k^2}}.$$

Based on the characteristic function of the loss in standard CreditRisk$^+$ the distribution will be determined numerically in the next section. To get a first impression of it, one may recall from (2.13) and (2.26) that

$$\mathbb{E}[X] = \sum_A p_A \nu_A, \tag{8.6}$$

$$\text{var}[X] = \sum_A p_A \nu_A^2 + \sum_{k=1}^{N} \sigma_k^2 \left(\sum_A w_{Ak} p_A \nu_A \right)^2. \tag{8.7}$$

8.4 Fourier Inversion in CreditRisk$^+$

Theorem 6 provides the characteristic function of the credit loss X in CreditRisk$^+$. In the first part of this chapter, general properties of characteristic functions were presented and two Fourier inversion methods were provided to obtain the distribution of a random variable from its characteristic function. Now Fourier inversion techniques will be applied to the characteristic function Φ_X.

The first method is based on the special algorithm of the fast Fourier transform (FFT). The general idea has been presented in Section 8.2.1 and now some hints on the specific use of FFT to determine the distribution of X will be given. The result is an approximative density of X.

It is obvious from its definition that X does not have a continuous density. In some cases, especially if the number of obligors is small, one would like to obtain the distribution without the assumption that an approximative density exists. Theorem 5 provides a method that applies in this context and that requires additional input, the first two moments and a distribution that is similar to the distribution of X. The extra information is available, since the expectation and variance of X are known and the gamma distribution is recommended as a first approximation of the loss.

8.4.1 Fourier inversion by FFT

Although the credit loss is a discrete random variable, it is intuitively obvious that there is an approximative density for the distribution if the number of obligors is large. This density can be determined using the FFT algorithm.

The loss of a loan portfolio lies naturally between 0, the case that no obligor defaults, and $\sum_A \nu_A$, which corresponds to the worst case that all obligors default. Hence the density f resides on $[0, \sum_A \nu_A]$. Let Z be the number of sample points; then the distances between two adjacent sampling points are given by

$$\Delta x = \frac{1}{Z-1} \sum_A \nu_A \quad \text{and} \quad \Delta z = \frac{2\pi}{Z\Delta x}.$$

The choice of Z depends on the accuracy one needs for further computations using the density. If g is a smooth function, more precisely if the first derivative of g is bounded, the error of the expectation of $g(X)$ is of order Δx:

$$\left| \mathbb{E}[g(X)] - \sum_{k=0}^{Z-1} g(k\Delta x)f(k\Delta x) \right| = \mathcal{O}(\Delta x).$$

In practice, Z will be taken as a power of 2, hence $Z = 2^n$ and typical values for n lie round about 10. If a proper value of Z has been chosen, one has to evaluate the characteristic function at $z_j = (j - \frac{Z}{2})\Delta z$, since these values are the input for the FFT algorithm. To compute $\Phi_X(z)$ one may adapt the formula given in Theorem 6:

$$\Phi_X(z) = \exp\left(\xi_0(z)\right) \prod_{k=1}^{N} \left(\frac{1}{1-\sigma_k^2 \xi_k(z)}\right)^{\frac{1}{\sigma_k^2}} \tag{8.8}$$

$$\text{where } \xi_k(z) := \sum_A p_A T w_{Ak}(e^{i\nu_A z} - 1) \quad k = 0, \dots, N.$$

Using the main branch of the natural logarithm one can rewrite the power expression:

$$\left(\frac{1}{1 - \sigma_k^2 \xi_k(z)}\right)^{\frac{1}{\sigma_k^2}} = \exp\left(-\frac{1}{\sigma_k^2}\ln(1 - \sigma_k^2 \xi_k(z))\right)$$

and hence the characteristic function can be evaluated using the equation

$$\Phi_X(z) = \exp\left(\xi_0(z) - \sum_{k=1}^{N} \frac{1}{\sigma_k^2}\ln\left(1 - \sigma_k^2 \xi_k(z)\right)\right).$$

This formula is numerically more stable than the representation (8.8), since one has to evaluate a sum of complex numbers, each with negative real part instead of a product of complex numbers. As shown in Section 8.2.1, the Fourier inversion of this function using the FFT algorithm yields the density of X.

To save computation time one should recall that $\Phi_X(-z) = \overline{\Phi_X(z)}$. The effort to compute the $\xi_K(z)$ is given by $\mathcal{O}(ZNK)$ and thereafter the computation of $\Phi_X(z)$ takes $\mathcal{O}(ZN)$ operations. The Fourier inversion done by FFT costs $\mathcal{O}(Z\ln Z)$ operations. Hence the overall computation time is given by

$$\text{Computational Effort} = \mathcal{O}(ZKN + Z\ln Z).$$

8.4.2 Fourier inversion using mean and variance

Another method to obtain the distribution of X from its characteristic function was presented in Section 8.2.2. Now the application of Theorem 5 to the specific situation in the CreditRisk$^+$ model is discussed. For an efficient use of that theorem it is necessary to find an approximative distribution with the same mean and variance.

The approximative distribution must have three moments and the first two moments have to be adjusted to the expectation and the variance of the loss X, which are given by (8.6) and (8.7). Since X is a non-negative random variable, the desired candidate should also reside on $\mathbb{R}^+$. Examples of such distributions are the lognormal distribution or the gamma distribution and their generalizations.

In [2] it is shown that the limiting distribution of a large loan portfolio is gamma distributed. Since the number of obligors is usually high, an approximation by the gamma distribution is reasonable. Additionally, all expressions that occur in the application of the gamma distribution in Theorem 5 can be expressed in closed form. Therefore the approximation of X by the gamma distribution is recommended.

Recall from (2.1) and (2.2) that the gamma distribution with density function is given by

$$f^{(\alpha,\beta)}(x) = \frac{x^{\alpha-1}}{\beta^\alpha \Gamma(\alpha)}e^{-x/\beta} \qquad x \geq 0$$

and a gamma-distributed random variable Y satisfies

$$\mathbb{E}[Y] = \alpha\beta, \qquad \text{var}[Y] = \alpha\beta^2.$$

Moreover, one has

$$\Phi_Y(z) = (1 - iz\beta)^{-\alpha}.$$

Additionally, the following statements also hold for $x \geq 0$:

$$\mathcal{H}(x) := \int_0^x f^{(\alpha,\beta)}(y)\, dy = \Upsilon(\alpha, x/\beta),$$

$$\hat{\mathcal{H}}(x) := \int_0^x G(y)\, dy = x\Upsilon(\alpha, x/\beta) - \alpha\beta\Upsilon(\alpha + 1, x/\beta)$$

where Υ denotes the incomplete gamma function (see e.g. [1, 7]).

Since the expectation and the variance of the loss X are known by (8.6) and (8.7), the parameters α and β of the gamma distribution with the same first two moments can be determined by

$$\beta = \frac{\text{var}[X]}{\mathbb{E}[X]} \qquad \text{and} \qquad \alpha = \frac{\mathbb{E}[X]^2}{\text{var}[X]}.$$

Let $F(x)$ denote the distribution function of X and $\hat{F}$ the integral of the distribution function. Then Theorem 5 yields

$$\hat{F}(x) = x\Upsilon(\alpha, x/\beta) - \alpha\beta\Upsilon(\alpha + 1, x\beta) - \frac{1}{2\pi}\int_{-\infty}^{\infty} e^{-izx}\frac{\Phi_X(z) - (1 - iz\beta)^{-\alpha}}{z^2}\, dz.$$

Due to the special choice of α and β, the integrand becomes zero for $z \to 0$. Hence the integral is well defined and converges, because the integrand decays with z^{-2} as $z \to \pm\infty$. In contrast to the FFT method, this integral may be evaluated by any integration method. Not having to depend on a particular integration algorithm gives the advantage that one can use more sophisticated integration tools, which may have a better rate of convergence. Hence the computation time will be reduced if the integration algorithm requires a smaller number of sample points.

Even by theory, the function $\hat{F}$ is continuous and so the numerical valuation of $\hat{F}$ will be stable. To obtain the distribution function of X, one has to differentiate $\hat{F}(x)$ with respect to x.

8.5 Conclusion

The CreditRisk$^+$ model has been presented in terms of characteristic functions instead of the usual approach by probability-generating functions. One advantage of this description is that no basic loss unit has to be introduced, since the proper choice of the basic loss unit may be critical, if the exposures have different orders of magnitude. Based on general results concerning the Fourier inversion of characteristic functions, alternative numerical algorithms to obtain the distribution of the credit loss in CreditRisk$^+$ have been provided.

The numerical algorithms are easy to implement since the main ingredients are basic complex arithmetic and a Fourier integration method. The fast Fourier transform is well known and is used in several applications, hence one might expect that this algorithm is already implemented in most banks and insurance companies. The algorithms presented are numerically stable and quite fast, so the idea of Fourier inversion applies also for large loan portfolios. Since one obtains the whole loss distribution, the Fourier inversion approach enables the efficient computation of some functionals like risk measures (e.g. value-at-risk, expected shortfall) or in the area of assets securitizations (e.g. asset-backed securities).

The Fourier inversion techniques presented are valid in more general contexts. Starting from a generalized CreditRisk$^+$ model, which provides an expression for the characteristic function, the presented Fourier inversion methods can be used again to obtain the corresponding distribution function.

Acknowledgements. The author thanks Hermann Haaf (Commerzbank AG) for raising interesting questions regarding the CreditRisk$^+$ model, and John Schoenmakers (Weierstrass Institute) for many valuable comments and suggestions. The research was done in cooperation with the bmb+f project "Effiziente Methoden zur Bestimmung von Risikomaßen", which is also supported by Bankgesellschaft Berlin AG.

References

1. M. Abramowitz and I. A. Stegun (eds.). *Handbook of Mathematical Functions: with Formulas, Graphs, and Mathematical Tables.* Dover Publications, Inc., New York, ninth printing, 1972.
2. P. Bürgisser, A. Kurth, and A. Wagner. Incorporating severity variations into credit risk. *Journal of Risk,* 3(4):5–31, 2001.
3. Credit Suisse Financial Products. CreditRisk$^+$: A credit risk management framework. London, 1997. Available at `http://www.csfb.com/creditrisk`.
4. W. Feller. *An Introduction to Probability Theory and Its Applications,* Wiley Series in Probability and Mathematical Statistics, John Wiley & Sons, Inc. New York. Volume I, Third Edition, 1968 – Volume II, 1966.
5. G. Giese. Enhancing CreditRisk$^+$. *Risk,* 16(4):73–77, 2003.
6. M. B. Gordy. Saddlepoint approximation of CreditRisk$^+$. *Journal of Banking & Finance,* 26:1335–1353, 2002.
7. W. H. Press, S. A. Teukolsky, W. T. Vetterling and B. P. Flannery. *Numerical Recipes in C: The Art of Scientific Computing.* Second Edition, Cambridge University Press, Cambridge, 1992.

8. O. Reiß. *Fourier inversion algorithms for generalized CreditRisk⁺ models and an extension to incorporate market risk*. WIAS–Preprint No. 817, 2003.

9

Incorporating Default Correlations and Severity Variations

Nese Akkaya, Alexandre Kurth and Armin Wagner

Summary. The original CreditRisk$^+$ methodology does not allow for incorporation of industry, geographical or other segment correlations in modelling default events. We provide an extension that enables modelling of default correlations among segments while preserving the analytical solution for the loss distribution. Moreover, the proposed methodology can consistently be extended to independently (of default events) model stochastic severities in collateral devaluation. This extension imposes further distributional assumptions on the model. Nevertheless, it is shown that in the limit of a large portfolio the loss distribution is determined solely by the systematic components of default and severity risk. Finally, the relevance of this observation to portfolio management and stress loss analysis is motivated.

Introduction

In this chapter we focus on the extension of CreditRisk$^+$ accounting for default correlations between different industries and the inclusion of severity risk, proceeding in the following way. We start with the original set-up of the model where the standard deviation of the loss distribution is derived by the use of probability-generating functions (PGF). Based on this technique we extend this concept to the incorporation of correlations between the risk factors. We then outline the drawback of the sometimes-mentioned concept of orthogonalization, resp. principal component transformation, in the context of the presented model. For the inclusion of severity risk we redefine the model set-up in the language of stochastic variables, introducing variables for default and severity risk. It is worthwhile noting that the newly derived standard deviation does not require the Poisson approximation for the individual default variable. The result shows that the original standard deviation is indeed an excellent approximation for small probabilities of default. The derivation is based on the decomposition into systematic and idiosyncratic risk components. The model just outlined makes the following three major assumptions: (i) systematic changes in default probabilities due to the economy can be described by scaling factors specific to industry segments, (ii) systematic changes in severities due to the economy can be described by scaling factors specific to the collateral segments, and (iii) the systematic risk factors for severities vary independently of those for default.

In the situation of a large portfolio the corresponding loss distribution fulfils specific properties that have been proven in [2, Chapter 6] on a rather abstract level. Here we derive these properties by an explicit construction using characteristic functions. The presented concept so far covers a rigorous quantification of the portfolio loss distribution respecting the drivers on default risk and those on the collateral devaluation. This is useful and necessary for the quantification of various risk measures, such as standard deviation, unexpected loss, value-at-risk, economic capital, expected shortfall, etc. For the management of a loan portfolio, the decomposition of such risk measures to contributors (e.g. individual entities or segments) is crucial. The determination of these contributions to the standard deviation is straightforward [3, Eq. 121]. Tasche and other authors [7, 8, 11] provided a rigorous calculation method for the contributions to value-at-risk and expected shortfall in the CreditRisk$^+$ environment.

9.1 Default Correlations

In its original form CreditRisk$^+$ starts off with modelling default events from a single segment. The introduction of a common risk scaling factor to describe default levels allows for modelling the common default behavior of obligors from a single segment. However, if we consider loans to corporates, the business activities of which extend over a variety of segments, we need to further enhance CreditRisk$^+$ by incorporating default correlations among different industries. One approach to handling the problem of correlations among obligors in CreditRisk$^+$ is to apportion the credit exposure of obligors among independent segments. Unfortunately, this approach is associated with practical problems at the implementation stage, since the correlations among certain industries are rather strong, and would therefore not be accounted for in the set-up outlined above. We will briefly revisit this standard CreditRisk$^+$ approach in Section 9.1.1. A praised remedy to the problem of industry correlations, the orthogonalization of the industry correlation matrix, will be discussed in detail in Section 9.1.4.

Another approach to consistently modelling default correlations among segments is introduced by Bürgisser et al. [1]. In this approach, the analytical formula for the variance of the portfolio loss, σ_X^2, is extended to incorporate the covariance information on default events from different industries. We present this extension in Section 9.1.3.

The current literature on CreditRisk$^+$ lacks contributions on modelling segment default correlations via copulas, which describe the dependence structure among the marginal distributions of a multivariate loss distribution, mainly for the reasons that:

a. the dependency among the lower-dimensional marginal distributions of a multivariate distribution are in general not completely described by the correlation matrix alone (exceptions like the multivariate normal distribution exist);
b. the scarcity of data would not support the estimation of the parameters of any postulated copula function.

In the presence of correlated segment defaults the assumption that the segment loss distributions are independent no longer holds. The dependence structure among the marginal distributions of a multivariate distribution is given by a so-called copula function, which together with the univariate marginal distributions uniquely defines the multivariate distribution. This implies that the loss distribution of a portfolio with obligors from a variety of correlated segments would be uniquely defined by the loss distribution for individual segments and the copula function describing their interdependency. This would require that the parameters of a postulated copula function need to be estimated besides the parameters of the individual segment loss distributions from the available portfolio loss data. Moreover, the observation that the functional forms of copulae can be quite general suggests that the predominant advantage of CreditRisk$^+$, the closed-form solution of the analytical presentation of risk, is no longer warranted.

9.1.1 Standard CreditRisk$^+$ revisited

Before proceeding to the extensions of CreditRisk$^+$, we briefly mention the main issues of the standard CreditRisk$^+$ model, which are relevant for this presentation. A detailed elaboration is found in Chapter 2. Recall that for a single-segment portfolio $\mathcal{S}$, the random variable for portfolio loss X is

$$X = \sum_{A \in \mathcal{S}} \mathbf{1}_A \nu_A \tag{9.1}$$

where $\mathbf{1}_A$ is the default indicator for obligor A with $\mathbb{P}[\mathbf{1}_A = 1] = p_A$, and ν_A is his exposure net of recovery.

As shown in Chapter 2, the first two moments of the portfolio loss distribution can be derived by utilizing the PGF $G(z) = \mathbb{E}[z^X] = \sum_{n \geq 0} p(n) z^n$ where $p(n)$ is the probability that the loss is of magnitude n and z is a formal variable. The application of PGFs is based on the following assumptions:

A1 the recovery rates are constant;
A2 the exposure nets of recovery are satisfactorily approximated by integer multiples of a fixed loss unit, which is necessary for the discrete presentation of losses (Section 2.2 shows that this is an unproblematic assumption);
A3 the distribution of the number of defaults can be approximated by the Poisson distribution (this is a valid approximation as long as the obligor default probabilities p_A are small).

In Section 9.2 we will weaken assumption A1. In Section 2.5 it is seen that in the single-segment case where we have a single risk factor S with mean 1 and variance σ_S^2, the PGF $G(z)$ has the following shape:

$$G(z) = \left(\frac{1}{1 - \sigma_S^2 \mu(\mathcal{P}(z) - \mathcal{P}(1))} \right)^{1/\sigma_S^2} \tag{9.2}$$

where $\mathcal{P}(z) = \mu^{-1} \sum_A p_A z^{\nu_A}$ is the portfolio polynomial with $\mu = \sum_A p_A$.

Using the first two moments of the PGF (9.2) we obtain the portfolio's expected loss, EL, and the variance, σ_X^2, of the loss distribution based on the single risk factor S (see Section 2.4 and [2]):

$$EL = \sum_{A \in S} p_A \nu_A, \qquad \sigma_X^2 = \sigma_S^2 EL^2 + \sum_{A \in S} p_A \nu_A^2. \tag{9.3}$$

The derivation based on the probability-generating function is not necessary, as we will show in Section 9.2.2.

As outlined in Section 2.4 the first component of σ_X^2 in (9.3) represents the risk due to systematic changes in the economy (as reflected by the relative default variance σ_S^2), whereas the second component reflects the risk from randomness of default events and is of importance in the case of small portfolios or when the systematic risk is low. For large and homogeneous portfolios and values of σ_S of order one, σ_X^2 is mainly driven by its first component. Setting $\nu_A = 1$ we get the variance of the number of default events in the portfolio to be $\sigma_S^2 \mu^2 + \mu$, which is for large portfolios roughly equal to the variance of default events observed in the economy, $\sigma_S^2 \mu^2$.

The assumption A3 and the choice of a gamma random variable for the risk scaling factor S (state of the economy) results in a loss distribution that is a mixture of a compound Poisson with a gamma distribution. This specific choice of random variables leads to the PGF $G(z)$ in (9.2), which is in fact the PGF of the negative binomial distribution with a closed analytical form that is amenable to an efficient computation of $\pi(\nu)$, the probability that portfolio loss is equal to ν. Utilizing a recursive relationship due to Panjer [10] one obtains (we here choose a different presentation of the same recursion (2.25))

$$\pi(\nu) = \frac{1}{\nu(1 + \mu \sigma_S^2)} \sum_{j=1}^{\min(m,\nu)} [\sigma_S^2 n\nu + (1 - \sigma_S^2)j] \mu_j \pi(\nu - j), \tag{9.4}$$

where $m = \deg(\mathcal{P})$ denotes the largest exposure in the portfolio and $\mu_j = \sum_{\nu_A = j} p_A$ (also cf. [3, Eq. (77)] or [2, Eq. (14)]). The starting value for the recursion is chosen as $\pi(0) = (1 + \mu \sigma_S^2)^{-1/\sigma_S^2}$.

9.1.2 Standard CreditRisk$^+$ with independent segments

In the next step we can extend the single-segment setting to a multi-segment setting where defaults from N different segments $S_1, \ldots S_N$ are independent, but are subject to segment specific default variations, i.e. we assume that the systematic risk factors $S_1, \cdots, S_N$ are mutually independent. In analogy to (9.1), the loss X in an N-segment setting is given by

$$X = \sum_{k=1}^{N} X_k = \sum_{k=1}^{N} \sum_{A \in \mathcal{S}_k} \mathbf{1}_A \nu_A.$$

The expected loss and variance of X is then given by

$$EL = \sum_{k=1}^{N} \sum_{A \in \mathcal{S}_k} p_A \nu_A, \quad \sigma_X^2 = \sum_{k=1}^{N} \sigma_{X_k}^2 = \sum_{k=1}^{N} \left(\sigma_{S_k}^2 EL_k^2 + \sum_{A \in \mathcal{S}_k} p_A \nu_A^2 \right). \quad (9.5)$$

Note that the mean of the individual risk scaling factor S_k, $k = 1, \ldots, N$, is normalized to 1. Under the assumption of independent segments, σ_X^2 is simply the sum of the variances of the segment losses $\sigma_{X_k}^2$. As we will see below, this statement is not true for the general case of correlated segments.

The extension of (9.4) for independent segments with systematic variation of defaults can be found in [3, Eq. (77)].

9.1.3 Moments of loss under correlated segments

We now go a step further and drop the assumption of independence among segment defaults. Indeed, for the rest of this section we assume that the defaults from different segments are correlated with each other via their systematic risk factors (specifically via the economic environment in which these segments are altogether embedded).

Since the expected value of a sum of a finite number of random variables (may they be independent or not) is the sum of their expected values, the expected loss of the portfolio EL is still the same as in (9.5).

However, when deriving the variance of the sum of dependent random variables we now need to consider their covariance. The variance σ_Z^2 of the random variable $Z = U + Y$, where U and Y are two (not necessarily independent) random variables for which the variances exist, is

$$\sigma_Z^2 = \sigma_U^2 + \sigma_Y^2 + 2\operatorname{cov}(U, Y) \qquad (9.6)$$

where $\operatorname{cov}(U, Y) = 2\mathbb{E}[UY] - 2\mathbb{E}[U]\mathbb{E}[Y]$. If $\operatorname{cov}(U, Y) = 0$ then $\sigma_Z^2 = \sigma_U^2 + \sigma_Y^2$. The covariance being zero does not imply that the random variables U and Y are independent, but $\operatorname{cov}(U, Y) = 0$ whenever U and Y are independent.

If U and Y are non-negative integer-valued random variables then we can use the moments from the PGF $G(z, w)$ to obtain $\operatorname{cov}(U, Y)$. Let

$$G(u, y) = \sum_{m,n} p(m, n) u^n y^m \qquad (9.7)$$

be the PGF of Z, where $p(m, n) = \mathbb{P}[U = m, Y = n]$ and u, y are formal variables. The covariance of U and Y is then given by

$$\operatorname{cov}(U, Y) = \left(\frac{\partial^2 G}{\partial z \partial w} - \frac{\partial G}{\partial z} \frac{\partial G}{\partial w} \right) (1, 1). \tag{9.8}$$

As the next step we derive the formula for the covariance of the segment losses X_1 and X_2 of a two-segment portfolio. For this, we need to rewrite (9.7) for the distribution of $X = X_1 + X_2$.

$$G(u, y) = \int_0^\infty \int_0^\infty G^{(s_1, s_2)}(u, y) f(s_1, s_2) ds_1 ds_2$$

where $G^{(s_1, s_2)}(u, y)$ is the probability-generating function of the portfolio loss conditional on $\{S_1 = s_1, S_2 = s_2\}$ and $f(s_1, s_2)$ is the joint density of the risk scaling factors S_1 and S_2. Employing (9.8) we get

$$\frac{\partial^2 G}{\partial u \partial y} = \int_0^\infty \int_0^\infty \frac{\partial^2 G^{(s_1, s_2)}}{\partial u \partial y}(1, 1) f(s_1, s_2) ds_1 ds_2 \tag{9.9}$$

$$= \int_0^\infty \int_0^\infty s_1 EL_1 s_2 EL_2 G^{(s_1, s_2)}(1, 1) f(s_1, s_2) ds_1 ds_2 = EL_1 EL_2 \mathbb{E}[S_1 S_2]$$

with EL_1 and EL_2 being the expected loss from portfolio segment 1 and portfolio segment 2 respectively.

Similarly it can be shown that $\frac{\partial G}{\partial z}(1, 1) = EL_1 \mathbb{E}[S_1]$ and $\frac{\partial G}{\partial w}(1, 1) = EL_2 \mathbb{E}[S_2]$. These together with (9.9) result in the covariance

$$\operatorname{cov}(X_1, X_2) = EL_1 EL_2 (\mathbb{E}[S_1 S_2] - \mathbb{E}[S_1] \mathbb{E}[S_2]) = EL_1 EL_2 \operatorname{cov}(S_1, S_2).$$

The covariance of the segment losses translates into the covariance of the risk scaling factors. We can now state the variance of the loss distribution of a two-segment portfolio with correlated risk scaling factors as

$$\sigma_X^2 = \sigma_{X_1}^2 + \sigma_{X_2}^2 + 2\operatorname{cov}(X_1, X_2) = \sigma_{X_1}^2 + \sigma_{X_2}^2 + 2EL_1 EL_2 \operatorname{cov}(S_1, S_2). \tag{9.10}$$

If the default events are not correlated among segments, then $\operatorname{cov}(S_1, S_2)$ in (9.10) vanishes. Thus, for independent segments the variance of the portfolio loss, σ_X^2, is simply the sum over segment loss variances $\sigma_{X_k}^2$ as in (9.5).

Using (9.5), (9.6), and $\operatorname{cov}(S_k, S_\ell) = \operatorname{corr}(S_k, S_\ell) \sigma_{S_k} \sigma_{S_\ell}$ we can generalize (9.10) for N segments with correlated defaults (cf. [1, Eq. (12)]):

$$\sigma_X^2 = \sum_{k=1}^N \sigma_{S_k}^2 EL_k^2 + \sum_{\substack{k, \ell \\ k \neq \ell}} \operatorname{corr}(S_k, S_\ell) \sigma_{S_k} \sigma_{S_\ell} EL_k EL_\ell + \sum_{k=1}^N \sum_{A \in S_k} p_A \nu_A^2. \tag{9.11}$$

Note that as in the single-segment portfolio case the first two sums of (9.11) incorporate the risk due to systematic changes in the default levels (as implied by the relative default variances $\sigma_{S_k}^2$ and their correlations $\mathrm{corr}(S_k, S_\ell)$. The third sum incorporates the risk due to the randomness of individual default events and is only of importance in the case of a small portfolio or when the systematic risk is low. In the sequel we will refer to the first two components of (9.11) as $\sigma_{X_{\mathrm{syst}}}^2$ to indicate that it incorporates the variability due to systematic effects and the third sum will be referred to as $\sigma_{X_{\mathrm{div}}}^2$ to indicate that it incorporates the variance due to the diversifiable nature of defaults.

To calculate the full loss distribution, we equate the single segment σ_X^2 of CreditRisk$^+$ (cf. Eq. (9.3)) to the σ_X^2 of (9.11) and get

$$\sigma_S^2 EL^2 = \sum_{k=1}^{N} \sigma_{S_k}^2 EL_k^2 + \sum_{\substack{k,\ell \\ k \neq \ell}} \mathrm{corr}(S_k, S_\ell) \sigma_{S_k} \sigma_{S_\ell} EL_k EL_\ell. \qquad (9.12)$$

The corresponding single-segment variance σ_S^2 of (relative) default rates is then derived by evaluating (9.12). This parameter is used thereafter in (9.4) as the variance of the gamma distribution describing the single-segment risk scaling factor. We call this procedure *moment matching*.

This mapping procedure is equivalent to an implicit choice of a dependency structure for the default correlations between the segments, which in full generality is done by copulae.[1] If compared with standard CreditRisk$^+$ with independent sectors, the moment-matching technique may lead to even lower loss quantiles in the tail in the case of low sector correlations, see Chapter 10 by Giese and [8, Table B]. However, this does not qualify one model over the other, since the dependency structure is not given from first principles and mostly chosen by convenience of computation.

9.1.4 A note on the orthogonalization of segments

A sometimes-mentioned work-around in handling correlations or covariances between segments is the orthogonalization or principal component transformation of the corresponding covariance matrix; see e.g. [9]. We briefly outline in this section that this is only of particular use and may lead to fallacies.

Introducing the notation $\mathcal{EL} := (EL^{(1)} \ldots EL^{(N)})^t$ for the transposed row vector of segment ELs and $\Sigma := (\mathrm{corr}(S_k, S_\ell)\sigma_{S_k}\sigma_{S_\ell})_{1 \le k,\ell \le N}$, we can write for the systematic term of the portfolio loss variance in (9.11)

$$\sigma_{X_{\mathrm{syst}}}^2 = \sum_{k,\ell=1}^{N} \mathrm{corr}(S_k, S_\ell)\sigma_{S_k}\sigma_{S_\ell} EL^{(k)} EL^{(l)} = \mathcal{EL}^t \Sigma \mathcal{EL}.$$

Since Σ is symmetric and positive (semi-)definite there is a non-singular $N \times N$ matrix Ψ such that $\Psi^t \Sigma \Psi = \mathrm{diag}(d_1, \ldots, d_N)$ is diagonal.

---

[1] The dependency structure of moment matching may not imply a unique choice of copula.

To explain a choice for Ψ we recall the factor-loading concept from Section 2.2 or [3, A12.3]. Each obligor A can be assigned to several industry segments by apportioning its default probability via factor loadings w_{Ak}, i.e.,

$$p_A^S = p_A \left(1 + \sum_{k=1}^{N} w_{A,k}(S_k - 1) \right) \quad \text{with } 0 \le w_{A,k} \le 1, \sum_{k=1}^{N} w_{A,k} \le 1.$$

Note that $w_{A,k} = 1$ corresponds to assigning obligor A to industry segment S_k. The weights $w_{A,k}$ may be obtained by an industry analysis of each individual firm, e.g., using the composition of the turnovers. For the remainder of this subsection we assume $\sum_{k=1}^{N} w_{A,k} = 1$; the following can be easily extended to the case of a reduced sensitivity to the N segments by introducing another industry with zero variance.

Coming back to the construction of a choice for Ψ, let $\Sigma = TT^t$ be the Cholesky decomposition of Σ. Then the row vectors of T correspond to the portfolio segments since the covariances of the segments are given by Σ. In particular, the squared lengths of the vectors correspond to the variances of the segments; more precisely, the system of N vectors is (uniquely) defined up to an orthogonal transformation. The set of obligors corresponds in this geometrical setting to an $(N-1)$-simplex Δ^{N-1} (cf. [12, 4.6]), the convex hull of these N vectors being a subset of a hyperplane H^{N-1} in the real space $\mathbb{R}^N$:

$$\Delta^{N-1} = \left\{ \sum_{k=1}^{N} w_{A,1} v_k \mid 0 \le w_{A,k} \le 1 \text{ and } \sum_{k=1}^{N} w_{A,k} = 1 \right\}.$$

Now Ψ can be chosen by $T^t \Psi$ being diagonal and each column adding up to one. This guarantees that for an obligor A with factor loadings $w_{A,1}, \ldots, w_{A,N}$, the transformed entries of $\Psi^{-1}(w_{A,1} \ldots w_{A,N})^t$ also add up to one.[2] Note that the transformed segments are uncorrelated.

Applying this diagonalization process to the systematic term of the portfolio loss variance one obtains:[3]

$$
\begin{aligned}
\sigma_{X_{\text{syst}}}^2 &= \mathcal{EL}^t (\Psi^t)^{-1} \operatorname{diag}(d_1, \ldots, d_N) \Psi^{-1} \mathcal{EL} \\
&= (\Psi^{-1} \mathcal{EL})^t \operatorname{diag}(d_1, \ldots, d_N)(\Psi^{-1} \mathcal{EL})
\end{aligned}
\tag{9.13}
$$

where

$$\Psi^{-1} \mathcal{EL} = \Psi^{-1} \sum_A \begin{pmatrix} w_{A,1} \\ \vdots \\ w_{A,N} \end{pmatrix} p_A \nu_A = \sum_A \Psi^{-1} \begin{pmatrix} w_{A,1} \\ \vdots \\ w_{A,N} \end{pmatrix} p_A \nu_A.$$

---

[2] In general Ψ may not be assumed to be orthogonal since an orthogonal transformation matrix may not necessarily preserve this hyperplane structure of segment loadings adding up to one.

[3] This orthogonalization does not influence the diversifiable term of (9.11). However, the second-order term of the generalization of the diversifiable part of the portfolio variance (Eq. (9.16)) is not reconcilable with such an orthogonalization.

This also shows that the simplex of obligors must be transformed under Ψ into a hyperplane, where the transformed components also add up to one. Note that the condition of non-negative components is not necessarily required for the transformed components.

For calculating the portfolio loss variance this diagonalization process is independent of the choice of Ψ as long as the described hyperplance structure is preserved. However, when it comes to calculating the loss distribution, resp. quantiles thereof (using (9.4) or for the multi-segment case [3, Eq. (77)]), the distribution indeed depends on the choice of Ψ. The multivariate distribution of the default rates of the correlated segments is described by the marginal gamma distributions and the correlation matrix, which does not specify the distribution in a unique way. Moreover, by considering a choice of a transformation matrix to uncorrelated segments does not necessarily imply independence of the segments. In fact, there are examples where different orthogonalizations, as described above, may lead to different results in quantile losses,[4] which in turn also lead to different portfolio management decisions according to contributions to value-at-risk or expected shortfall; cf. [8].

Example 1. Let us now give a numerical example for the choice of various orthogonalizations. Say, we have a two-segment portfolio with a covariance matrix $\Sigma = \begin{pmatrix} 2 & -1 \\ -1 & 1 \end{pmatrix}$. The Cholesky matrix of Σ is $T := \begin{pmatrix} 1 & -1 \\ 0 & 1 \end{pmatrix}$; i.e., $TT^t = \Sigma$. Hence the row vectors $v_1 = (1, -1)$, $v_2 = (0, 1)$ of T correspond to the segments whose variances are given by the squared lengths of v_1, v_2. It follows that the obligors of the portfolio can be seen as the 1-simplex $\Delta^1 = \{(x, y) \mid y = 1 - 2x \text{ and } 0 \le x \le 1\}$. Following the construction described above, we find two orthogonalizations of this system that preserve the line (hyperplane) $y = 1 - 2x$ containing the simplex Δ^1. The diagonal covariance matrices of these orthogonalizations are given by:

- $w : w_1 = (1, -1)$, $w_2 = (\frac{1}{3}, \frac{1}{3})$. $\Sigma_w = \begin{pmatrix} 2 & 0 \\ 0 & \frac{2}{9} \end{pmatrix}$.
- $w' : w'_1 = (\frac{1}{2}, 0)$, $w'_2 = (0, 1)$. $\Sigma_{w'} = \begin{pmatrix} \frac{1}{2} & 0 \\ 0 & 1 \end{pmatrix}$.

Note again that the orthogonal vectors correspond to the uncorrelated (hypothetical) segments, which are no longer pure industry segments. These orthogonalizations are constructed in such a way that all vectors v_1, v_2, w_1, w_2, and w'_1, w'_2 are on the line $y = 1 - 2x$ so that any obligor can be written as a linear combination of each pair where the components add up to one.

We now show that the portfolio loss quantiles are different in both orthogonalizations for the following sample portfolio, where we have two types of loans. All loans in segment 1 have an exposure net of recovery of 2 (monetary units), and all loans in segment 2 have an exposure net of recovery of 1. The relevant portfolio parameters are given in the following table:

segment	# obligors	exposure per obligor	PD	LGD
1	100	2	1%	100%
2	500	1	0.5%	100%

---

[4] By applying [3, Eq. 77] for independent segments after transformation.

The portfolio with its segmentation is given with respect to the covariance matrix Σ_w. Then the corresponding transformation matrix $\Psi^{-1} = \begin{pmatrix} 2 & -\frac{2}{3} \\ -1 & \frac{1}{3} \end{pmatrix}$ translates the segment coordinates with respect to w to those with respect to w'. The calculated portfolio loss quantiles are listed in the following table:

	w	w'
total exposure	700	700
EL	4.5	4.5
standard deviation	3.99	3.99
90%-quantile	9	9
99%-quantile	19	16

The standard deviation coincides for both calculations according to (9.13). Recall that the determination of the loss distribution, and therefore the quantiles, requires that all exposures be given as multiple of a loss unit, which is here chosen to be 1. So the quantiles are also derived as multiples of this unit. The difference of both 99%-quantile calculations amounts to roughly 19%. □

To emphasize this shortcoming from another practical point of view, we cite Embrechts et al. [5, 1.2] (short version [4]). It strikingly shows that based on correlations only, there are quite different ways to define a multivariate distribution with marginal gamma distributions. The multivariate structure with maximal correlations does not even produce the highest quantile losses, see [5, 5. Remark 2]. We may conclude that the multivariate recursion for several independent segments as in [3, A10] is therefore inappropriate for the situation of correlated industry segments, since an orthogonalization imposes a particular multivariate structure which may misspecify the risk structure.

We conclude that for variance considerations, orthogonalization is an appropriate tool to simplify the presentation to uncorrelated segments. But going into the kernel of CreditRisk$^+$ by analytically calculating the loss distribution with its quantile loss or expected shortfall orthogonalization is of no use.

9.1.5 Linking loss volatility to obligor default correlation

The CreditRisk$^+$ model provides a link to the default correlations among obligors in case of independent segments, cf. [3, Eq. 38]. Here, we want to derive a simple relation of the model presented so far with a meaningful average correlation among obligors.

Suppose we computed the portfolio variance by (9.11). Assuming that the correlation between any two (non-equal) obligors is given by ϱ, we have

$$\sigma_X^2 = \sum_{A \neq B} \varrho \sqrt{p_A(1-p_A)} \sqrt{p_B(1-p_B)} \nu_A \nu_B + \sum_A p_A(1-p_A)\nu_A^2$$

$$= \varrho \Big(\underbrace{\sum_{A,B} \sqrt{p_A(1-p_A)p_B(1-p_B)}\nu_A\nu_B}_{\sigma_X^2_{\{\varrho=1\}}} - \underbrace{\sum_A p_A(1-p_A)\nu_A^2}_{\sigma_X^2_{\{\varrho=0\}}} \Big) + \underbrace{\sum_A p_A(1-p_A)\nu_A^2}_{\sigma_X^2_{\{\varrho=0\}}}$$

where $\sigma^2_{X_{\{\varrho=0\}}}$ is the portfolio variance with the assumption of no correlation among obligors, and $\sigma^2_{X_{\{\varrho=1\}}}$ with all correlations among obligors set to one. Hence we obtain

$$\varrho = \frac{\sigma^2_X - \sigma^2_{X_{\{\varrho=0\}}}}{\sigma^2_{X_{\{\varrho=1\}}} - \sigma^2_{X_{\{\varrho=0\}}}}.$$

This gives a simple cross-check opportunity for the level of an average correlation for all obligors in the portfolio.

9.2 Introducing Stochastic Variation of Severity

Collateral risk is, besides counterparty risk, an important component in modelling credit risk. Modelling collateral risk is especially crucial to financial institutions operating in the retail and the middle market loan segments, whose loan portfolio is to a high degree secured by pledged real estate or other illiquid assets. For such business a sudden shift in the collateral value (e.g. as a result of natural catastrophes such as flood or earthquake or as a result of the prevailing market conditions for the specific collateral) can significantly alter portfolio risk.

9.2.1 Portfolio loss extended by systematic severity variations in the single-segment case

The systematic risk inherent to collateral value can be modelled, in analogy to modelling systematic variation of default levels, by introducing a random severity scaling factor, Λ. The scaling factor Λ models the systematic severity variation inherent to a specific collateral type. Indicating the obligor specific severity variation[5] by Λ_A, we can write the portfolio loss for a single-segment portfolio with one collateral segment as

$$X = \sum_A \mathbf{1}_A \nu_A \Lambda_A \Lambda \tag{9.14}$$

where $\mathbf{1}_A$ is the default indicator and ν_A is the exposure net of recovery. In the sequel we assume that Λ_A and Λ are independent random variables with means standardized to 1. It is further assumed that both Λ_A and Λ are independent of the default scaling factor S introduced in Section 9.1.1. We will denote the volatilities of Λ_A and Λ by σ_{Λ_A} and σ_Λ respectively. Equipped with these assumptions we can derive the EL and σ^2_X of the loss distribution of a single-segment portfolio as follows.

The expected value EL of X in (9.14) is given by

---

[5] Gordy in Chapter 7 also introduces idiosyncratic severity risk in his extension of the saddlepoint approximation.

$$EL = \mathbb{E}[\mathbb{E}[X \mid S, \Lambda]] = \mathbb{E}[S]\mathbb{E}[\Lambda]\sum_A p_A \nu_A \mathbb{E}[\Lambda_A] = \sum_A p_A \nu_A. \qquad (9.15)$$

Note that the second equality of (9.15) holds by the mutual independence of S, Λ and Λ_A, whereas the third equality is implied by their normalized means. The volatility of portfolio loss, on the other hand, is given by

$$\sigma_X^2 = \sigma_S^2 EL^2 + \sigma_S^2 \sigma_\Lambda^2 EL^2 + \sigma_\Lambda^2 EL^2$$
$$+ (1+\sigma_\Lambda^2)\sum_A [(1 + \sigma_{\Lambda_A}^2)p_A - (1+\sigma_\Lambda^2)p_A^2]\nu_A^2. \qquad (9.16)$$

The incorporation of the severity variations into the loss distribution $\pi(\nu)$ requires a distributional assumption on Λ. Distinct internal loss measurement processes, such as handling of extra recovery costs (administrative costs, cost of carry), different types of collateral, and changing economic environment may lead to different shapes of the distribution of severity variable. Typical choices for the distribution of Λ are beta distributions or lognormal distributions, but the results presented in the sequel remain valid for any distribution defined on the positive axis. Note that the unconditional loss distribution can be obtained by averaging the conditional loss distributions over all states $\{S = s, \Lambda = \lambda\}$ using the probability densities $f_S(s)$ and $f_\Lambda(\lambda)$ as weights (compare with (9.19) in the multiple segment setting). Since the random variables S and Λ are assumed to be independent, this averaging process can be performed in two steps: first over $\{S = s\}$ and then over $\{\Lambda = \lambda\}$. Note that the averaging over $\{S = s\}$ has readily been performed in (9.4), resulting in the discrete distribution $\pi(\nu)$, where only at this point the presentation as integer multiples of a fixed loss unit is necessary (cf. assumption A2 in Section 9.1.1). The unconditional loss distribution would thus be given by the product $\nu \cdot \Lambda$, where ν and Λ are independent. In other words, the unconditional loss distribution is the multiplicative convolution of the distributions of ν and Λ and its cumulative distribution $F(n)$ can be expressed by

$$F(n) = \mathbb{P}[\nu \cdot \Lambda \le n] = \mathbb{P}[\nu = 0] + \sum_{\nu > 0} \pi(\nu)\mathbb{P}[\Lambda \le n/\nu]$$
$$= \pi(0) + \sum_{\nu > 0} \pi(\nu) F_\Lambda(n/\nu). \qquad (9.17)$$

This last expression shows that the presentation of the losses as integer multiples is also unproblematic for the extension to random systematic severities (independent of the default scaling variables), since F_Λ can be assumed to be continuous.

In order to compute the probabilities $F(n)$ and corresponding percentiles, the infinite sum in (9.17) is replaced by a finite summation over ν with the upper summation bound depending on n. For the resulting approximation error we refer the reader to [2, Appendix B]. Moreover, note that in (9.17), the inclusion of the obligor specific severities Λ_A are omitted since they are only of practical relevance for small portfolios (cf. [2, Section 8]).

9.2.2 Portfolio loss extended by systematic severity variations in the multi-segment case

We can now proceed to generalize (9.15) and (9.16) to derive the EL and the σ_X^2 of the loss distribution of a portfolio with N industry segments, $\mathcal{S}_k, k = 1, \ldots, N$, and M collateral segments, $\mathcal{L}_r, r = 1, \ldots, M$.

In this general set-up we assume that each obligor can be assigned to exactly one of the industries $\mathcal{S}_1, \ldots, \mathcal{S}_N$ and exactly one collateral segment from $\mathcal{L}_1, \ldots, \mathcal{L}_M$ such that $A \in \mathcal{S}_k \cap \mathcal{L}_r$ (compare Figure 9.1).

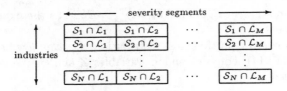

Fig. 9.1. Segmentation of a portfolio.

Hence, for obligor $A \in \mathcal{S}_k \cap \mathcal{L}_r$, the systematic risk is modelled by segment-specific scaling factors: $\mathcal{S}_k$, describing the default level in industry segment $\mathcal{S}_k$, and Λ_r, describing the severity variation in collateral segment $\mathcal{L}_r$. These random scaling variables are all normalized to mean equal to one.

The loss of obligor $A \in \mathcal{S}_k \cap \mathcal{L}_r$ in this framework is now given by the variable $\mathbf{1}_A \nu_A \Lambda_A \Lambda_r$, where the indicator variable $\mathbf{1}_A$ of default has conditional expectation $\mathbb{E}[\mathbf{1}_A \mid S_k = s_k] = p_A s_k$. The portfolio loss X conditional on $S := (S_1, \ldots, S_N)$ and $\Lambda := (\Lambda_1, \ldots, \Lambda_M)$ is

$$\mathbb{E}[X \mid S = s, \Lambda = \lambda] = \sum_{k=1}^{N} \sum_{r=1}^{M} s_k \lambda_r \sum_{A \in \mathcal{S}_k \cap \mathcal{L}_r} p_A \nu_A \qquad (9.18)$$

by our assumption of independence.

Let $f_X(x \mid S = s, \Lambda = \lambda)$ denote the density function of the random variable X given the states of the economy $\{S, \Lambda\}$ with realizations $s := (s_1, \ldots, s_N)$ and $\lambda := (\lambda_1, \ldots, \lambda_M)$, and let $f_{S,\Lambda}(s, \lambda)$ be the corresponding multi-dimensional density function of the scaling factors. The *unconditional* probability density function $f_X(x)$ of the portfolio losses is then obtained by averaging the conditional densities $f_X(x \mid S = s, \Lambda = \lambda)$ over all possible states of the economy $\{S = s, C = c\}$. Hence, we have

$$f(x) = \int \int f_X(x \mid S = s, \Lambda = \lambda) f_{S,\Lambda}(s, \lambda) ds d\lambda, \qquad (9.19)$$

in the most general case, since a priori severity and default risk are driven by the same macro-economic factors, and are therefore not necessarily independent from each other. Here, to simplify the model, we assume that S and

Λ are independent, and therefore $f_{S,\Lambda}(s, \lambda)$ decomposes into the product of densities $f_S(s)$ and $f_\Lambda(\lambda)$, specific to each risk factor.

The mean of the loss distribution, EL, is calculated as the expectation of the conditional expectations of (9.18) (compare (9.15)):

$$EL = \mathbb{E}[\mathbb{E}[X \mid S, \Lambda]] = \sum_{k=1}^{N}\sum_{r=1}^{M} \mathbb{E}[S_k]\mathbb{E}[\Lambda_r] \sum_{A \in S_k \cap \mathcal{L}_r} \mathbb{E}[\Lambda_A]p_A\nu_A = \sum_A p_A\nu_A.$$

Similarly, $EL_r^{(k)} = \sum_{A \in S_k \cap \mathcal{L}_r} p_A\nu_A$ is the expected loss in industry segment S_k and collateral segment $\mathcal{L}_r$. Moreover, we define $EL^{(k)} = \sum_{r=1}^{M} EL_r^{(k)}$ as the expected loss in segment S_k, and $EL_r = \sum_{k=1}^{N} EL_r^{(k)}$ as the expected loss in collateral segment $\mathcal{L}_r$.

The variance of the portfolio loss distribution, σ_X^2, can now be determined analogously to (2.14): For two random variables Z and Θ, the variance of Z can be expressed as

$$\text{var}[Z] = \text{var}[\mathbb{E}[Z \mid \Theta]] + \mathbb{E}[\text{var}[Z \mid \Theta]]. \qquad (9.20)$$

Thus, the unconditional variance is the sum of the conditional mean's variance (between variance) and the conditional variance's mean (within variance).

We will see below that the term $\mathbb{E}[\text{var}[Z \mid \Theta]]$ of (9.20) represents the diversifiable risk when Z describes the portfolio loss. The first term, $\text{var}[\mathbb{E}[Z \mid \Theta]]$, is the systematic risk that cannot be eliminated by diversification.

Before introducing Theorem 1, stating the variance of the portfolio loss distribution, we present the relevant notation on variances and correlations of the systematic default and severity scaling factors:

$$\sigma_{S_k}^2 := \text{var}[S_k], \quad \rho_{k\ell} := \text{corr}(S_k, S_\ell),$$

$$\sigma_{\Lambda_r}^2 := \text{var}[\Lambda_r], \quad \psi_{rs} := \text{corr}(\Lambda_r, \Lambda_s).$$

Theorem 1. *The standard deviation σ_X of the unconditional loss distribution is given by the following relation:*

$$\sigma_X^2 = \sigma_{X_{syst}}^2 + \sigma_{X_{div}}^2 \qquad (9.21)$$

where

$$\sigma_{X_{syst}}^2 = \sum_{k,\ell=1}^{N} \rho_{k\ell}\sigma_{S_k}\sigma_\ell EL^{(k)} EL^{(\ell)} + \sum_{k,\ell=1}^{N}\sum_{r,s=1}^{M} \rho_{k\ell}\sigma_{S_k}\sigma_\ell\psi_{rs}\sigma_{\Lambda_r}\sigma_{\Lambda_s} EL_r^{(k)} EL_s^{(\ell)}$$

$$+ \sum_{r,s=1}^{M} \psi_{rs}\sigma_{\Lambda_r}\sigma_{\Lambda_s} EL_r EL_s$$

is the variance due to systematic risk (default and loss severity risk), and

$$\sigma^2_{X_{div}} = \sum_{r=1}^{M}(1+\sigma^2_{\Lambda_r}) \sum_{k=1}^{N} \sum_{A \in \mathcal{S}_k \cap \mathcal{L}_r} \left((1+\sigma^2_{\Lambda_A})p_A - (1+\sigma^2_{S_k})p_A^2\right)\nu_A^2$$

is the variance due to the diversifiable (statistical) nature of default events and obligor-specific severity risk.

Proof. See Bürgisser et al. [2, Appendix A]. □

For the single-segment case with $N = M = 1$ the formulae for $\sigma^2_{X_{syst}}$ and $\sigma^2_{X_{div}}$ reduce to

$$\sigma^2_{X_{syst}} = \sigma^2_S EL^2 + \sigma^2_S\sigma^2_\Lambda EL^2 + \sigma^2_\Lambda EL^2,$$
$$\sigma^2_{X_{div}} = (1+\sigma^2_\Lambda)\sum_A \left((1+\sigma^2_{\Lambda_A})p_A - (1+\sigma^2_S)p_A^2\right)\nu_A^2, \qquad (9.22)$$

which are identical to the two components of σ^2_X in (9.16). The first term of the systematic variance $\sigma^2_{X_{\{syst\}}}$ in (9.22) represents the systematic default risk. The third term represents the systematic severity risk, and the second one is due to the mixed systematic risk of default and severity variation.

The decomposition of σ^2_X into $\sigma^2_{X_{syst}}$ and $\sigma^2_{X_{div}}$ is motivated by separating the risk into the component solely determined by systematic risk parameters, and the risk diversified away for a large portfolio with many obligors. The fact that $\sigma^2_{X_{div}}$ is negligible for a large portfolio is easily seen as follows: $\sigma^2_{X_{syst}}$ consists of K^2 terms when EL is written as the sum over all K obligors. On the other hand, $\sigma^2_{X_{div}}$ only has K terms and is therefore of relatively minor importance for portfolios with many obligors. This would suggest that in the limit of a large portfolio, the loss distribution does not depend upon any obligor-specific quantities. This statement will be formalized in Theorem 2 in Section 9.4.

9.2.3 Computation of the loss distribution for multiple segments

In the single-segment case we have assumed that S follows a gamma and Λ e.g. a lognormal distribution to derive the portfolio loss distribution. It would be natural to extend these assumptions to the situation of many segments. However, whereas a multivariate lognormal distribution is fully specified by its marginal distributions and the correlations, there is no canonical way to define a multivariate distribution whose marginals are gamma distributed. Note that the assumption of a gamma distribution is crucial for the calculation of the loss distribution with the Panjer recursion. Moreover, there are few data to determine the shape of a multidimensional default or severity distribution.

As mentioned in Section 9.1.4, Embrechts et al. [4, §1.2] show that, based on correlations only, there are quite different ways to define a multivariate distribution with marginal gamma distributions. The multivariate structure with maximal correlations does not even produce the highest quantile losses. The multivariate recursion for several independent segments as done in [3, §A10] is therefore inappropriate for our situation of correlated industry segments, since an orthogonalization (as explained in Section 9.1.4) imposes a particular multivariate structure that may misspecify the risk structure.

As a consequence, we take a pragmatic approach in which we reduce a multivariate structure to a single-segment model with the same EL and σ_X. The calculation procedure is as follows. First, we calculate the standard deviation of the portfolio by Theorem 1, thereby taking into account the segment structure. Then we estimate single systematic default and severity volatilities σ_S and σ_A such that the standard deviation of the portfolio, computed with the single-segment formula (9.22), matches the standard deviation computed before. Finally, the loss distribution is calculated as in the single-segment situation, where the systematic default behaviour is gamma and the systematic severity variation is lognormally distributed.[6]

There are two natural possibilities to estimate the implied overall systematic volatilities in the number of defaults σ_S and in the severities σ_A:

1. We may determine σ_S by moment matching for the case of pure default risk (see (9.12)), i.e., by setting $\sigma_{A_r} = 0$, $\sigma_{A_A} = 0$, cf. [1, Eq. (13)].

2. Alternatively, we may focus on pure severity risk and determine σ_A by equating the σ_X-formulae of the single and multi-segment situation by setting $\sigma_{S_k} = 0$.

Because of the non-uniqueness in the parameter estimation, one has to assess the impact of the choice of parameters on the overall loss distribution. We have performed various experiments but found no indication of a problem resulting from this issue. Even in the rather risky case of volatilities $\sigma_{S_k}, \sigma_{A_r}$ equal to one, the shape of the loss distribution hardly depends on this choice: in extreme cases the relative difference in the 99.98% percentile was less than 3% in the examples we considered.

9.3 Active Portfolio Management and Stress Analyses

In a way, the extensions presented so far give us more insight on the level of risk adequate for a portfolio. This is important for the knowledge of the total capital to be allocated, based on σ_X (Eq. (9.21)) or quantile losses (derived by the loss distribution, cf. Subsection 9.2.3). But in order to know the decomposition of the portfolio in terms of which position (or obligor) adds how much to the total capital, we need a notion of contribution. It is important to specify the level of contributors. For obligors, the contributions may be given on the level of legal entities or of economic groups or holdings of legal entities.

An immediate approach is to define the contribution of a position as the difference of capital of the total portfolio minus the capital of the total portfolio reduced by this position. However, this definition does not fulfil the

---

[6] Any other severity distribution works as well for this computation.

property of additivity, in the sense that the sum of these contributions does not result in the total capital of the portfolio.

To define the contribution to portfolio risk, the only way compatible with portfolio optimization is given by partial derivatives; cf. [11, Theorem 4.4]. Hence for the standard deviation of the portfolio losses as risk measure the contribution of an obligor is defined by

$$r_A(\sigma_X) = \nu_A \frac{\partial \sigma_X}{\partial \nu_A} . \tag{9.23}$$

Clearly by Euler's equation, these contributions add up to σ_X. This means that the σ_X-contribution is computed by the first-order sensitivity of the portfolio standard deviation σ_X with respect to small changes in the exposures ν_A of obligor A. Note that variations in severity are properly included in (9.23) following the assumptions made in Section 9.2.

If a financial institution wants to manage, and thus optimize its loan portfolio on the level of the standard deviation, it has to do so by allocating capital and charging its costs on the σ_X-contributions given by (9.23). More precisely, the total capital would be determined by a constant multiple of σ_X; and the contributions are thus given by (9.23) times this constant multiple. However, due to the limitations to small exposures, pricing a sale or purchase of risks or assets in the secondary market would not be based on (9.23), but rather on the difference of the total capital (based on σ_X) minus the capital of the resulting portfolio excluding the risks or assets to be securitized (resp. including for the purchase).

Ignoring the severity risk, the introduced model set-up allows an analytical definition of contributions to a quantile loss (value-at-risk, *VaR*) and expected shortfall (*ES*), which is defined by the conditional loss expectation above a threshold, see also Chapter 3, here specified by a quantile q:

$$ES_q = \mathbb{E}[X \mid LX > VaR_q(X)].$$

As suggested by Tasche and others ([7, 8, 11]), the contribution to ES_q for an obligor A can be defined by

$$ES_{q,A} = \nu_A \, \mathbb{E}[\mathbf{1}_A \mid X > VaR_q(X)]. \tag{9.24}$$

It can be shown ([8, Section 6]) that $ES_{q,A}$ can be analytically computed for N independent segments. Using the moment-matching method these contributions can also be calculated for N correlated segments. Note that this concept of contribution also applies to VaR_q; see [8, Section 4] or Section 3.4.

It is interesting to note that generally the expected shortfall computed from the loss distribution requires infinitely many calculation steps. However, in this model set-up, expected shortfall to any quantile is derived in finitely many steps by the equation $ES_q = \sum_A ES_{q,A}$. It is even seen [8, Section 5] that ES_q is given by the tail information up to the quantile loss VaR_q.

Now what is the practical rather than conceptual difference of contributions to σ_X and to ES (resp. to VaR)? The rapid availability of the cost of capital for pricing new loans is very important. A full run of the entire portfolio calculation may be too time consuming. Contributions to standard deviation provide analytical approximations when adding a new loan that is small in the portfolio. Moreover, it may be a matter of personal opinion whether rare and therefore extreme deviations from the mean should influence the pricing of a loan. So, to base the capital to be priced on the appropriate multiple of the standard deviation σ_X, and its contributions on this multiple times the σ_X-contributions appears to be a meaningful solution. Note that (9.23) covers severity risk, whereas (9.24) does not.

However, for an independent risk control, analysis of contributions to extreme portfolio losses is indispensable. Particularly, contributions to VaR and ES are a powerful notion in quantifying stress losses.[7] Conceptually, contributions to ES are to be preferred, since all losses above threshold are taken into account. We want to mention two examples of stress-testing applications based on ES-contributions (or VaR-contributions).

(I) *Single name stress*: The contributions to ES on an obligor level may be used to obtain a measure of the lumpiness of certain exposures, especially if the relative ES-contribution is significantly higher than exposure contribution in the portfolio.[8]

(II) *Segment contribution*: The portfolio is apportioned into segments, e.g. industries, client segment types, and countries. Segment contributions to ES are controlled with respect to some predefined levels derived by the risk appetite.

Both examples may be used for setting limits – stress loss limits or exposure limits – on a single name level in (I) and on segment level in (II).

9.4 Loss Distribution in the Limit of a Large Portfolio

Loan portfolios of banks active in the middle market segment usually consist of a large number of loans (of order 10,000 or more). In this section we show that the loss distribution of such large portfolios approaches a limit distribution, which can be explicitly described. The point is that this limit loss distribution depends on the portfolio only through the segment expected losses, while the obligor-specific severity variables and default probabilities are not relevant. In addition, the limit distribution directly mirrors the assumed behaviour of

---

[7] For quantifying stress losses the probability occurrence is quite irrelevant since stress testing refers to the analysis "what if...".

[8] Here, results and their interpretation for a specified portfolio may significantly differ on the level of legal entities, in comparison with the level of economic groups or holdings of legal entities.

the economy, modelled by the distributions f_S and f_Λ describing the systematic default and severity impact. Moreover, the limit distribution is invariant under the exchange of the distributions f_S and f_Λ. These insights reveal the importance of the economy and the associated distributional assumptions for quantifying credit risk.

9.4.1 Loss distribution

Before stating the result we have to explain what we mean by the limit of a large portfolio. We start with a loan portfolio $\mathcal{PF}$ with expected loss EL, which has relative expected losses $el_r^{(k)} := EL_r^{(k)}/EL$ in segment $\mathcal{S}_k$ and collateral segment $\mathcal{L}_r$. Now we replicate each position of this portfolio n times by taking identical risk parameters. The resulting hypothetical portfolio is denoted by $\mathcal{PF}_n$. We intend to study the limit of the loss distribution of the portfolio $\mathcal{PF}_n$ as $n \to \infty$. Clearly, the expected loss nEL of $\mathcal{PF}_n$ tends to infinity as $n \to \infty$. Therefore, in order to study the limit, we normalize the loss of $\mathcal{PF}_n$ to the expected loss of one by scaling down the loss with $1/(nEL)$.

The following intuitive result gives an explicit representation of the normalized loss distribution in the limit.

Theorem 2. *The limit of the normalized loss distribution of the portfolio $\mathcal{PF}_n$ as $n \to \infty$ has the following cumulative distribution function:*

$$F_{lim}(x) = \int_{E(x)} f_S(s) f_\Lambda(\lambda) ds d\lambda,$$

where the integration is over the set

$$E(x) := \left\{ (s, \lambda) \in \mathbb{R}_+^N \times \mathbb{R}_+^M \mid \sum_{k,r} el_r^{(k)} s_k \lambda_r \leq x \right\},$$

which is bounded by a quadratic hypersurface (compare Figure 9.2).

In [2, Appendix C] there is an indirect proof of the theorem using stochastic calculus based on the strong law of large numbers and the theorem of dominated convergence. In the next section we give a more constructive proof using characteristic functions.

The theorem states that in the limit, the probability of losing at most xEL is obtained by integrating over all states of the economy (s, λ) having the property that the sum of the segment expected losses $EL_r^{(k)}$, weighted with $s_k \lambda_r$, is less than xEL. In the single-segment case, the integration set is bounded by a hyperbola as illustrated in Figure 9.2.

It is worthwhile noting that in the limit, we have symmetry with regard to the systematic default variable S and severity variable Λ. In general, there

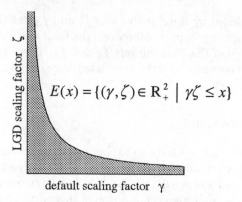

Fig. 9.2. The hyperbola bounding the integration set in the (s, λ)-state space for one segment.

is no such symmetry, which can be seen for instance by looking at the σ_X-formula in Theorem 1. Only the systematic part of the σ_X is symmetric, but the statistical term vanishes in the limit.

Next we are going to have a closer look at the special case of one segment $N = M = 1$. In this situation we have the cumulative loss distribution

$$F_{lim}(x) = \int_0^\infty \left(\int_0^{x/\lambda} f_S(s)ds \right) f_\Lambda(\lambda)d\lambda.$$

The corresponding density of the loss distribution $f_{lim}(x) = F'_{lim}(x)$ is obtained by taking the first derivative:

$$f_{lim}(x) = \int_0^\infty \frac{1}{\lambda} f_S \left(\frac{x}{\lambda} \right) f_\Lambda(\lambda)d\lambda.$$

This density $f_{lim}(x)$ allows an important interpretation. By substituting $\lambda = e^\eta$, $x = e^y$ one observes that

$$f_{lim}(e^y) = \int_{-\infty}^\infty f_S(e^{y-\eta}) f_\Lambda(e^\eta)d\eta$$

which is the convolution with respect to the additive group of real numbers. Thus we can think of f_{lim} as the multiplicative convolution, also called Mellin convolution, of the functions f_S and f_Λ defined on the group of positive real numbers.

Since the (additive) convolution of the densities of normal distributions is normal, the multiplicative convolution of the densities of lognormal distributions is also lognormal. This implies that if we assumed that the systematic default and severity variables have lognormal distribution, then the limit of the loss distribution for large portfolios would be lognormal too.

If we do not impose systematic severity variations, the limit of the loss distribution of a large portfolio is the assumed distribution f_S for the relative number of defaults. In standard CreditRisk$^+$, the limit distribution is therefore a gamma distribution.

9.4.2 Characteristic function of the loss distribution

We will provide the proof of Theorem 2 using the technique of characteristic functions. These form an important and flexible tool in probability theory (see their properties below and in Chapter 8 as well as for instance [6]). They uniquely determine probability distributions and therefore provide an equivalent way of describing them. The characteristic function of the distribution of a random variable Y is defined by

$$\Phi_Y(u) := \mathbb{E}[e^{iuY}]$$

where $i = \sqrt{-1}$. Note that if the distribution of X has a density $f(x)$, then the characteristic function is the Fourier transform of $f(x)$, that is,

$$\Phi_f(u) = \int e^{iux} f(x)dx.$$

The relation between the PGF $G(z)$ (cf. Section 9.1.1 and Section 2.3) and the characteristic function ϕ of Y is realized by a change of variable $z = e^{it}$:

$$\Phi(t) = \mathbb{E}[e^{itY}] = \mathbb{E}[e^{(it)^Y}] = G(e^{it}).$$

We mention some of the important properties of characteristic functions that will be needed in the following:

1. The characteristic function of the sum of independent random variables Y and Z equals the product of the characteristic functions of the random variables: $\Phi_{Y+Z}(u) = \Phi_Y(u)\Phi_Z(u)$.
2. We have $\Phi_{\alpha Y}(u) = \Phi_Y(\alpha u)$ for a constant α and random variable Y.
3. A sequence of probability distributions converges (in distribution) to a limit distribution if and only if we have pointwise convergence of the corresponding characteristic functions (provided the characteristic function of the limit is continuous).

In the first step, we determine the characteristic function of the final loss distribution.

Lemma 1. *The loss distribution of the portfolio $\mathcal{PF}$, incorporating systematic default and severity risk as well as obligor-specific severity risk, is described by the characteristic function $\Phi(u) = \int \int \Phi_{S,\Lambda}(u) f_S(s) f_\Lambda(\lambda) ds d\lambda$, where*

$$\Phi_{S,\Lambda}(u) = \prod_{k,r} \prod_{A \in \mathcal{S}_k \cap \mathcal{L}_r} \left(1 + s_k p_A \left(\int e^{iu\nu_A \lambda_A \lambda_r} f_{\Lambda_A}(\lambda_A) d\lambda_A - 1 \right) \right).$$

Note that λ_A denotes the realization of the obligor-specific severity variable Λ_A, and f_{Λ_A} denotes the corresponding density function.

Proof. The loss of obligor $A \in \mathcal{S}_k \cap \mathcal{L}_r$ in this framework is given by the variable $\mathbf{1}_A \nu_A \Lambda_A \Lambda_r$, where the indicator variable $\mathbf{1}_A$ of default has conditional expectation $\mathbb{E}[\mathbf{1}_A \mid S_k] = p_A S_k$. For the portfolio loss X conditional on $S :=$ $(S_1, \ldots, S_N)$ and $\Lambda := (\Lambda_1, \ldots, \Lambda_M)$ we have that

$$\mathbb{E}[X \mid S, \Lambda] = \sum_{k=1}^{N} \sum_{r=1}^{M} S_k \Lambda_r \sum_{A \in \mathcal{S}_k \cap \mathcal{L}_r} p_A \nu_A$$

by our assumption of independence of S and Λ; cf. (9.18).

Let $f_X(x \mid S = s, \Lambda = \lambda)$ denote the density function of the random variable X given the states of the economy S and Λ with realizations $s = (s_1, \ldots, s_N)$ and $\lambda = (\lambda_1, \ldots, \lambda_M)$. The density $f(x)$ of the final loss distribution is obtained by averaging the conditional densities $f_X(x \mid S = s, \Lambda = \lambda)$ over all possible states of the economy $S = s, \Lambda = \lambda$:

$$f(x) = \int \int f_X(x \mid S = s, \Lambda = \lambda) f_{S,\Lambda}(s, \lambda) ds d\lambda,$$

The conditional characteristic function of the random variable $\mathbf{1}_A \nu_A \Lambda_A \Lambda_r$ equals by definition

$$\Psi_{S_k, \Lambda_r, A}(u) := 1 - s_k p_A + s_k p_A \int e^{iu \nu_A \lambda_A \lambda_r} f_{\Lambda_A}(\lambda_A) d\lambda_A.$$

By conditional independence we conclude with the above property 1 of characteristic functions that

$$\Phi_{S,\Lambda}(u) = \prod_{k,r} \prod_{A \in \mathcal{S}_k \cap \mathcal{L}_r} \Psi_{S_k, \Lambda_r, A}(u) \tag{9.25}$$

is the characteristic function of $\sum_A \mathbf{1}_A \nu_A \Lambda_A \Lambda_r$ conditional on S, Λ. Since the characteristic function of the final loss distribution is obtained by averaging over the $\Phi_{S,\Lambda}(u)$ the assertion follows.   □

In the second step, we derive the limit of the characteristic function of the normalized loss distribution of the portfolio $\mathcal{PF}_n$.

Lemma 2. *The characteristic function of the normalized loss distribution of the portfolio $\mathcal{PF}_n$ has the following limit as $n \to \infty$:*

$$\Phi_{lim}(u) = \int_0^\infty \int_0^\infty \exp\left(iu \sum_{k,r} e l_r^{(k)} s_k \lambda_r \right) f_S(s) f_\Lambda(\lambda) ds d\lambda.$$

Proof. We first claim that by Lemma 1, the characteristic function $\Phi_n(u)$ of the loss distribution of $\mathcal{PF}_n$ has the form

$$\Phi_n(u) = \int \int (\Phi_{S,\Lambda})^n(u) f_S(s) f_\Lambda(\lambda) ds d\lambda.$$

This follows from the above-mentioned property 1 of characteristic functions, since each position of the portfolio $\mathcal{PF}$ is replicated n times by taking identical risk parameters. This means that we have to replace the random variable X of portfolio losses describing the losses conditional on the state of the economy by the sum of n independent copies. The corresponding characteristic function is therefore the n-th power $(\Phi_{S,\Lambda})^n(u)$ as claimed.

By property 2 above, the characteristic function of the normalized loss distribution is given by $\Phi_n(u/(nEL))$. For the proof of the lemma, it is therefore sufficient to show that

$$\lim_{n\to\infty} (\Phi_{S,\Lambda})^n(u/(nEL)) = \exp\left(iu \sum_{k,r} el_r^{(k)} s_k \lambda_r\right).$$

Taking into account that by (9.25)

$$\Phi_{S,\Lambda}(u/(nEL)) = \prod_{k,r} \prod_{A\in\mathcal{S}_k\cap\mathcal{L}_r} \Psi_{S_k,\Lambda_r,A}(u/(nEL)),$$

we see that it is enough to show that

$$\lim_{n\to\infty} (\Psi_{S_k,\Lambda_r,A}(u/(nEL)))^n = \exp(ius_k\lambda_r p_A \nu_A/EL). \qquad (9.26)$$

We write $\Psi_{S_k,\Lambda_r,A}(u/(nEL)) = 1 + s_k p_A f_n(u)$, where

$$f_n(u) = \int e^{iu\nu_A\lambda_A\lambda_r/(nEL)} f_{\Lambda_A}(\lambda_A)d\lambda_A \; - \; 1$$

$$= \frac{iu\nu_A c_r}{nEL} \int \lambda_A f_{\Lambda_A}(\lambda_A)d\lambda_A + O\left(\frac{1}{n^2}\right) = \frac{iu\nu_A\lambda_r}{nEL} + O\left(\frac{1}{n^2}\right),$$

using the first-order Taylor expansion of exp. By the expansion $\log(1 + x) = x + O(x^2)$ we get

$$n\log\Psi_{S_k,\Lambda_r,A}(u/(nEL)) = ius_k\lambda_r p_A \nu_A/EL + O(1/n),$$

which indeed implies (9.26). □

Proof (Proof of Theorem 2). From Lemma 2 we conclude that

$$\Phi_{lim}(u) = \int e^{iux} F'_{lim}(x)dx,$$

since

$$\int_0^\infty e^{iux} F'_{lim}(x)dx = \int_0^\infty e^{iux} \left(\int_{E(x)} f_S(s)f_\Lambda(\lambda)dsdc\right)' dx$$

$$= \int_0^\infty \int_0^\infty \exp\left(iu \sum_{k,r} el_r^{(k)} s_k\lambda_r\right) f_S(s)f_\Lambda(\lambda)dsd\lambda$$

which is due to the definition of $E(x)$. This implies that Φ_{lim} is the characteristic function of F_{lim}. Therefore, by the above-mentioned property 3 of characteristic functions, the normalized loss distribution of $\mathcal{PF}_n$ converges to the distribution function F_{lim}. This completes the proof of Theorem 2.   □

Acknowledgements. Opinions expressed herein are the authors' and do not necessarily reflect the opinions of UBS. The authors thank Peter Bürgisser and Isa Cakir for helpful suggestions and valuable remarks.

References

1. P. Bürgisser, A. Kurth, A. Wagner, and M. Wolf. Integrating correlations. *Risk*, 12(7):57–60, 1999.
2. P. Bürgisser, A. Kurth, and A. Wagner. Incorporating severity variations into credit risk. *Journal of Risk*, 3(4):5–31, 2001.
3. Credit Suisse Financial Products. CreditRisk$^+$: A credit risk management framework. London, 1997. Available at http://www.csfb.com/creditrisk.
4. P. Embrechts, A. McNeil, and D. Straumann. Pitfalls and alternatives. *Risk*, 12(5):69–71, 1999.
5. P. Embrechts, A. McNeil, and D. Straumann. Correlation and dependency in risk management: properties and pitfalls.
http://www.math.ethz.ch/~mcneil/pub_list.html, 1999.
6. M. Fisz. *Probability Theory and Mathematical Statistics.* John Wiley & Sons, Inc., New York, London, 1963.
7. H. Haaf and D. Tasche. Credit portfolio measurements. *GARP Risk Review*, (7):43–47, 2002.
8. A. Kurth and D. Tasche. Contributions to credit risk. *Risk*, 16(3):84–88, 2003.
9. M. Lesko, F. Schlottmann, and S. Vorgrimler. Die Datenqualität entscheidet. *Schweizer Bank* 2001/4:54–56, 2001.
10. H: Panjer. Recursive evaluation of a family of compound distributions. *ASTIN Bulletin*, 12:22–26, 1981.
11. D. Tasche. Risk contributions and performance measurement. Working paper, Technische Universität München.
http://citeseer.nj.nec.com/tasche99risk.html, 1999.
12. W. Warner. *Foundations of Differentiable Manifolds and Lie Groups.* Springer-Verlag, Berlin, Heidelberg, New York, 1983.

10

Dependent Risk Factors

Götz Giese

Summary. As an extension to the standard CreditRisk$^+$model we discuss two multivariate factor distributions, which include factor correlations. The moment-generating functions (MGFs) of both distributions have a simple analytical form, which fits into the framework of Chapter 6 so that the nested evaluation recursion scheme can be applied. We show how the parameters of the new distributions can be fitted to an externally given covariance matrix for the risk factors. With the example of a test portfolio we compare the new models with a single-factor approach to correlation, which has been proposed in [1].

Introduction

In this chapter we draw on the results of Chapter 6, where the general relationship (6.7) between the probability-generating function (PGF), $G_X(z)$, of the loss distribution and the MGF, $M_{\mathbf{S}}(\mathbf{t})$, of the factor distribution has been discussed. For MGFs that can be written as a combination of simple algebraic operations and (nested) exp and ln functions, we have proposed a simple, but general recursion algorithm to calculate the loss distribution. In the following we will discuss two multivariate distributions, which have sufficiently simple MGFs and allow representing factor dependence. The first is a well-known multivariate gamma distribution, whereas the second one (which we call "compound gamma") has gamma-distributed marginals only in two trivial limiting cases. It has to our knowledge not yet been discussed in the literature.

Before we look at these distributions in more detail, let us briefly recapitulate the relationship between factor distribution and obligor default correlation in CreditRisk$^+$. Consider two Poisson default indicator variables D_A and D_B, which – conditional on the risk factors $\mathbf{S}$ – are independent and have conditional intensities p_A^S and p_B^S, subject to the linear relationship (2.9). Applying the standard formula for conditional random variables we obtain for the covariance of the default variables

$$\text{cov}[D_A, D_B] = \mathbb{E}\big[\text{cov}[D_A, D_B \mid S]\big] + \text{cov}\big[\mathbb{E}[D_A|S], \ \mathbb{E}[D_B|S]\big]$$

$$= \mathbb{E}\big[\delta_{AB}\, p_A^S\big] + \text{cov}\big[p_A^S, \ p_B^S\big]$$

$$= \delta_{AB}\, p_A + p_A p_B \sum_{k,l=1}^{N} w_{Ak} w_{Bl}\, \text{cov}[S_k, S_l].$$

(As usual, δ_{AB} denotes Kronecker's delta.) Keeping only terms of first order in the default rates and denoting the covariance matrix of the risk factors by σ_{kl} we can therefore write for the default correlation (see also (2.23))

$$\rho_{AB}^{Def} = \frac{\text{cov}[D_A, D_B]}{\text{var}[D_A]\text{var}[D_B]} \approx \sqrt{p_A\, p_B} \sum_{k,l=1}^{N} w_{Ak} w_{Bl}\, \sigma_{kl} \quad \text{for } A \neq B. \qquad (10.1)$$

Note that the deterministic risk factor $S_0 \equiv 1$, which represents the obligor-specific component of default risk, does not appear in the above equations. Another interesting point is that in contrast to other portfolio models (e.g. those based on asset correlations) the default correlation in this framework does *not* depend on the particular form of the factor distribution, but only on the covariance matrix σ_{kl}. In the standard model with independent factors, the only way to introduce default correlation between obligors in different sectors is to set the factor loadings w_{Ak} accordingly. There are good reasons to assign an obligor to more than one sector, but surely this should not be misused to mimic a dependence structure that in reality arises on the level of the risk factors.

Our aim in the following is to investigate dependent factor distributions that reproduce the same default correlation as standard CreditRisk+ within sectors, i.e. we require

$$\sigma_{kk} = \sigma_k^2 \qquad (10.2)$$

on the diagonal of the covariance matrix. In addition, we will allow for default correlation between sectors, which is generated by non-vanishing off-diagonal elements σ_{kl}. As before, the class of admissible multivariate distributions is restricted by the requirements of normalization and positivity:

$$\mathbb{E}[S_k] = 1, \qquad S_k \geqslant 0. \qquad (10.3)$$

10.1 Multivariate Gamma Distribution

The first candidate is a well-known (see e.g. [2]) N-dimensional gamma distribution, which is constructed from a linear superposition of $N+1$ independent standard gamma variables $\hat{Y}, Y_1, \ldots, Y_N$. (A gamma-distributed random variable is called "standard gamma" if its scale parameter β is equal to 1.) Let us denote the shape parameters of the $N+1$ variables by $\hat{\alpha}$ and $\alpha_1, \ldots, \alpha_N$, respectively. The key point for the following is that the sum of two independent standard gamma variables with shape parameters α_1 and α_2 is also standard gamma with shape parameter $\alpha = \alpha_1 + \alpha_2$. On the other hand, in order to satisfy (10.2) and (10.3) the k-th factor has to have a scale parameter $\beta_k = \sigma_{kk}$. This motivates the following linear representation of the factors:

$$S_k = \sigma_{kk} \left\{ Y_k + \hat{Y} \right\} \quad \text{for } k = 1, \ldots, N. \tag{10.4}$$

Note that the dependence between the S_k results entirely from the common dependence on the same background variable $\hat{Y}$, which is why we call the distribution "hidden gamma" distribution. (10.4) leads to the following covariance structure:

$$\text{cov}[S_k, S_l] = \sigma_{kk}\, \sigma_{ll} \left\{ \delta_{kl}\, \alpha_k + \hat{\alpha} \right\}. \tag{10.5}$$

The requirements (10.2) for the diagonal elements result in N parameter restrictions

$$\frac{1}{\sigma_{kk}} = \alpha_k + \hat{\alpha}. \tag{10.6}$$

These conditions also define an upper bound for the parameter $\hat{\alpha}$,

$$0 \leq \hat{\alpha} \overset{!}{\leq} \min_k \left\{ \frac{1}{\sigma_{kk}} \right\}, \tag{10.7}$$

which means that the degree of dependence that can be modelled with this distribution is limited by the largest sector variance. Another disadvantage of the hidden gamma distribution is that its covariance matrix is of a very particular form. For $N = 3$ we have for example from (10.5) and (10.6)

$$\text{cov}[S] = \begin{pmatrix} \sigma_{11} & \hat{\alpha}\,\sigma_{11}\sigma_{22} & \hat{\alpha}\,\sigma_{11}\sigma_{33} \\ \hat{\alpha}\,\sigma_{11}\sigma_{22} & \sigma_{22} & \hat{\alpha}\,\sigma_{22}\sigma_{33} \\ \hat{\alpha}\,\sigma_{11}\sigma_{33} & \hat{\alpha}\,\sigma_{22}\sigma_{33} & \sigma_{33} \end{pmatrix}, \tag{10.8}$$

which imposes via (10.1) a very heterogenous structure on the default correlation between obligors in different sectors. There will be only very rare cases where the above matrix provides a good approximation for the true covariance matrix to be modelled. Noting that the correlation between two different sectors is

$$\rho_{kl}^S = \hat{\alpha}\sqrt{\sigma_{kk}\sigma_{ll}}$$

we see that the hidden gamma distribution can provide a good fit only if sectors with high variance are also significantly stronger correlated than other sectors.

In any case, the MGF of the distribution can be easily obtained by averaging over the standard gamma variables:

$$M_S^{HG}(t) = \mathbb{E}[e^{t_0 S_0} e^{\sum_{k=1}^N t_k \sigma_{kk}(Y_k+\hat{Y})}] = e^{t_0}\mathbb{E}[e^{\sum_{k=1}^N t_k \sigma_{kk}\hat{Y}}]\prod_{k=1}^N \mathbb{E}[e^{t_k \sigma_{kk} Y_k}],$$

where for completeness we have also allowed for the deterministic factor $S_0 \equiv 1$. Applying the general formula ((6.8) for the MGF of a gamma-distributed variable and using (10.6) we end up with

$$M_{\mathbf{S}}^{HG}(\mathbf{t}) = \exp\left[t_0 + \sum_{k=1}^{N} \left(\hat{\alpha} - \frac{1}{\sigma_{kk}} \right) \ln\left(1 - \sigma_{kk} t_k\right) - \hat{\alpha} \ln\left(1 - \sum_{k=1}^{N} \sigma_{kk} t_k \right) \right].$$

Plugged into the general equation (6.7) this MGF can be easily computed with the nested evaluation algorithm proposed in that chapter. A more complex covariance structure than (10.5) can be obtained if the distribution is constructed from $N + M$ standard gamma variables (with M rather than only one background factor). However, the limitations imposed by (10.7) become in this case even more complicated and render the calibration to a target covariance matrix more difficult. The main shortcoming of this model, the rather unintuitive heterogeneity in the covariance structure, cannot be remedied by these means.

10.2 Compound Gamma Distribution

We will now consider a multivariate factor distribution with a more convenient covariance structure. As before, we want to constrain the diagonal elements of the covariance matrix to the same values as in standard CreditRisk$^+$, $\sigma_{kk} = \sigma_k^2$. In addition, the new distribution has constant, non-zero factor covariance, which is determined by a parameter $\hat{\sigma}$. This modelling leaves the level of default correlation within sectors unchanged, but introduces additional correlation between obligors in different sectors as given by (10.1).

Being a mixture distribution, the distribution is constructed rather similarly to the way the PGF of standard CreditRisk$^+$ has been derived. We consider factor variables S_k, which are independently gamma distributed conditional on a positive random variable T. Let the respective conditional shape parameters be

$$\hat{\alpha}_k(T) = \alpha_k T, \quad \alpha_k > 0 \qquad (10.9)$$

and assume that the scale parameters β_k are constants. (Note that T uniformly multiplies the shape parameters rather than the variables S_k themselves.) We model the background variable T also as gamma-distributed with shape $\alpha = \frac{1}{\hat{\sigma}^2}$ and scale parameter $\beta = \hat{\sigma}^2$ such that

$$\mathbb{E}[T] = 1 \quad \text{and} \quad \mathrm{var}[T] = \hat{\sigma}^2.$$

As usual, the N-dimensional distribution of the S_k is obtained by averaging over T, which is why we call the distribution "compound gamma". We emphasize that we do not associate any economic interpretation with T – we simply describe a way to construct a multivariate distribution that is a reasonable candidate for modelling the factor distribution.[1] Formally, we can write for the joint probability density function of $\mathbf{S}$

$$f_{\mathbf{S}}^{CG}(\mathbf{s}) = \int\limits_{0}^{\infty} dt\, f_{T}^{\left(\hat{\sigma}^{-2},\hat{\sigma}^{2}\right)}(t) \prod_{k=1}^{K} f_{S_k}^{(\hat{\alpha}_k(t),\beta_k)}(s_k),$$

where $f_{Y}^{(\alpha,\beta)}(y)$ denotes the density function of the univariate gamma distribution with parameters α and β as introduced in Chapter 2. Although it seems difficult to further simplify the above expression, calculating the moments and the MGF of the distribution is straightforward. For the first moment we have

$$\mathbb{E}[S_k] = \mathbb{E}[\mathbb{E}[S_k\,|T]] = \mathbb{E}[\hat{\alpha}_k(T)\,\beta_k] = \alpha_k\beta_k.$$

From the normalization condition (10.3) it follows that

$$\alpha_k = \frac{1}{\beta_k}. \tag{10.10}$$

Combining this equation with the covariance formula for conditional random variables we get

$$\sigma_{kl} = \mathbb{E}\left[\mathrm{cov}\left[S_k, S_l\,|T\right]\right] + \mathrm{cov}\left[\mathbb{E}[S_k\,|T], \mathbb{E}[S_l\,|T]\right] = \delta_{kl}\beta_k + \hat{\sigma}^2. \tag{10.11}$$

The distribution thus has $N+1$ free parameters ($\hat{\sigma}$ and the β_k), with the β_k contributing only to the diagonal of the covariance matrix. Using the diagonal elements as input for calibration (cf. (10.2)) we immediately see that

$$\beta_k = \sigma_{kk} - \hat{\sigma}^2 \overset{!}{\geq} 0. \tag{10.12}$$

In this setting, the degree of correlation attainable via the parameter $\hat{\sigma}^2$ is therefore limited by the smallest sector variance. For $N = 3$, the final form of the covariance reads

$$\mathrm{cov}[\mathbf{S}] = \begin{pmatrix} \sigma_{11} & \hat{\sigma}^2 & \hat{\sigma}^2 \\ \hat{\sigma}^2 & \sigma_{22} & \hat{\sigma}^2 \\ \hat{\sigma}^2 & \hat{\sigma}^2 & \sigma_{33} \end{pmatrix}. \tag{10.13}$$

From (10.1) we see that the compound gamma distribution generates uniform default correlation between sectors. This is of course a less artificial correlation structure than the one produced by the hidden gamma distribution (cf. (10.8)). The MGF of the compound gamma distribution can be calculated by iterative application of the MGF formula (6.8) for the univariate gamma distribution:

---

[1] We should also point out that there is no connection between this distribution and distributions of random sums, where the term "compound" is sometimes also used.

$$M_{\mathbf{S}}^{CG}(\mathbf{t}) = \mathbb{E}\left[\mathbb{E}[e^{\mathbf{t}\cdot\mathbf{S}}|\,T]\right] = \mathbb{E}\left[\exp\left(-T\sum_{k=1}^{N}\alpha_k\ln\left(1-\beta_k t_k\right)\right)\right]$$

$$= \exp\left\{-\frac{1}{\hat{\sigma}^2}\ln\left(1+\hat{\sigma}^2\sum_{k=1}^{K}\frac{1}{\beta_k}\ln\left(1-\beta_k t_k\right)\right)\right\},$$

where the linearity (10.9) of the shape parameters $\hat{\alpha}(T)$ has been used in the second step. Plugging this MGF into the general formula (6.7) and including the deterministic sector we obtain the PGF of the compound gamma model,

$$G_X^{CG}(z) = \exp\left\{P_0(z) - \frac{1}{\hat{\sigma}^2}\ln\left(1+\hat{\sigma}^2\sum_{k=1}^{K}\frac{1}{\beta_k}\ln\left(1-\beta_k P_k(z)\right)\right)\right\}, \quad (10.14)$$

where the $P_k(z)$ denote the sector polynomials defined in (6.4). For completeness, we also provide the "modified PGFs" $G_{Xk}^{CG}(z) := \frac{\partial}{\partial t_k}M_{\mathbf{S}}^{CG}(\mathbf{P}(z))$, which play the key role for calculating exact risk contributions (see (6.18)):

$$G_{Xk}^{CG}(z) = G_X^{CG}(z) \quad \text{for } k = 0$$
$$G_{Xk}^{CG}(z) = \exp\left\{\left(1+\hat{\sigma}^2\right)\ln G_X^{CG}(z) - \hat{\sigma}^2 P_0(z)\right. \quad (10.15)$$
$$\left. - \ln\left(1-\beta_k P_k(z)\right)\right\} \quad \text{for } k \geq 1.$$

Note that $G_X^{CG}(z)$ as well as the $G_{Xk}^{CG}(z)$ is still only a result of nested ln and exp functions and therefore amenable to the nested evaluation recursion scheme, while the Panjer algorithm can no longer be used.

An interesting way to look at the compound gamma distribution is to consider it as an "interpolation" between the independent gamma distribution ($\hat{\sigma} = 0$) and a single-factor gamma distribution (all $\beta_k = 0$).

In the latter case the multivariate distribution simply consists of N replicas of a single gamma-distributed variable with variance $\hat{\sigma}^2$. For intermediate parameter values, the marginal distribution is non-gamma and has to be computed numerically by Fourier inversion of the characteristic function.[2] For the k-th factor, the characteristic function reads

$$C_k^{CG}(t) = M_{\mathbf{S}}^{CG}(t_1 = 0, \ldots, t_k = it, \ldots, t_N = 0) = \left[1+\frac{\hat{\sigma}^2}{\beta_k}\ln\left(1-i\beta_k t\right)\right]^{-\frac{1}{\hat{\sigma}^2}}.$$

The differences between the gamma and the compound gamma distribution are rather small for $\sigma_{kk} \leq 1$, which is the typical parameter range in practical applications. Figure 10.1 provides an example for $\sigma_{kk} = 0.5$. In general, the compound gamma distribution is slightly more fat-tailed. Test calculations comparing standard CreditRisk$^+$ with a model with independent compound gamma-distributed factors did not show a significant effect on the portfolio

---

[2] See for example [4, Chapter 12] or Chapter 8 for an introduction to the Fourier transformation and its numerical calculation.

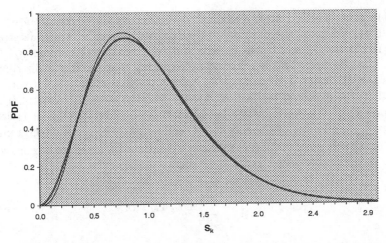

Fig. 10.1. Comparison of the probability density functions of the gamma (thin line) and the compound gamma distribution (thick line) with mean 1. Both distributions have the same standard deviation $\sigma_{kk} = 0.5$. The parameters of the compound gamma distribution are $\hat{\sigma} = 0.3$ and $\beta_k = 0.16$.

loss distribution, when both distributions were calibrated to the same variance structure. This is in line with the findings of Koyluoglu and Hickman [3] and demonstrates that the specific choice of the marginal factor distribution is mainly a matter of mathematical convenience.

Let us conclude this section by noting that the model can be further generalized by constructing the compound gamma distribution from an M-dimensional vector of independent background factors. In this case one replaces (10.9) by

$$\hat{\alpha}_k(\mathbf{T}) = \boldsymbol{\alpha}_k \cdot \mathbf{T}, \quad \alpha_{km} > 0 \text{ for } m = 1, \ldots, M,$$

which allows modelling of different (positive) levels of correlation between sectors. A higher degree of differentiation is particularly desirable if dependencies between countries (and continents) are to be represented in the model.

10.3 Fitting to an Externally Given Covariance Structure

In practical applications, one typically estimates a covariance matrix V_{kl} from time series data, e.g. from a regression analysis of historical sector default rates or from bond default data. The two distributions discussed above have only $N + 1$ parameters, which need to be fitted to this (generally more complex) covariance structure (with obviously $N(N + 1)/2$ degrees of freedom). We illustrate the fitting procedure with the example of the compound gamma distribution.

The general idea is easily explained. As before, we aim to retain the original values $\sigma_{kk} = V_{kk}$ on the diagonal of the covariance matrix, which determines via (10.12) the β_k and leaves us with only one remaining parameter $\hat{\sigma}$ that has to be adjusted. This is achieved by requiring that the factor distribution generates the same loss variance σ_X^2 as the original matrix V_{kl}. (Note that by (10.3) the expected loss will always be exactly matched.) As explained in Chapter 2, the loss variance can be split into a portfolio-specific, diversifiable part, σ_{div}^2, and a factor-dependent, systematic part, σ_{syst}^2. By definition, we only have to fit the distribution to the systematic variance contribution, which reads in the presence of a general covariance matrix

$$\sigma_{\text{syst}}^2 = \sum_{k,l=1}^{N} EL_k \, V_{kl} \, EL_l. \tag{10.16}$$

Here $EL_k = \frac{d}{dz} P_k(z = 1)$ is the expected loss of sector S_k. If we equate this target equation with the corresponding expression for the covariance matrix σ_{kl} of the compound gamma model and assume for the moment that the diagonal of V_{kl} can be matched exactly, we get the simple equation for $\hat{\sigma}$

$$\hat{\sigma}^2 = \frac{\sum\limits_{k \neq l} EL_k \, V_{kl} \, EL_l}{\sum\limits_{k \neq l} EL_k \, EL_l}.$$

The potential problem with this approach is that the parameters of the compound gamma distribution cannot be chosen freely, but have to be non-negative. Non-negativity of $\hat{\sigma}^2$ is in practice not a constraint, since the average correlation between credit risk factors is usually positive. For a large number of factors, however, and depending on the particular form of V_{kl}, the above equation may lead to a $\hat{\sigma}^2$ larger than some of the diagonal variances. This would in turn require negative β_k for these factors (see again (10.12)). To circumvent these problems we suggest the following iterative algorithm.

Step 1: Calculate $\hat{\sigma}$ from

$$\hat{\sigma}^2 = \frac{\sum\limits_{k \neq l} EL_k \, V_{kl} \, EL_l + \sum\limits_{k \in \mathcal{M}} V_{kk} EL_k^2}{\sum\limits_{k \neq l} EL_k \, EL_l + \sum\limits_{k \in \mathcal{M}} EL_k^2},$$

where we set the index set $\mathcal{M} = \emptyset$ at the start. We also define a second index set $\mathcal{I}$, which represents those diagonal elements of the original matrix that could be successfully matched and which is initialized as $\mathcal{I} = \{1, \ldots, N\}$.

Step 2: Calculate

$$\beta_k = V_{kk} - \hat{\sigma}^2 \quad \text{for all } k \in \mathcal{I}.$$

Step 3: If all $\beta_k \geq 0$ the fitting procedure is completed. Otherwise find the index i of the smallest β_k, i.e.

$$i = \arg\min_{k \in \mathcal{I}}(\beta_k),$$

and set

$$\beta_i = 0$$
$$\mathcal{M} = \mathcal{M} \cup \{i\}, \qquad \mathcal{I} = \mathcal{I} \setminus \{i\}.$$

Go back to step 1.

The resulting covariance matrix σ_{kl} contains $\hat{\sigma}^2$ on the off-diagonal and in those places on the diagonal, where the original values were smaller than $\hat{\sigma}^2$. At the same time, the loss variance matches the original value. The fitting algorithm for the hidden gamma distribution works rather similarly except that here we have to diminish the largest elements on the diagonal first (cf. (10.7)), thereby lessening the default correlation within those sectors that are likely to be the main risk drivers.

10.4 Model Comparison

Let us now compare the two distributions introduced in this chapter with another correlation approach, which has been proposed by Bürgisser, Kurth, Wagner and Wolf [1]. These authors suggest calculating an equivalent single-factor CreditRisk$^+$ model where the only free parameter, the sector variance σ_{sf}^2, is also calibrated by matching the loss variance. (As pointed out above, the single-factor model is in fact a special case of the compound gamma model.)

To highlight the differences, we consider a test portfolio, which is made up of $N = 12$ sectors, each containing 3,000 obligors. The obligors in a sector belong to equal parts to one of three homogenous classes with identical adjusted exposures E_I, E_{II} and E_{III} and respective default probabilities p_I, p_{II} and p_{III}, as detailed in Table 10.1. As can be seen, all sectors are nearly identical, except that sectors $\mathcal{S}_{11}$ and $\mathcal{S}_{12}$ contain larger exposures. In addition, sector $\mathcal{S}_{12}$ is characterized by a much larger factor variance. We assume a uniform sector correlation $\rho^S = 10\%$ such that on the off-diagonal we have $V_{kl} = 0.1\sqrt{V_{kk} V_{ll}} = 0.014$, if k or $l = 12$, and 0.004 otherwise.

In this set-up, $\mathcal{S}_{11}$ and $\mathcal{S}_{12}$ carry the dominating part of the portfolio loss variance (9% and 58%, respectively), whereas the other ten sectors contribute only 3.3% each. Table 10.2 compares the statistics of the loss distribution from the standard CreditRisk$^+$ model with the three alternative models incorporating correlation. By construction, the introduction of sector correlation equally increases the standard deviation of the models with factor dependence. The single-factor model, however, produces implausible tail behaviour:

162   Götz Giese

	Sectors 1, ..., 10	Sector 11	Sector 12
p_I	5.5%	5.5%	5.5%
p_{II}	0.8%	0.8%	0.8%
p_{III}	0.2%	0.2%	0.2%
E_I	10	20	20
E_{II}	25	50	50
E_{III}	50	100	100
Sector variance V_{kk}	0.04	0.04	0.49
Sector correlation ρ^S	0.1		

Table 10.1. Parameter summary for the test portfolio used for model comparison. The portfolio consists of 36,000 obligors uniformly distributed over 12 sectors, the first 10 sectors being identical. In each sector, there are three different obligor classes (each comprising 1,000 identical obligors) with the above default rates and exposures (in arbitrary units).

For confidence levels larger than 99% the loss quantiles of this distribution are even smaller than those of the standard model. By replacing all elements of the original covariance matrix with one single value $\sigma^2_{sf} = 0.019$, the single-factor model cannot adequately capture the concentration risk in sector S_{12} (large exposures and large factor variance). The hidden gamma model, on the other hand, fully preserves the correct default correlation within the sectors by using the original V_{kk} (both the hidden gamma and the compound gamma model can be fitted exactly to the original diagonal elements in this set-up). However, the heterogeneous covariance structure generated by the model overemphasizes the weight of the dominating sector S_{12}, producing off-diagonal terms $\sigma_{k12} = 0.029$ (instead of the original value 0.014). This leads to a significantly fatter tail distribution than that of the compound gamma model, which employs a uniform average inter-sector covariance $\hat{\sigma}^2 = 0.007$. The compound gamma model seems to be a reasonable compromise, displaying consistently fatter tails than standard CreditRisk+ and the single-factor

	Standard CR+	Single factor	Hidden gamma	Compound gamma
Inter-sector covariance	–	0.019	0.002 / 0.029	0.007
Expected loss	1.00%	1.00%	1.00%	1.00%
Std. deviation	0.12%	0.14%	0.14%	0.14%
99% quantile	1.36%	1.37%	1.45%	1.40%
99.5% quantile	1.42%	1.41%	1.52%	1.46%
99.9% quantile	1.55%	1.50%	1.70%	1.60%

Table 10.2. Comparison of loss distributions from standard CreditRisk+, the single-factor, the hidden gamma and the compound gamma model for the test portfolio described in Table 10.1. All loss statistics are quoted as percentages of the maximum loss $\sum_A \nu_A$.

model, but smoothing rather than amplifying the heterogeneity of the original covariance matrix.

With its uniform sector correlation ρ^S the test portfolio might appear tailored to a successful application of the compound gamma model. As discussed above, the hidden gamma distribution is better suited to modelling situations where the systematic risk is dominated by a number of highly correlated sectors with large variances. In practice, however, the differences between sectors are rarely that pronounced. In our experience the compound gamma model provides in most cases the best fit to the empirical covariance structure.

It is also interesting to analyse individual risk contributions in this portfolio. We concentrate on the compound gamma model for this purpose. Figure 10.2 provides a comparison of exact UL contributions in different sectors, expressed in terms of penalty factors, i.e. we essentially plot the ratio of individual risk contribution and expected loss. As discussed in Chapter 6, the penalty factors are determined by the tail behaviour of the modified PGFs $G_{X_k}^{CG}(z)$ (cf. (6.19) and (10.15)).

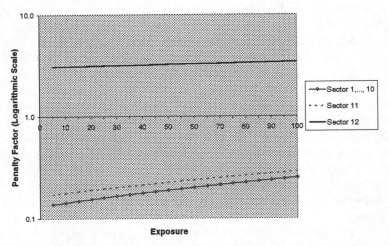

Fig. 10.2. Comparison of the penalty factors for exact UL contributions for the test portfolio using the compound gamma model. By construction, the curves for the first ten sectors are identical. The confidence level has been set to $\epsilon = 0.01\%$.

Two immediate observations can be made. As expected, penalty factors are increasing functions of exposure size. In fact, due to the extreme granularity of the test portfolio, the slope of the curves is much smaller than what one usually sees for "real life" portfolios, where the exponential increase from smallest to largest exposure size is more pronounced.

Note, however, that the exponential growth constants vary among sectors (Figure 10.2 is a logarithmic plot), which already indicates that even for a very homogeneous portfolio the saddlepoint approximation may not be suitable for calculating risk contributions. Another observation is that the average magnitude of the penalty factors differs from sector to sector, reflecting the fact that sectors have significantly different contributions to total UL. These differences do not only arise from the correlation structure, but are also driven by size concentrations (compare the contributions of sectors S_1 and S_{11}).

	Sectors 1, ..., 10	Sector 11	Sector 12
Variance	3.3%	9.3%	58.1%
Unexpected loss (UL)	1.8%	5.0%	76.6%
UL saddlepoint	1.6%	4.4%	79.5%
Unexpected shortfall (US)	1.7%	4.6%	78.7%

Table 10.3. Risk contributions in percent for different risk measures aggregated to sector level. UL and US contributions have been calculated on an $\epsilon = 0.01\%$ confidence level.

This can also be seen from Table 10.3, where we have aggregated the risk contributions of all obligors in a sector. We compare the contributions to different risk measures (variance, unexpected loss and unexpected shortfall (US), cf. definition (6.21)). Not surprisingly, the impact of large exposures (in S_{11} and S_{12}) on tail risk is significantly underestimated by a variance-based contribution method.[3] We also check the validity of the saddlepoint approximation for UL contributions, see (6.20). In this approximation, the contribution of the dominating sector S_{12} is significantly overestimated, which leads to a relatively high error for the smaller sectors $S_1, \ldots, S_{10}$. Here the saddlepoint result is more than 13% below the true value. Conversely, the saddlepoint approximation is nearly perfect for estimating the tail probability, where we typically observe less than $3 \cdot 10^{-6}$ relative error. As pointed out before, the method is tailor-made for fitting the loss distribution, but tends to distort the conditional probability measure, under which the risk contributions are calculated.

As it turns out, the saddlepoint result for S_{12} is even larger than the US contribution, which naturally puts a stronger weight on contributions to tail risk. This interpretation is also supported by Figure 10.3, which displays the penalty factor curve for S_{12} for the different contribution methods. In order to allow comparison we have rescaled variance and US contributions by $\frac{UL}{\sigma_X^2}$ and $\frac{UL}{US}$, respectively.

Again, level and slope of variance-based penalty factors are much smaller than those from risk measures that focus on worst-case events. We emphasize that variance-based factors are only linear in the exposure size, whereas the other factor curves increase exponentially (which is difficult to discern in this plot due to the strong granularity of the portfolio).

---

[3] We use the generalization of formula (2.27) to an arbitrary covariance matrix V_{kl}.

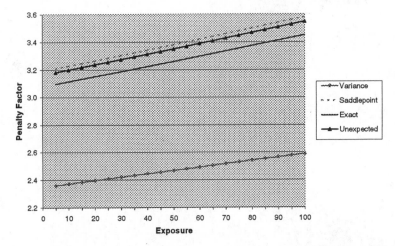

Fig. 10.3. Penalty factors for different risk measures and the saddlepoint approximation for sector S_{12}. This sector is the main risk driver in the test portfolio. The factors for variance and US contributions have been rescaled to allow comparison with the curves for UL.

This concluding section may have demonstrated the power of the framework proposed in Chapter 6. We have been able to calculate the loss distribution for a more realistic factor distribution than that of the standard model without losing analytical tractability. By calculating risk contributions we have obtained the main tool for identifying the risk drivers in a portfolio. We have introduced size- and sector-dependent penalty factors, which provide a very condensed and convenient means to represent the risk profile of a portfolio and to include economic capital costs into pricing. They also allow estimating the potential impact of new transactions on portfolio risk. All these results can be obtained with high accuracy within a few minutes on a standard PC, even for very large and heterogeneous portfolios.

References

1. P. Bürgisser, A. Kurth, A. Wagner, and M. Wolf. Integrating correlations. *Risk*, 12(7):57–60, 1999.
2. N. L. Johnson and S. Kotz. *Distributions in Statistics: Continuous Multivariate Distributions*. Wiley, 1972.
3. H. Koyluoglu and A. Hickman. Reconcilable differences. *Risk* 11(10):56–62, 1998.
4. W. H. Press, S. A. Teukolsky, W. T. Vetterling and B. P. Flannery. *Numerical Recipes in C: The Art of Scientific Computing*. Second Edition, Cambridge University Press, Cambridge, 1992.

11

Integrating Rating Migrations

Frank Bröker and Stefan Schweizer

Summary. While CreditRisk$^+$ is generally regarded as a powerful, fast and easy-to-use credit portfolio model, there is often criticism that its definition of credit loss according to the default mode approach is inferior to the more comprehensive mark-to-market approach used in other credit portfolio models like CreditMetrics. In this chapter we present a practical, "easy to implement" procedure that allows us to integrate the rating migration concept – an important advantage of CreditMetrics – into CreditRisk$^+$. Rating-driven changes in market value that are characteristic of liquid portfolios are included in this model without losing the benefits of CreditRisk$^+$.

Introduction

Generally, CreditRisk$^+$ focuses on the modelling and managing of credit default risk. Therefore a separate model for the incorporation of credit spread (or rating migration) risk is advocated by experts [5], if these risks are relevant to the portfolios analysed. Like other default mode (DM) approaches CreditRisk$^+$ differentiates only between two different states of a credit exposure: performing (i.e. debt service as contractually agreed) or non-performing.

On the other hand, mark-to-market (MtM) credit portfolio models like CreditMetrics rely on a more comprehensive definition of credit risk.[1]

Contrary to CreditRisk$^+$ the MtM modelling methodology incorporates:

- several different future states of creditworthiness (rating grades) instead of a binary performing/non-performing approach;
- future net present values determined with a mark-to-model approach (potential changes in market value caused by rating upgrades and downgrades) instead of book or nominal values.

---

[1] CreditMetrics offers no real MtM, because basically only rating changes drive the valuation, in fact spreads move more often and especially much earlier than ratings do.

Therefore, MtM credit portfolio models are usually advantageous if mark-to-market valuation or accounting is relevant, typically in trading book portfolios of bonds or credit derivatives, for credit default adjustments of over-the-counter (OTC) products etc. The more extensive credit risk definition is also useful for modelling credit risk impacts on typical bank balance sheets in which accrual accounting and mark-to-market accounting elements exist side by side (e.g. banking and trading book exposures). Furthermore, instruments that transfer credit risk in the banking book to the capital markets with its MtM valuation like asset-backed securities (ABS), credit-linked obligations (CLO) or basket credit derivatives structures are gaining importance; see Chapters 18 and 19 for the pricing of such products.

Practical experiences in working with CreditRisk$^+$ suggest that any potential modification of the CreditRisk$^+$ model that tries to integrate the effect of rating changes should retain the main advantages of CreditRisk$^+$: the relatively low data requirements, the analytical tractability and the high flexibility.

This chapter is organized as follows. In the first section we present the basic idea of a practical, "easy to implement" procedure that allows the integration of the rating migration concept into CreditRisk$^+$. In Section 11.2 we will show how correlations can be integrated in this concept. Section 11.3 is devoted to the topic of calculating risk contributions in the migration mode. In Section 11.4 we will give some numerical results by analysing a real banking book portfolio with both approaches. Section 11.5 concludes with some final remarks.

11.1 Integrating Rating Migrations into CreditRisk$^+$

11.1.1 Some basic notations

We consider a loan portfolio with K different obligors. Every obligor A belongs to a specific non-default rating category $i \in \{1, ..., R\}$, which determines its probability of default p_A. Rating 1 is considered to be the "best" rating grade with the lowest default probability. Rating R is the worst possible "performing" rating grade carrying the highest default probability. The probabilities of rating grade changes are given by the migration rates $m_{i,j}$, with $i \in \{1, ..., R\}$ and $j \in \{1, ..., R+1\}$. The migration rate $m_{i,j}$ indicates the probability that an obligor changes its rating from currently i to j within a specific time period, usually one year. The migration rate $m_{i,R+1}$ is therefore the default rate of an obligor with a current rating i.

11.1.2 The general idea of integrating rating migrations into CreditRisk$^+$

The integration of rating migrations into the basic CreditRisk$^+$ framework can be accomplished in five consecutive steps (see Figure 11.1).

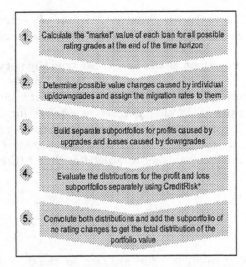

Fig. 11.1. Steps to follow for the integration of rating migrations into CreditRisk$^+$

In the beginning all potential future loan values are evaluated depending on the possible rating categories at the end of the chosen time horizon. Typically one chooses a time horizon of one year but other time horizons are equally manageable by the following procedure. Following the CreditMetrics concept the loan cash flows are discounted using forward zero curves by rating categories. Subsequently, the potential profits caused by upgrades and the losses caused by downgrades (default is considered as a special case of a downgrade) are evaluated by subtracting that future loan value from the one under the assumption of no rating change. The central idea is to model possible profits and losses due to rating changes in the same way as default events are modelled in CreditRisk$^+$. This idea was first published in [9].

Hence, the potential value changes of all loans and their corresponding migration rates must be assigned separately to a subportfolio of profits due to upgrades and a subportfolio of losses due to downgrades. This separation is inevitable, because CreditRisk$^+$ cannot handle exposures with different signs. Furthermore, the total portfolio value with unchanged ratings is just the sum of all future loan values that are associated with their initial rating categories. The fourth step comprises the separate calculation of profit and loss distributions of both subportfolios using the usual recurrence relation of the CreditRisk$^+$ methodology as presented in [5] and Chapter 2. The absolute amounts of profits and losses are used as net exposures and the default rates correspond to the migration rates.

For an alternative approach to rating migrations and further theoretical aspects of our approach we refer to Chapter 12. Our contribution is focused on the presentation of an "easy to implement" approach.

In the sequel we will take a closer look at the five different steps of the implementation procedure to integrate rating migrations into CreditRisk$^+$.

11.1.3 Step 1: Calculate the "market" value of each loan for all possible rating categories at the end of the time horizon

For the loans in our portfolio, market prices are typically not available. Therefore we calculate the value of the loans at the end of the time horizon by discounting all future cash flows (CF_t is the cash flow at time t) with the risk-adjusted forward rates. This value is taken as a proxy for the market value ("mark-to-model"). Knowing the term structures of the rating specific credit spread curves we can calculate – based on a rating grade i_A at the current point in time τ – the loan value for a rating j at time t (the end of the time horizon) according to[2]

$$V_A^j(\tau, t) = \sum_{l=1}^{L} \frac{CF_{t_l}^A}{(1 + f^j(\tau, t, t_l))^{t_l - t}}$$

with $f^j(\tau, t, t_l)$ being the time τ forward rate for the time interval $[t, t_l]$ depending on the future rating grade j. We can see that the loan value depends on the possible rating migrations from currently i_A to j. Therefore, we have $R + 1$ possible loan values at the end of the time horizon for each exposure: one for each of the R possible rating categories plus one recovery value RV_A in the case of a default before t.

$$V_A^j = \begin{cases} \sum_{l=1}^{L} \dfrac{CF_{t_l}^A}{(1 + f^j(\tau, t, t_l))^{t_l - t}} & \text{if } j \leq R \\ RV_A & \text{if } j = R + 1 \end{cases} \tag{11.1}$$

RV_A depends on the recovery assumption. It is often modelled as a fraction of the face value, while others use a fraction of the market value prior to default.[3]

11.1.4 Step 2: Determine possible value changes caused by individual rating up- or downgrades and assign the migration rates to them

From Step 1 we know that according to (11.1) each exposure in our portfolio will have one out of $R + 1$ different loan values at the end of the time horizon t. The different values stem from possible rating up- or downgrades (rating migration) during the time period looked at. In order to compare the different values we define the value change at the end of the time horizon

$$\Delta V_A = V_A^j - V_A^{j_A}.$$

---

[2] For simplification we neglect potential cash flows between τ and t.

[3] Recovery rates can also be modelled as stochastic variables. See [2, 4] and Chapters 9, 7 and 3 for possible approaches to integrating stochastic recovery rates into CreditRisk+.

We get a positive value change in the case of a rating upgrade $(j < i_A)$, a negative value change when we have a rating downgrade $(j > i_A)$ and a value change of zero when the rating remains constant $(j = i_A)$. The possibility of each value change is simply the one-year migration rate $m_{i_A,j}$ to the respective rating category. Applying these migration rates we can calculate the expected loan value at the end of the time horizon t as the probability weighted sum of all possible value changes plus the loan value in case of no rating change

$$\mathbb{E}[V_A] = \underbrace{\sum_{j=1}^{i_A-1} [V_A^j - V_A^{i_A}]m_{i_A,j}}_{\Delta V_A^+} + \underbrace{\sum_{j=i_A+1}^{R+1} [V_A^j - V_A^{i_A}]m_{i_A,j}}_{\Delta V_A^-} + \underbrace{V_A^{j=i_A}}_{\Delta V_A^0}$$

The expected value of the total portfolio can be calculated by simply summing up all expected loan values

$$\mathbb{E}[V_{PF}] = \underbrace{\sum_{A=1}^{K} \sum_{j=1}^{i_A-1} [V_A^j - V_A^{i_A}]m_{i_A,j}}_{\Delta V_{PF}^+}$$

$$+ \underbrace{\sum_{A=1}^{K} \sum_{j=i_A+1}^{R+1} [V_A^j - V_A^{i_A}]m_{i_A,j}}_{\Delta V_{PF}^-} + \underbrace{\sum_{A=1}^{K} V_A^{j=i_A}}_{\Delta V_{PF}^0}.$$

11.1.5 Step 3: Build up separate subportfolios for profits caused by upgrades and losses caused by downgrades

Since we know the potential loan values at the end of the time horizon and their probabilities, we can build up three different subportfolios: one labelled "+" for the portfolio of profits due to possible rating upgrades, one labelled "0" for the portfolio without rating changes and one labelled "–"for the portfolio of losses due to possible rating downgrades. For every loan we come up with $R + 1$ different possible loan values and therefore $R + 1$ different "virtual" loans. The number of "virtual" loans in the three subportfolios is given by

$$K^+ = \sum_{A=1}^{K}(i_A - 1), \quad K^- = \sum_{A=1}^{K}(R - i_A + 1) \quad \text{and} \quad K^0 = K \tag{11.2}$$

with $K^+ + K^- + K^0 = K(R+1)$.

Before we continue our implementation procedure, Figure 11.2 illustrates for a sample loan with $R = 7$ what has already been worked out in Steps 1 to 3.

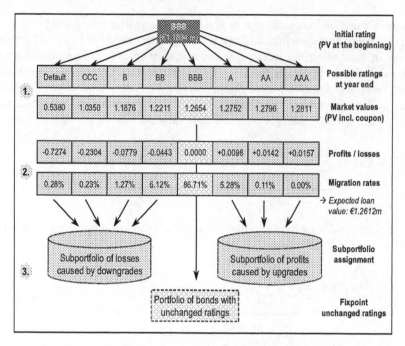

Fig. 11.2. Steps 1 to 3 of the implementation procedure

11.1.6 Step 4: Evaluate the distributions for the profit and loss subportfolios separately using CreditRisk$^+$

The fourth step comprises the separate calculation of both subportfolios using the usual recurrence relation of the CreditRisk$^+$ methodology. The absolute amounts of profits and losses ΔV_A^j are used as net exposures and the default rate corresponds to the migration rate $m_{i_A,j}$. The same loss unit L_0 has to be chosen for the profit distribution and the loss distribution. In Figure 11.3 results are presented for a illustrative portfolio of 20 loans with varying face values, ratings and maturities and an initial total market value of EUR 67.54 m (see Table 11.1, taken from [7], for details).

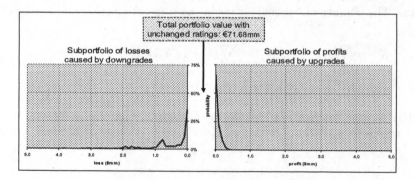

Fig. 11.3. Profit and loss distributions according to CreditRisk$^+$

a: Sample portfolio at the beginning of the time horizon

Obligor A	Rating	Notional	Rem. maturity	Seniority	Coupon	Market value
1	AAA	7.000.000	3 years	Sen. Sec.	8.95%	7.821.049
2	AA	1,000,000	4 years	Sen. Sec.	10.07%	1,177,268
3	A	1,000,000	3 years	Sen. Sec.	9.29%	1,120,831
4	BBB	1,000,000	4 years	Sen. Sec.	10.93%	1,189,432
5	BB	1,000,000	3 years	Sen. Sec.	12.54%	1,154,641
6	B	1,000,000	4 years	Sen. Sec.	16.29%	1,263,523
7	CCC	1,000,000	2 years	Sen. Sec.	22.88%	1,127,628
8	A	10,000,000	8 years	Sen. Sec.	11.87%	14,229,071
9	BB	5,000,000	2 years	Sen. Sec.	10.22%	5,386,603
10	A	3,000,000	2 years	Sen. Sec.	7.52%	3,181,246
11	A	1,000,000	4 years	Sen. Sec.	10.35%	1,181,246
12	A	2,000,000	5 years	Sen. Sec.	10.85%	2,483,322
13	B	600,000	3 years	Sen. Sec.	14.66%	705,409
14	B	1,000,000	2 years	Sen. Sec.	11.82%	1,087,841
15	B	3,000,000	2 years	Sen. Sec.	11.82%	3,263,523
16	B	2,000,000	4 years	Sen. Sec.	16.29%	2,527,046
17	BBB	1,000,000	6 years	Sen. Sec.	11.88%	1,315,720
18	BBB	8,000,000	5 years	Sen. Sec.	11.46%	10,020,611
19	BBB	1,000,000	3 years	Sen. Sec.	9.54%	1,118,178
20	AA	5,000,000	5 years	Sen. Sec.	10.56%	6,181,784
Σ		55,600,000				67,535,972

b: One-year transition matrix

	AAA	AA	A	BBB	BB	B	CCC	D
AAA	90.81%	8.33%	0.68%	0.06%	0.12%	0.00%	0.00%	0.00%
AA	0.70%	90.65%	7.79%	0.64%	0.06%	0.14%	0.02%	0.00%
A	0.09%	2.27%	91.05%	5.52%	0.74%	0.26%	0.01%	0.06%
BBB	0.02%	0.33%	5.95%	86.93%	5.30%	1.17%	0.12%	0.18%
BB	0.03%	0.14%	0.67%	7.73%	80.53%	8.84%	1.00%	1.06%
B	0.00%	0.11%	0.24%	0.43%	6.48%	83.46%	4.07%	5.21%
CCC	0.22%	0.00%	0.22%	1.30%	2.38%	11.24%	64.86%	19.78%
D	0.00%	0.00%	0.00%	0.00%	0.00%	0.00%	0.00%	100.00%

c: One-year forward discount factors by rating category

	Year 1	Year 2	Year 3	Year 4	Year 5	Year 6	Year 7	Year 8
AAA	96.53%	92.15%	87.05%	81.90%	77.91%	74.11%	70.50%	67.07%
AA	96.48%	92.07%	86.93%	81.74%	77.72%	73.90%	70.27%	66.81%
A	96.41%	91.89%	86.56%	81.27%	77.17%	73.27%	69.57%	66.06%
BBB	96.06%	91.28%	85.77%	80.33%	76.04%	71.99%	68.15%	64.52%
BB	94.74%	88.97%	82.14%	75.52%	70.41%	65.63%	61.19%	57.04%
B	94.30%	87.31%	79.32%	72.10%	66.44%	61.23%	56.42%	51.99%
CCC	86.92%	75.59%	67.44%	60.22%	53.04%	46.73%	41.16%	36.26%

d: Average recovery rates by seniority

Seniority	Average recovery rate (with respect to the notional)
Senior secured	53.80%

Table 11.1. Sample portfolio at the beginning of the time horizon and calculation parameters

Table 11.2 shows the portfolio at the end of the time horizon and gives some details of how Steps 1 to 3 are performed. The expected portfolio market value at the end of the time horizon is EUR 71.09 m. The expected change in market value due to rating changes is EUR 0.59 m with expected losses due to downgrades contributing EUR 0.64 m and expected profits due to upgrades contributing EUR 0.05 m. Seventy-eight percent of the expected losses are due to defaults, which demonstrates already that the distribution of the total portfolio value will be dominated by the loss distribution.

1	2	3	4	7	8	9	10	11	12	13	14	15	16	17
Virtual Obligor C	Corresponding "real" obligor A	Rating	Potential new rating at end of time horizon	Potential market value	Potential profit/ losses	Migration rate	Portfolio of no rating changes (9)*()	Portfolio of profits (b')	Portfolio of losses (b')	Portfolio of losses (only defaults)	Weighted market value (9)*(7)	Weighted profit/loss (9)*(8)	Weighted profit (9)*(11)	Weighted loss (9)*(12)
1	1	AAA	AAA	8,259,166		90.81%	8,259,166	0		0	7,500,166		0	0
2	1	AAA	AA	8,252,153	-7,034	8.33%	0	0	-7,034	0	687,404	-586	0	-586
3	1	AAA	A	8,238,290	-20,896	0.66%	0	0	-20,896	0	56,020	-142	0	-142
4	1	AAA	BBB	8,189,297	-69,889	0.06%	0	0	-69,889	0	4,914	-42	0	-42
5	1	AAA	BB	8,004,882	-254,305	0.12%	0	0	-254,305	0	9,606	-305	0	-305
6	1	AAA	B	7,875,878	-383,308	0.00%	0	0	-383,308	0	0	0	0	0
7	1	AAA	CCC	6,935,610	-1,323,576	0.00%	0	0	-1,323,576	0	0	0	0	0
8	1	AAA	D	3,766,000	-4,493,186	0.00%	0	0	-4,493,186	-4,493,186	0	0	0	0
9	2	AA	AAA	1,249,066	1,507	0.70%	0	1,507	0	0	8,743	11	11	0
10	2	AA	AA	1,247,559		90.65%	1,247,559	0	0	0	1,130,912	0	0	0
11	2	AA	A	1,243,218	-4,341	7.79%	0	0	-4,341	0	96,847	-338	0	-338
12	2	AA	BBB	1,233,581	-13,978	0.64%	0	0	-13,978	0	7,895	-89	0	-89
13	2	AA	BB	1,189,920	-57,638	0.06%	0	0	-57,638	0	714	-35	0	-35
14	2	AA	B	1,155,781	-90,777	0.14%	0	0	-90,777	0	1,619	-127	0	-127
15	2	AA	CCC	1,006,846	-240,713	0.02%	0	0	-240,713	0	201	-48	0	-48
16	2	AA	D	536,000	-709,559	0.00%	0	0	-709,559	-709,559	0	0	0	0
⋮	⋮		⋮		⋮			⋮			⋮		⋮	
145	19	BBB	AAA	1,196,875	10,063	0.02%	0	10,063	0	0	239	2	2	0
146	19	BBB	AA	1,195,862	9,050	0.33%	0	9,050	0	0	3,946	30	30	0
147	19	BBB	A	1,193,867	7,056	5.95%	0	7,056	0	0	71,035	420	420	0
148	19	BBB	BBB	1,186,812		86.93%	1,186,812	0	0	0	1,031,695	0	0	0
149	19	BBB	BB	1,160,253	-26,559	5.30%	0	0	-26,559	0	61,493	-1,406	0	-1,406
150	19	BBB	B	1,141,700	-45,111	1.17%	0	0	-45,111	0	13,356	-528	0	-528
151	19	BBB	CCC	1,006,252	-180,560	0.12%	0	0	-180,560	0	1,206	-217	0	-217
152	19	BBB	D	536,000	-648,812	0.18%	0	0	-648,812	-648,812	968	-1,168	0	-1,168
153	20	AA	AAA	6,510,260	9,973	0.70%	0	9,973	0	0	45,572	70	70	0
154	20	AA	AA	6,500,267		90.65%	6,500,267	0	0	0	5,892,509	0	0	0
155	20	AA	A	6,471,361	-28,926	7.79%	0	0	-28,926	0	504,119	-2,253	0	-2,253
156	20	AA	BBB	6,409,602	-90,685	0.64%	0	0	-90,685	0	41,021	-580	0	-580
157	20	AA	BB	6,105,687	-394,400	0.06%	0	0	-394,400	0	3,664	-237	0	-237
158	20	AA	B	5,890,863	-609,424	0.14%	0	0	-609,424	0	8,247	-853	0	-853
159	20	AA	CCC	5,070,206	-1,430,061	0.02%	0	0	-1,430,061	0	1,014	-286	0	-286
160	20	AA	D	2,690,000	-3,810,287	0.00%	0	0	-3,810,287	-3,810,287	0	0	0	0
Σ				71,684,644			3,782,495		-63,164,602	-41,771,644	71,093,217	-591,426	47,094	-638,520

Table 11.2. Portfolio at the end of the time horizon

The CreditRisk⁺ concept of default rate volatility and sector analysis is not necessary at this stage, as we assume independent obligors here. However, one has to look at other implications closely: due to the application of the Poisson approximation in CreditRisk⁺ the migration rates have to be small to obtain correct results. Transition matrices usually show quite small upgrade and downgrade rates. In our example the only migration rates that exceed 10% p.a. belong to the upgrade of a CCC obligor to the B category (11.24%) and to the default of a CCC obligor (19.79%). Modest approximation errors are likely to occur for large and risky exposures of speculative grade bonds.

11.1.7 Step 5: Convolute the two distributions and add the value of the portfolio with no rating changes to get the total distribution of the portfolio value

The last step is to combine the profit and the loss distribution and to add the total portfolio value with unchanged ratings. From Step 4 we get two probability distributions, one profit distribution due to upgrades $p^+(V) = p^+(V_{PF}^+)$ and one loss distribution due to downgrades $p^-(V) = p^-(V_{PF}^-)$. Assuming independence[4] the final distribution of portfolio values $p(V_{PF})$ is given by the convolution of both distributions shifted by the portfolio value V_{PF}^0 in the case of no rating changes.

The profit and loss distributions are given as discrete probabilities for multiples of the banding width L_0. We use the notation G_i for the discrete probability of a gain iL_0 with $0 \le i \le N_u$ and L_j for the corresponding probability of a loss jL_0 with $0 \le j \le N_d$.[5] Based on (11.3) the final probability p_k for a portfolio value of $V_{PF}^0 + kL_0$ is given by

$$
p_k = \begin{cases}
\sum_{l=0}^{N^+} G_{l+k}L_l & \text{with } N^+ = \min(N_u - k; N_d) \ \forall \ 0 \le k \le N_u, \\
\sum_{l=0}^{N^-} L_{l-k}G_l & \text{with } N^- = \min(N_d + k; N_u) \ \forall \ 0 > k \ge -N_d, \\
0 & \text{otherwise.}
\end{cases}
\tag{11.3}
$$

In our example the resulting distribution of the portfolio value is presented in Figure 11.4. Using the CreditMetrics methodology and a Monte Carlo simulation with 20,000 simulation runs a nearly identical histogram can be obtained with excessively higher effort.

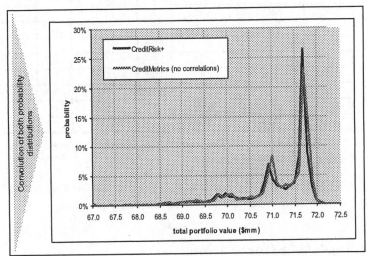

Fig. 11.4. Distribution of the total portfolio value by CreditRisk$^+$ & CreditMetrics

---

[4] In reality the two distributions are not mutually independent, but exhibit some kind of correlation. It is fair to assume that the correlation is small and negative, as massive upgrades will usually not go hand in hand with massive downgrades.

[5] $N_u * Q_0$ and $N_d * Q_0$ represent the maximum portfolio profit and the maximum portfolio loss.

A detailed account of the deviations is given in Columns 3 and 4 of Table 11.3. This table contains the amounts the portfolio value will not fall below with a probability given by the percentile level. Due to the mentioned Poisson approximation and the independence assumption, the use of the CreditRisk+ model conducts a very small overestimation of risk. Chapter 12 by Binnenhei shows how our independence assumption could be relaxed.

Column 6 of Table 11.3 additionally shows the distribution characteristics if we simplify our calculation further and just consider the loss distribution. As we can see by comparing Columns 3 and 6 in Table 11.3, the effect of the profit distribution on the percentile levels is rather small.

1	2	3	4	5	6	7	8
		CreditRisk+ (full calculation)	CreditMetrics (no correlations)	Δ 3, 4 in %	CreditRisk+ (only loss distr.)	Δ 3, 6 in %	CreditRisk+ (only defaults)
Expected Portfolio Value		71.09	71.09		71.05		71.09
Standard Deviation		0.91	0.88		0.91		0.87
Percentiles	90.00%	69.76	69.88		69.71		70.01
	95.00%	69.19	69.29		69.13		69.72
	97.50%	68.61	68.85		68.53		69.00
	99.00%	67.84	68.12		67.74		68.10
	99.50%	67.07	67.51		67.00		67.36
	99.75%	65.35	66.07		65.27		65.44
	99.90%	64.29	64.79		64.20		64.59
Value-at-Risk (=Expected Value- Percentile)	90.00%	-1.32	-1.21	-8.47%	-1.34	0.90%	-1.08
	95.00%	-1.90	-1.79	-5.59%	-1.92	1.16%	-1.37
	97.50%	-2.48	-2.24	-9.66%	-2.52	1.69%	-2.09
	99.00%	-3.24	-2.97	-8.58%	-3.31	1.91%	-2.99
	99.50%	-4.01	-3.58	-10.88%	-4.05	0.80%	-3.73
	99.75%	-5.74	-5.02	-12.56%	-5.78	0.73%	-5.65
	99.90%	-6.80	-6.30	-7.38%	-6.85	0.76%	-6.50

Table 11.3. Expected value – percentile and value-at-risks of future portfolio values (million $) using different methods

Figure 11.5 illustrates the value-at-risk results for the CreditRisk+ (full calculation) and the CreditMetrics (no correlations) calculation.

11.2 Integrating Correlations

The consideration of dependencies between obligor rating migrations generally results in a widening of the distribution and higher portfolio percentiles. The integration of default correlation does not affect the general procedure of the first three steps. With regard to the fourth step our experience has shown that for most portfolios the parameterisation of the correlation between default events is far more important than correlation values between

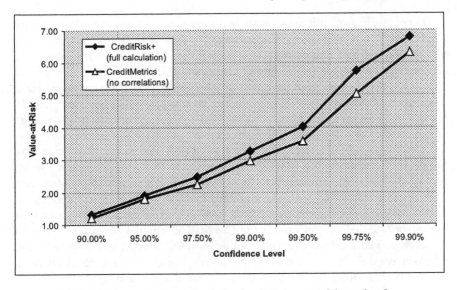

Fig. 11.5. Value-at-risk results for different confidence levels

default and non-default migrations events or non-default migrations events only.[6] Referring to the example calculation in the previous chapter this finding is not surprising, as 78% of the expected negative value change was due to real default events and only 22% due to (non-default) downgrade events. Accordingly, the portfolio distribution and the risk figures are mainly driven by the parameterization for default events. This clear domination of default events is typical for most portfolios (highest-quality investment grade portfolios in a risk-averse market sentiment being an exception) although several factors can influence it significantly. The share of expected value in the loss distribution that is due to default events rather than (non-default) credit migration events increases with decreasing credit quality. Furthermore, low-level credit spreads stress the importance of defaults, while the correlation modelling of non-default events is more important in a high-level credit spread environment (in which those non-default downgrades can result in significant losses in market value). Taking into account the general problems of assessing default correlation values it seems an acceptable assumption for many practical purposes to parameterize non-default migration events in the same way as real default events.

If the fifth step is performed in the same manner as before, the profit and loss distributions as a whole will be regarded as independent, while the dependencies in each subportfolio still remain. Consequently, risk will be underestimated marginally.

---

[6] Our experience is mainly based on reference results with CreditMetrics (correlation modelling based on the Merton approach) determined for several large bond and loan portfolios.

There are many different ways of integrating default correlations into CreditRisk$^+$ and calibrating the necessary parameters for the CreditRisk$^+$ sector analysis and the default rate volatilities in practice. One way is to choose only one sector (representing the global economy) and to allocate all obligors to this sector. Another possibility is to use industries as sectors. For a description of different possible parameterizations of CreditRisk$^+$ see [1, 3, 7]. For our purpose we will follow the correlation approach proposed by [3], mainly due to reasons of robustness and simplicity. We estimate the correlation of default events using industries as the drivers of risk. Following [3] we calculate the portfolio variance in a way consistent with the observed correlation structure of the different industries of the obligors and try to match this variance with the one calculated with the standard one-sector CreditRisk$^+$ model. As a result we will get the relative default variance[7] $\hat{\sigma}_{PF}$ as the parameter that has to be applied in the one-sector CreditRisk$^+$ model in order to match the observed correlation structure as well as possible with the means chosen. The relative default variance can then be used for the gamma distribution in CreditRisk$^+$ in order to calculate the portfolio distribution.

11.2.1 Illustrative Example

To illustrate the approach we provide a simple example. We choose a portfolio of 500 obligors. Each obligor belongs to one of eight different industries in Germany. For the calibration of the relative default variance $\hat{\sigma}_{PF}^2$ we use the correlation of the default rates of the different industries, which we can get from industry-specific time series of historically observed default events.[8] Figure 11.5 shows the correlations structure and the result for $\hat{\sigma}_{PF}^2$ as well as the results for the calculation of the loss distribution with our modelling approach (labelled "Special CR+*") and the basic CreditRisk$^+$ model.

From Figure 11.6 we can see that by setting all industry correlation to 100% the results with our model match those of the basic one-sector CreditRisk$^+$ model. We can also see that if we apply the one-sector model and neglect the real correlations structure we would overestimate the value-at-risk of the portfolio.

The concept of modelling correlation in CreditRisk$^+$ presented above is independent of our approach of integrating rating migrations. Therefore the correlation model can easily be replaced by other models, respectively calibration processes.

In order to apply the presented correlation procedure within our approach we use the relative default variance for the relative migration rate variances of both the profit and the loss distribution. It appears intuitive that the economic dependencies are the same for either positive or negative changes in

---

[7] The relative default volatility of the portfolio $\hat{\sigma}_{PF}$ equals the sum of all default rate volatilities of all obligors in the portfolio divided by the sum of all default rates of all obligors in the portfolio.

[8] Here we used data from the Statistisches Bundesamt (www.destatis.de).

Industry		Correlations							
		1	2	3	4	5	6	7	8
1 Agriculture, Hunting, Forestry	1	100.00%	72.03%	95.25%	87.70%	82.21%	64.54%	85.26%	57.52%
2 Manufacturing	2	72.03%	100.00%	79.32%	93.20%	82.21%	91.02%	93.39%	58.74%
3 Construction	3	95.25%	79.32%	100.00%	91.19%	87.95%	74.22%	89.80%	63.71%
4 Wholesale and Retail Trade	4	87.70%	93.20%	91.19%	100.00%	90.12%	84.71%	97.02%	70.00%
5 Hotels and Restaurants	5	82.21%	82.21%	87.95%	90.12%	100.00%	90.82%	95.99%	89.11%
6 Transport, Storage and Communication	6	64.54%	91.02%	74.22%	84.71%	90.82%	100.00%	93.47%	74.65%
7 Real Estate, Renting and Business Activities	7	85.26%	93.39%	89.80%	97.02%	95.99%	93.47%	100.00%	77.13%
8 Other Community, Social and Personal Service Activities	8	57.52%	58.74%	63.71%	70.00%	89.11%	74.65%	77.13%	100.00%

$$\hat{\sigma}_{PF}=0.65$$

Portfolio model	CR+*		CR+ original
Industry correlations	From time series	100%	All obligors in one sector
Modelling approach	Default mode		Default mode
Number of obligors	500		500
Exposure	315,104,735		315,104,735
VAR (99%)	14,152,572	14,844,297	14,844,297

Fig. 11.6. Sample calculations of the correlation approach

the obligors, expected default rates expressed by their rating grades. Further research is necessary to base this intuition on empirical evidence.

After our short digression to the topic of modelling default correlations we are now able to feed CreditRisk$^+$ with the necessary data and we can evaluate the two subportfolios with correlations.

All obligors are allocated to one sector and the correlation model presented above gives us the relative default variances $\hat{\sigma}^{+^2}$ for the portfolio of profits and $\hat{\sigma}^{-^2}$ for the portfolio of losses to account for the correlation structure of the portfolio.

Now we apply this correlation approach to the sample calculation from Section 11.1. In Table 11.4 we compare again our results with CreditMetrics.[9] Unfortunately, the different modelling techniques concerning correlation effects restrict the possibilities for a meaningful comparison of CreditMetrics with the approach presented here. For a valid comparison the parameterizations of both correlation approaches have to be set consistently.[10] For the sake of simplicity we choose a parameterization in our example that leads us to comparable standard deviations in both approaches.

---

[9] In CreditMetrics we use a constant asset correlation of 20% for all obligors.

[10] See [6] for a detailed description of consistent parameterizations of asset value models like CreditMetrics and CreditRisk$^+$.

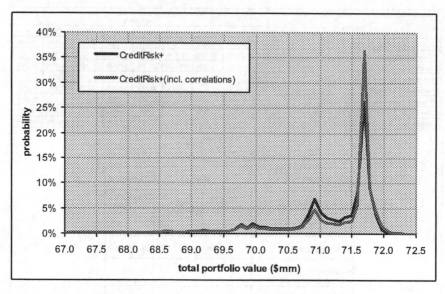

Fig. 11.7. Distribution of the total portfolio value using CreditRisk⁺ with and without correlations

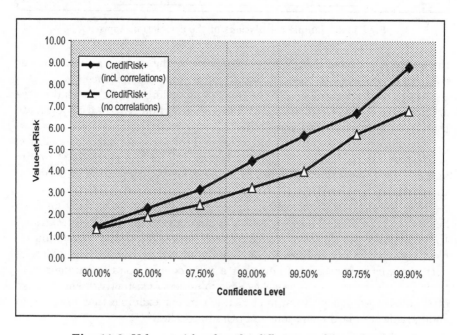

Fig. 11.8. Value-at-risk values for different confidence levels

1	2	3	4	5	5	7
		CreditRisk+ (incl. correlations)	CreditRisk+ (no correlations)	Δ 3, 4 in %	CreditMetrics (incl. asset correlations 20%)	Δ 3, 6 in %
Expected portfolio value		71.09	71.09		71.09	
Standard deviation		1.11	0.91		1.11	
Percentiles	90.00%	69.67	69.76		69.77	
	95.00%	68.80	69.19		68.95	
	97.50%	67.94	68.61		68.12	
	99.00%	66.59	67.84		66.88	
	99.50%	65.44	67.07		65.68	
	99.75%	64.39	65.35		64.38	
	99.90%	62.27	64.29		61.66	
Value-at-Risk (=Expected Value-Percentile)	90.00%	-1.42	-1.32	-6.76%	-1.32	-7.31%
	95.00%	-2.28	-1.90	-16.82%	-2.14	-6.48%
	97.50%	-3.15	-2.48	-21.35%	-2.97	-5.67%
	99.00%	-4.49	-3.24	-27.79%	-4.21	-6.26%
	99.50%	-5.65	-4.01	-28.92%	-5.41	-4.14%
	99.75%	-6.70	-5.74	-14.33%	-6.71	0.12%
	99.90%	-8.81	-6.80	-22.88%	-9.43	6.96%

Table 11.4. Expected value – percentile and value-at-risks of future portfolio value using different methods (incl. correlations)

11.3 Risk Contributions in the Migration Mode

For risk management purposes beyond the pure measurement of the total portfolio risk it is desirable to calculate the risk contributions of parts of the portfolio or even single obligors to the total risk of the portfolio. As in [5] we follow the way that we calculate the risk contributions as the contribution to the standard deviation of the portfolio. The formulas presented in [5] can be applied in a straightforward manner in order to calculate the risk contributions in our migration mode model.

According to (2.16) the risk contribution of obligor A to the standard deviation can be written as

$$r_A(\sigma_{PF}) = \nu_A \frac{\partial \sigma_{PF}}{\partial \nu_A} \quad \text{with} \quad \sum_{A=1}^{K} r_A(\sigma_{PF}) = \sigma_{PF}.$$

In our case we get for the two subportfolios

$$\sum_{A=1}^{K} r_A(\sigma^+) = \sigma^+ \quad \text{and} \quad \sum_{A=1}^{K} r_A(\sigma^-) = \sigma^-.$$

To get the obligor-specific risk contributions we have to sum up the risk contributions of the "virtual" obligors of the corresponding loan

$$r_A(\sigma^+) = \sum_{j<i_A} r_{A,j}(\sigma^+) \quad \text{and} \quad r_A(\sigma^-) = \sum_{j>i_A} r_{A,j}(\sigma^-).$$

Taking into account that the convolution of the profit and loss distribution assumes independence, we finally get the risk contribution for obligor A by

$$r_A(\sigma_{PF}) = r_A(\sigma^+)\frac{\partial\sigma_{PF}}{\partial r_A(\sigma^+)} + r_A(\sigma^-)\frac{\partial\sigma_{PF}}{\partial r_A(\sigma^-)} = \frac{1}{\sigma_{PF}}(r_A(\sigma^+)\sigma^+ + r_A(\sigma^-)\sigma^-).$$

Corresponding to the usual CreditRisk$^+$ methodology the risk contribution relevant for specific percentile values can be simply derived by scaling up the risk contributions calculated above. For further aspects about the calculation of risk contributions we refer to Chapter 3.

11.4 Numerical Example

In this section we give a numerical example of our migration mode approach. All subsequent calculations are based on a sample credit portfolio consisting of 500 small and medium-size enterprise loans. For the sake of simplicity we use a product specific subset of a credit portfolio with just plain vanilla bullet loans. All loans have different notionals, ranging from EUR 10,000 to EUR 15,000,000 and their time to maturity ranges from 1 to 5 years. All exposures are assumed to be net exposures so that $RV = 0$ for all obligors. The notional value of the whole portfolio is EUR 300,000,000. All obligors belong to one of eight different German industries. Each obligor within the portfolio has been assigned a rating grade by the bank's internal rating system. Using the one-year migration matrix of the internal rating system as shown in Table 11.5 we can evaluate the one-year default rate of each obligor.

		Rating at the end of the time horizon						
		1	2	3	4	5	6	7
Rating at the beginning of the time horizon	1	82.00%	16.00%	1.78%	0.10%	0.01%	0.01%	0.10%
	2	9.00%	75.50%	14.00%	0.75%	0.30%	0.10%	0.35%
	3	0.20%	8.70%	73.00%	12.50%	3.50%	0.50%	1.60%
	4	0.50%	2.50%	10.00%	69.00%	8.00%	3.50%	6.50%
	5	0.10%	1.30%	2.60%	14.00%	58.00%	10.00%	14.00%
	6	0.10%	1.00%	2.00%	12.00%	14.90%	50.00%	20.00%
	7	0.00%	0.00%	0.00%	0.00%	0.00%	0.00%	100.00%

Table 11.5. Migration matrix of the internal rating system

In Figure 11.9 we can see the results for the separate calculation of the two subportfolios (profits and losses) as well as the combined distribution of the portfolio market value at the end of the time horizon.

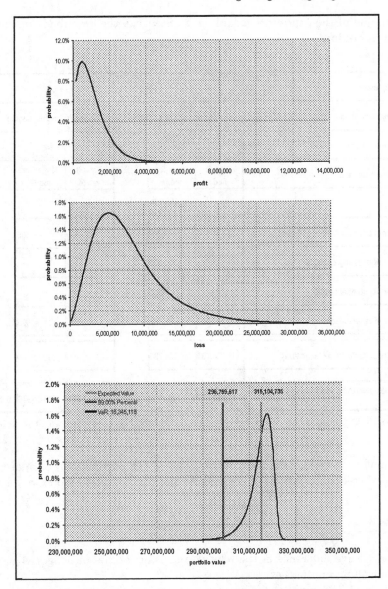

Fig. 11.9. Profit, loss and total portfolio value distributions

Table 11.6 shows the results both for the basic default mode case and for the migration mode case. It can easily be seen that the unexpected loss measured in the migration mode is about 15% higher as in the default mode. We underestimate the risk if we just measure the risk of pure defaults and neglect the risk of potential rating up- or downgrades. Note that the difference in the expected value or exposure between the extended or migration mode CreditRisk$^+$ and the default mode model stems from the market-value-based calculation of the exposures as laid out in Step 1 of the described implementation procedure.

Portfolio model	CR+*			CR+*		
Industry correlations	Yes (from time series)			Yes (from time series)		
Rating migrations	Yes			No		
Modelling approach	Default mode	Migration mode		Default mode	Migration mode	
Number of obligors (virtual obligors)	500	3,500		500	500	
Number of no rating changes	-	500				
Number of potential profits	-	852		-	-	
Number of potential losses	500	2,148		500	500	
Expected portfolio value (exposure)	315,104,735	315,104,735		320,112,700	320,112,700	
			Δ			Δ
Unexpected portfolio value change (value-at-risk), 99%	14,163,619	16,345,118	15.40%	14,407,970	14,407,970	0.00%

Table 11.6. Comparison of the sample calculation for the migration and the default mode

Finally, we look at the risk contributions of the portfolio. To compare the obligors' risk contributions in the default and the migration mode we define

$$\Delta_A = \frac{r_A(\sigma_{PF}^{Migration})}{\sigma_{PF}^{Migration}} / \frac{r_A(\sigma_{PF}^{Default})}{\sigma_{PF}^{Default}} - 1$$

as the relative difference in the relative risk contribution of obligor A in the two modelling approaches.

Table 11.7 shows $\bar{\Delta}$ as the average of Δ for the different rating categories. Consistent with our findings in Section 11.2 we can see that $\bar{\Delta}$ is positive for "good" and negative for "bad" rating grade. The relative risk contribution of an obligor measured in the migration mode is higher (lower) than the one measured in the default mode for obligors with "good" ("bad") ratings. For good ratings the potential losses due to downgrades usually exceed the potential profits due to rating upgrades. The opposite is true for bad ratings.

Rating	Average Δ
1	101.1%
2	86.0%
3	48.2%
4	-9.8%
5	-16.3%
6	-16.6%

Table 11.7. Average relative risk contributions by rating category

11.5 Conclusion

The motivation of this chapter is to provide a practical and "easy to implement" procedure that combines the rating migration concept of CreditMetrics with the actuarial approach of CreditRisk$^+$. The range of portfolios the latter model can be successfully applied to is extended from illiquid portfolios to liquid ones that demand a mark-to-market valuation that integrates potential rating migrations. It is demonstrated that this analytical approach can be used instead of computationally costly Monte Carlo simulations. Despite the simplification of the calculation procedure the results prove to be reliable and robust.

References

1. L. Boegelein, A. Hamerle, R. Rauhmeier, and H. Scheule. Modelling default rate dynamics in the CreditRisk$^+$ framework. *Risk*, 15(10):S24–S28, 2002.
2. M. Bolder, F. B. Lehrbass, M. Lesko, S. Vorgrimler. Modellierung von Netto-Exposures im Hypothekenbankengeschäft. *Die Bank*, 06/2002:405–408, 2002.
3. P. Bürgisser, A. Kurth, A. Wagner, and M. Wolf. Integrating correlations. *Risk*, 12(7):7–60, 1999.
4. P. Bürgisser, A. Kurth, and A. Wagner. Incorporating severity variations into credit risk. *Journal of Risk*, 3(4):5–31, 2001.
5. Credit Suisse Financial Products. CreditRisk$^+$: A credit risk management framework. London, 1997. Available at http://www.csfb.com/creditrisk.
6. M. B. Gordy. A comparative anatomy of credit risk models. *Journal of Banking & Finance*, 24:119–149, 2000.
7. M. Lesko, F. Schlottmann, and S. Vorgrimler. Fortschritte bei der Schätzung von Risikofaktorgewichten für CreditRisk$^+$. *Die Bank*, 6:436–441, 2001.
8. J. P. Morgan. CreditMetrics^{TM} – Technical Document. 1997.
9. B. Rolfes and F. Bröker. Good migrations. *Risk*, 11(11):72–73, 1998.

12

An Analytic Approach to Rating Transitions

Carsten Binnenhei

Summary. Extending CreditRisk$^+$ to a multi-state model allows one to incorporate credit quality changes into the calculation of portfolio credit risk. Several modifications to the original methodology are proposed to make the extension to a mark-to-market model tractable. The distribution of portfolio value changes is obtained analytically by a two-dimensional recursion algorithm.

Introduction

Current credit risk models such as J. P. Morgan's *CreditMetrics*, Moody's KMV's *Portfolio Manager* or McKinsey's *CreditPortfolio View*, which are capable of capturing the effect of credit quality changes on the value of a bond portfolio[1] rely heavily on the use of Monte Carlo techniques. It is the aim of the present chapter to outline a novel approach to rating transitions on the portfolio level that avoids the use of Monte Carlo simulations completely.

There have been attempts to treat rating transitions within the standard CreditRisk$^+$ framework in the past, notably by Rolfes and Bröker [4], see also Chapter 11. These authors propose of splitting each obligor into R "clones", where R is the number of rating categories under consideration. Each clone represents a proper rating transition[2] of the obligor at the time horizon: assigned to the clones are the corresponding rating transition probabilities and bond value changes, as analogues of default probabilities and net exposures in standard CreditRisk$^+$. Then, since the model cannot simultaneously handle exposures with opposite signs, CreditRisk$^+$ is applied separately to the "subportfolios" of profits (positive bond value changes) and losses (negative value changes). The total portfolio's profit and loss distribution is finally obtained by convoluting the distributions of the two subportfolios, neglecting all de-

---

[1] Bond valuation is a problem that will not be treated here. Bond value changes due to credit events are part of the model's input.

[2] Default is included, but unchanged rating at horizon is disregarded.

pendencies between profits and losses. Correlations between different obligors are not considered either.

The approach of Rolfes and Bröker, though easily implemented, has several shortcomings. The model does not account for the fact that transitions of one and the same obligor to different rating categories are mutually exclusive. In principle, an obligor can default and at the same time migrate to all rating categories.[3] Next, the aggregation of profits and losses by convolution is not satisfactory. It should be possible to calculate the total portfolio's profit and loss distribution more accurately. Furthermore, neglecting correlations between obligors leads to a significant underestimation of risk. Correlations cannot be introduced by choosing the *same* factor loadings (w.r.t. appropriate stochastic background factors) for all clones of a given obligor, since that would lead to high correlations between clones, i.e. to high probabilities of multiple transitions. Rather, correlations between upgradings and downgradings should be negative, a feature that is incompatible with standard CreditRisk$^+$. These remarks show that sector assignment has to be done with care.

We will address each of these topics in the sequel. The objective of the present chapter is to outline the theoretical framework of an analytic migration model. It is the author's intention to report on model implementations somewhere else.

The starting point of our analysis in Section 12.2 is a more realistic model of credit events. At the horizon the state of an obligor is described by an indicator random vector with $(R + 1)$ components. An obligor's rating at the horizon is well-defined. In the case of independent, statistically identical obligors the number of credit events then follows a multinomial distribution (generalizing the binomial distribution one encounters in the standard CreditRisk$^+$ case).

Replacing indicator functions by indicator vectors is a very natural choice for a model of rating transitions. It implies that *multivariate* probability-generating functions have to be considered if standard CreditRisk$^+$ techniques are to be adopted. The necessary mathematical background is developed in Section 12.1.

In Section 12.3 the aggregation of value changes over obligors is done by a Poisson approximation in R dimensions. This step seems to be inevitable if the computational efficiency of CreditRisk$^+$ is to be maintained. As a side effect, multiple transitions of single obligors are no longer excluded in the model. This phenomenon is analogous to the occurrence of multiple defaults in standard CreditRisk$^+$ and can be viewed as a partial justification of the approach of Rolfes and Bröker. What we have gained compared with [4] is the complete information on the joint distribution of profits and losses.

Stochastic migration rates are discussed in Section 12.4. Negative correlations between upgradings and downgradings can be realized by admitting *neg-*

---

[3] As mentioned in [4], this effect is small as long as all transition probabilities are small.

ative factor loadings. But negative factor loadings w.r.t. gamma-distributed background factors imply negative conditional transition probabilities, which can cause a breakdown of the model. This obstruction is overcome by adding one *uniformly* distributed factor to the model. Relative to this factor, which can be viewed as an indicator of the business cycle, negative sensitivities are admissible within certain limits. (Alternatively, if one prefers to stick to gamma-distributed factors only, then a conservative and pragmatic approach to correlations is to treat only rating deterioration as stochastic and to neglect correlations between rating upgradings.)

Exact algorithms for the calculation of the distribution of portfolio value changes are derived in Section 12.5. The basic idea is to split the portfolio value change variable into a positive and a negative part and to rearrange these parts into a bivariate random vector. The distribution of this vector, from which the distribution of portfolio value changes is readily obtained, can be calculated by two-dimensional recurrence schemes based either on standard CreditRisk$^+$ [1], or on adaptations of Giese's algorithm [2], see also Chapter 6.

Section 12.6 describes model variants based on Gaussian background factors. Linear dependence of migration rates on Gaussian factors leads to significant simplifications of the algorithms but suffers from the problem of negative conditional probabilities mentioned above. This defect is cured by using squares of Gaussians as background factors.

In Appendix 1 we make precise the concept (and implications) of "conditional independence" underlying the model. The model's computation time is roughly estimated in Appendix 2.

12.1 Preliminaries on Probability-Generating Functions

Some mathematical tools and notations are introduced here for later use. We discuss multivariate probability-generating functions, and we show how the distribution of a random variable with values in the (positive *and* negative) integers can be computed using probability-generating functions.

12.1.1 Multivariate probability-generating functions

Throughout this section let $X = (X_1, \ldots, X_N)$ be a random vector with components taking values in the set $\mathbb{N}$ of non-negative integers. We use the convention $0^0 \equiv 1$ and the abbreviation PGF for probability-generating function.

The generalization of univariate PGFs to higher dimensions is straightforward:

Definition 1. *(a) The PGF $G_X(z_1, \ldots, z_N)$ of X is defined as*

$$G_X(z_1, \ldots, z_N) \equiv \mathbb{E}(z_1^{X_1} \cdots z_N^{X_N}). \tag{12.1}$$

The function $G_X(z)$ exists for all $z = (z_1, \ldots, z_N) \in [-1,1]^N$. (By standard arguments from complex analysis, $G_X(z)$ is analytic in the ball $\{z \in \mathbb{C}^N \mid |z_i| < 1\}$.)

(b) The conditional PGFn $G_X(z_1, \ldots, z_N \mid \mathcal{B})$ of X under a hypothesis[4] $\mathcal{B}$ is a univariate random variable, defined by

$$G_X(z_1, \ldots, z_N \mid \mathcal{B}) \equiv \mathbb{E}\left(z_1^{X_1} \cdots z_N^{X_N} \mid \mathcal{B}\right). \tag{12.2}$$

Notation. *It is convenient to use multi-index notation henceforth. For $z = (z_1, \ldots, z_N)$ we set*

$$z^X \equiv z_1^{X_1} \cdots z_N^{X_N}.$$

Then the PGF of X can be written in the usual form:

$$G_X(z) = \mathbb{E}\left(z^X\right).$$

For a multi-index $n = (n_1, \ldots, n_N) \in \mathbb{N}^N$ with non-negative integer entries we use notations as follows:

$$z^n \equiv z_1^{n_1} \cdots z_N^{n_N} \tag{12.3a}$$

$$|n| \equiv n_1 + \cdots + n_N \tag{12.3b}$$

$$n! \equiv n_1! \cdots n_N! \tag{12.3c}$$

The following properties of multivariate probability-generating functions are readily checked, except for property (g) which is a consequence of Proposition 2 proved in the appendix.

Proposition 1. *The following properties hold:*

(a) The coefficients of the Taylor series expansion of G_X around the origin describe the joint distribution of the components of X:

$$G_X(z) = \sum_{n \in \mathbb{N}^N} \mathbb{P}[X = n] z^n.$$

These coefficients can be computed by taking partial derivatives at $z = 0$:

$$\mathbb{P}[X = n] = \frac{1}{n!} \left(\frac{\partial^{n_1}}{\partial z_1^{n_1}} \cdots \frac{\partial^{n_N}}{\partial z_N^{n_N}} G_X \right)(0).$$

(b) Let Y be another random vector with N components. If X and Y are independent, then the PGF of $X + Y$ equals the product of G_X and G_Y:

$$G_{X+Y} = G_X G_Y.$$

---

[4] Technically speaking, a hypothesis is a sub-σ-algebra on the given probability space. The reader may think of the sub-σ-algebra generated by some other random variables, e. g. the systematic risk factors.

(c) For all $n \in \mathbb{N}^N$ one has

$$G_{(n_1 X_1, \ldots, n_N X_N)}(z) = G_X(z^n).$$

(d) Let $\{i_1, \ldots, i_m\}$ be a subset of $\{1, \ldots, N\}$, and let $\{j_1, \ldots, j_{N-m}\} = \{1, \ldots, N\} \setminus \{i_1, \ldots, i_m\}$ be its complement. The probability-generating function of $\{X_{i_1}, \ldots, X_{i_m}\}$, which describes the marginal distribution of $X_{i_1}, \ldots, X_{i_m}$, is equal to the restriction of G_X to the m-dimensional plane given by $z_{j_1} = \cdots = z_{j_{N-m}} = 1$:

$$G_{(X_{i_1}, \ldots, X_{i_m})} = G_X|_{z_{j_1} = \cdots = z_{j_{N-m}} = 1}. \tag{12.4}$$

(e) The (univariate) PGF of the sum of the components of X is equal to the restriction of G_X to the diagonal $z_1 = \cdots = z_N$:

$$G_{X_1 + \cdots + X_N}(t) = G_X(t, \ldots, t), \quad t \in [-1, 1].$$

(f) The PGF of X can be obtained by averaging the conditional PGF under some hypothesis $\mathcal{B}$:

$$G_X(z) = \mathbb{E}(G_X(z \mid \mathcal{B})).$$

(g) Let $X_1, \ldots, X_M$ ($M < \infty$) be random vectors with values in $\mathbb{N}^N$. If $X_1, \ldots, X_M$ are conditionally independent[5] under some hypothesis $\mathcal{B}$, then

$$G_{X_1 + \cdots + X_M}(z \mid \mathcal{B}) = G_{X_1}(z \mid \mathcal{B}) \cdots G_{X_M}(z \mid \mathcal{B}) \quad \text{almost surely.}$$

12.1.2 The distribution of integer-valued random variables

Let X be an integer-valued random variable such that $\mathbb{P}[X < 0] > 0$. Then care has to be taken of the fact that the "would be" probability-generating function $t \mapsto \mathbb{E}[t^X]$ is singular at $t = 0$.

If there exists an integer n_0 such that $\mathbb{P}[X < n_0] = 0$ then the distribution of X can be obtained by computing the probability-generating function of the non-negative random variable $X - n_0$. Hence this case poses no substantial problems.

In the application we have in mind, where X represents the portfolio value change caused by rating transitions, there is, however – due to the Poisson approximation – no lower bound on the range of X so that $\mathbb{E}[t^X]$ is essentially singular at $t = 0$.

We propose to deal with the latter case in the following way. Select non-negative integer-valued random variables $X^\pm$ such that[6]

$$X = X^+ - X^-.$$

---

[5] See the appendix for a rigorous definition.
[6] A canonical choice would be to take the positive, resp. negative parts $X^\pm \equiv \frac{1}{2}(|X| \pm X)$ of X. This choice is distinguished by the condition that $X^+ X^- = 0$.

Then let the bivariate random vector $\tilde{X}$ be defined by

$$\tilde{X} = \begin{pmatrix} X^+ \\ X^- \end{pmatrix}. \tag{12.5}$$

Assume that the coefficients g_{mn} of the PGF of $\tilde{X}$,

$$G_{\tilde{X}}(z_1, z_2) = \sum_{m,n=0}^{\infty} g_{mn} z_1^m z_2^n,$$

are known. Then, by Proposition 1(a), the distribution of X is given by

$$\mathbb{P}[X = n] = \sum_{m \geq \max\{-n,0\}} g_{m+n,m}, \quad n \in \mathbb{Z}. \tag{12.6}$$

In practical applications one has of course to find appropriate cut-offs for the infinite sums appearing above.

12.2 Modelling Credit Events

We consider a rating system consisting of R categories plus a default state. Rating categories are labelled by an integer $r = 0, \dots R - 1$, where ordering is by deteriorating credit quality. The value $r = R$ is reserved for the default state.

Let a portfolio $\mathcal{P}_{\text{tot}}$ of size $|\mathcal{P}_{\text{tot}}|$ be given. The *initial rating* of an obligor $A \in \mathcal{P}_{\text{tot}}$ will be denoted by $r(A)$. The probability that obligor A will fall into category r at the given time horizon will be denoted by p_{Ar}. This probability typically depends only[7] on the initial rating $r(A)$ and the *final rating* r. Transition probabilities add up to one:

$$\sum_{r=0}^{R} p_{Ar} = 1. \tag{12.7}$$

We associate with each obligor A an indicator random vector $\mathbf{1}_A$ with $R + 1$ components,

$$\mathbf{1}_A = (\mathbf{1}_{A0}, \dots, \mathbf{1}_{AR}),$$

subject to the constraint

$$\sum_{r=0}^{R} \mathbf{1}_{Ar} = 1. \tag{12.8}$$

---

[7] We prefer to write p_{Ar} instead of $p_{r(A),r}$ because in Section 12.4, conditional transition probabilities will indeed depend on A and not only on initial and final ratings.

The components $\mathbf{1}_{Ar}$ are the indicator functions of the events "obligor A migrates to state r". The constraint (12.8) ensures that each obligor is in *exactly* one state at the horizon. The expectation value of $\mathbf{1}_A$ is determined by the rating transition probabilities:

$$\mathbb{E}(\mathbf{1}_A) = (p_{A0}, \ldots, p_{AR}).$$

Note that, by (12.8), this model of migrations contains a certain amount of redundancy. Given, e.g. the values of $\mathbf{1}_{A1}, \ldots, \mathbf{1}_{AR}$, the value of $\mathbf{1}_{A0}$ is known with certainty, reflecting the dependencies between credit events. This redundancy will be eliminated in Section 12.3.

It is instructive to look at the case of *independent* obligors. Let N_{red} be the sum of all indicator vectors:

$$N_{\text{red}} \equiv \sum_{A \in \mathcal{P}_{\text{tot}}} \mathbf{1}_A,$$

where the subscript "red" indicates that this description still contains redundancies. The components of N_{red} model the number of obligors belonging to the different rating categories at the horizon.

In case of independent indicator vectors the distribution of N_{red} can be obtained by combinatorial considerations. For $n \in \mathbb{N}^{R+1}$ one has

$$\mathbb{P}[N_{\text{red}} = n] = \begin{cases} \displaystyle\sum_{\substack{\mathcal{P}_0 \cup \cdots \cup \mathcal{P}_R = \mathcal{P}_{\text{tot}}, \\ |\mathcal{P}_r| = n_r}} p(\mathcal{P}_0, 0) \cdots p(\mathcal{P}_R, R), & |n| = |\mathcal{P}_{\text{tot}}| \\ 0, & |n| \neq |\mathcal{P}_{\text{tot}}| \end{cases} \tag{12.9}$$

where $|n|$ is defined in (12.3), the sum runs over all partitions of the total portfolio $\mathcal{P}_{\text{tot}}$ into subportfolios $\mathcal{P}_r$ (corresponding to final rating r) of size n_r, and, for a subportfolio $\mathcal{P} \subset \mathcal{P}_{\text{tot}}$, $p(\mathcal{P}, r)$ is defined by

$$p(\mathcal{P}, r) \equiv \prod_{A \in \mathcal{P}} p_{Ar}.$$

If in addition all obligors share the same initial rating, i.e. the indicator vectors are i.i.d., then N_{red} is *multinomially distributed*:

$$\mathbb{P}[N_{\text{red}} = n] = \frac{|\mathcal{P}_{\text{tot}}|!}{n!} p_0^{n_0} \cdots p_R^{n_R} \qquad (|n| = |\mathcal{P}_{\text{tot}}|), \tag{12.10}$$

where $n!$ is defined in (12.3), and the p_r are the common transition probabilities.

12.3 Poisson Approximation

To reduce computational complexity we seek approximations of the distributions of the indicator vectors $G_{\mathbf{1}_A}$ and of their sum by R-dimensional Poisson distributions, just as in the standard CreditRisk$^+$ case. It is convenient to do this in the framework of probability-generating functions.

12.3.1 Single obligors

The value of the probability-generating function G_A of an indicator vector $\mathbf{1}_A$ at $z = (z_0, \dots, z_R) \in [-1, 1]^{R+1}$ can be computed as follows:

$$
\begin{aligned}
G_A(z) &= \sum_{n_0 + \cdots + n_R = 1} \mathbb{P}[\mathbf{1}_A = n]\, z^n \\
&= \sum_{r=0}^{R} \mathbb{P}[\mathbf{1}_{Ar} = 1,\ \mathbf{1}_{As} = 0\ (s \neq r)]\, z_r = \sum_{r=0}^{R} p_{Ar} z_r.
\end{aligned}
\tag{12.11}
$$

Typically, the probability of "no migration" dominates the probabilities of proper migration events:

$$
p_{A, r(A)} \gg p_{Ar}, \qquad r \neq r(A).
$$

Thus the Poisson distribution, being adequate for low-probability events, appears to be a reasonable approximation for *proper* credit events only.

The probability-generating function for proper credit events is equal to the restriction of $G_{\mathbf{1}_A}(z)$ to the hyperplane $z_{r(A)} = 1$, cf. (12.4). Using (12.7), the R-dimensional Poisson approximation for a single obligor looks as follows:

$$
\begin{aligned}
G_A(z)|_{z_{r(A)}=1} &= p_{A, r(A)} + \sum_{r \neq r(A)} p_{Ar}\, z_r \\
&= 1 + \sum_{r \neq r(A)} p_{Ar}\, (z_r - 1) \approx \exp \sum_{r \neq r(A)} p_{Ar}\, (z_r - 1).
\end{aligned}
\tag{12.12}
$$

This is the natural generalization of the one-dimensional approximation in [1, Eq. (6)]. The R-dimensional approximation is simply a product of R univariate Poisson distributions. Note that, by restricting to proper credit events, the redundancy in the description of credit events has now been removed. This is in line with standard CreditRisk+, i.e. the case $R = 1$, where one restricts attention to the default indicator function.

Note further that, due to the nature of the Poisson distribution, and since product distributions are tantamount to independence, multiple credit events of single obligors now crop up in the model in any possible combination (but with low probabilities).

12.3.2 Homogeneous subportfolios

The Poisson approximation enables one to sum over obligors in a very efficient manner, by simply adding the respective Poisson parameters. This should be contrasted with the unwieldy combinatorics behind (12.9) and (12.10).

The sum of indicator vectors $\mathbf{1}_A$ is a random vector that models the occupation numbers of the various rating categories at the horizon. Knowledge of these occupation numbers alone does not suffice to determine the portfolio's value distribution; rather, one has to combine it with information on the value changes that the different positions incur in case of credit events.

On an aggregate level, this missing piece of information can be provided by partitioning the total portfolio $\mathcal{P}_{tot}$ into *homogeneous subportfolios* $\mathcal{P}_1, \ldots, \mathcal{P}_M$, the counterparts of the exposure bands in standard CreditRisk$^+$ [1]. A homogeneous subportfolio $\mathcal{P}_j$ is characterized by a fixed *initial rating* r_j,

$$r_j \in \{0, \ldots, R-1\},$$

together with a vector v_j of approximate *value changes*,

$$v_j = (v_{jr})_{r \neq r_j} \in \mathbb{Z}^R.$$

As in standard CreditRisk$^+$, value changes are expressed as integer multiples of a fixed positive *value unit*. We prefer to use opposite signs compared with [1]. We regard losses as negative and profits as positive:

$$v_{jr} \lesseqgtr 0 \quad \text{if} \quad r \gtreqless r_j. \tag{12.13}$$

See Chapter 11 by Bröker and Schweizer (e.g. Fig. 11.2) for examples of choices of the v_{jr}. One of the obstacles we have to overcome in the following is the simultaneous occurrence of positive and negative signs in (12.13).

Value changes are always considered relative to the reference state "final rating = initial rating". This reference state must correspond to zero value change with necessity because the information on the occupation number of this state has been lost during the approximation (12.12).

By definition, the subportfolio $\mathcal{P}_j$ comprises all obligors $A \in \mathcal{P}_{tot}$ with initial rating equal to r_j,

$$r(A) = r_j,$$

and such that the obligor's position incurs a value change approximately equal to v_{jr} should the obligor migrate to category r.

Remark 1. The grouping of obligors into subportfolios makes sense only if the number M of subportfolios required to match all value change vectors occurring in $\mathcal{P}_{tot}$ is significantly smaller than the portfolio size $|\mathcal{P}_{tot}|$. To keep the number M manageable we recommend using the following improved aggregation mechanism. This mechanism applies to standard CreditRisk$^+$ as well, so we will explain and illustrate it in the standard CreditRisk$^+$ framework. The generalization to R dimensions should be obvious.

Instead of choosing exposure bands with constant band width (the loss unit L_0) as done in [1], let the band width grow *exponentially* with some fixed rate. This means that the relative rounding error in the single bands is kept fixed, rather than the absolute error. In this way the number of exposure bands can be drastically reduced, especially in the case of strongly inhomogeneous portfolios, without significantly affecting the model's precision.

As an illustration, consider a large portfolio $\mathcal{P}$ consisting of a retail subportfolio $\mathcal{P}_1$, with net exposures in the range \$1000 to \$1,000,000, and a subportfolio $\mathcal{P}_2$ of corporate loans, with exposures ranging from $\$10^6$ to $\$10^9$. The number of obligors in $\mathcal{P}_1$ might amount to some $100,000$, the number of obligors in $\mathcal{P}_2$ to some 1000. A reasonable choice[8] of the loss unit L_0 might then be $L_0 = 100,000$. Forming exposure bands as suggested in [1], one would end up with several 1000 bands, with most of them sparsely filled. Working with an exponentially growing band width instead presumes the choice of an initial band width l_0 together with a growth rate $r > 0$. The width of the j-th band is then equal to $(1+r)^{j-1}l_0$ $(j = 1,\ldots,M)$. Put differently, the j-th band comprises all obligors A with exposure $\tilde{\nu}_A$ in the range

$$\tfrac{(1+r)^{j-1}-1}{r}l_0 < \tilde{\nu}_A \le \tfrac{(1+r)^j-1}{r}l_0.$$

The number of exposure bands needed to cover the portfolio $\mathcal{P}$ is the smallest integer M such that

$$M \ge \frac{\log(1 + r\frac{\tilde{\nu}_{\max}}{l_0})}{\log(1+r)}$$

where $\tilde{\nu}_{\max}$ denotes the maximum exposure in $\mathcal{P}$. In the situation envisaged above, an appropriate choice of the parameters could be $l_0 = 1.5L_0$, $r = 10\%$, leading to $M = 69$. One should assign to a band the quotient of its expected loss and Poisson parameter (rounded to the nearest integer multiple of L_0) as average exposure. The band's Poisson parameter can then be set to the quotient of expected loss and average exposure, so that the pre-rounding expected loss is *exactly* preserved.[9]

Let us finally remark that the model's computation time does not really depend on M. It depends to some extent on the number of different values of v_{jr} occurring in the portfolio. This contribution can in fact be neglected in view of the computational burden implied by the recurrence schemes that will be discussed in Section 12.5. □

Now let

$$N_j = (N_{jr})_{r\ne r_j} \tag{12.14}$$

be the sum of the approximate indicator vectors of proper credit events in $\mathcal{P}_j$ described by (12.12). Assuming independence of obligors, the distribution of N_j is a product of R univariate Poisson distributions with parameters

$$\mu_{jr} \equiv \sum_{A\in\mathcal{P}_j} p_{Ar}, \qquad r \ne r_j. \tag{12.15}$$

---

[8] With this choice, the rounding error in $\mathcal{P}_1$ and the number of CreditRisk$^+$ recurrence steps should both be acceptable.

[9] Notice that during the rounding process, the same average exposure can be assigned to adjacent bands. Such bands have to be suitably merged. Notice further that the average exposure of the lowest band can be equal to zero. It should then be set to L_0, and the Poisson parameter be correspondingly adjusted.

The probability-generating function of N_j is given by

$$G_{N_j}(z_0, \ldots, \widehat{z_{r_j}}, \ldots, z_R) = \exp \sum_{r \neq r_j} \mu_{jr}(z_r - 1). \qquad (12.16)$$

The notation "$\widehat{z_{r_j}}$" indicates that the variable z_{r_j} is omitted.

(12.16) is the R-dimensional analogue of (7) in [1]. Using multi-index notations as in (12.3), the distribution of N_j is explicitly given by

$$\mathbb{P}[N_j = n] = e^{-|\mu_j|} \frac{\mu_j^n}{n!}, \qquad n \in \mathbb{N}^R, \qquad (12.17)$$

where

$$\mu_j \equiv (\mu_{jr})_{r \neq r_j}.$$

The distribution of profits and losses of the subportfolio $\mathcal{P}_j$ is modelled by the integer-valued random variable V_j,

$$V_j \equiv \sum_{r \neq r_j} v_{jr} N_{jr}. \qquad (12.18)$$

Because of the complications implied by (12.13), the discussion of the distribution of profits and losses is postponed to Section 12.5.

12.4 Stochastic Migration Rates

Correlations between obligors are introduced by letting obligor-specific transition probabilities depend on a random vector S,

$$S = (S_0, \ldots, S_N),$$

consisting of independent non-negative risk factors with mean one:

$$S_k \geq 0, \quad \mathbb{E}(S_k) = 1 \qquad (k = 0, \ldots, N). \qquad (12.19)$$

We suggest specifying risk factors as follows:

S_0 uniformly distributed in $[0, 2]$ (with variance $\sigma_0^2 = \frac{1}{3}$),   (12.20a)

$S_1 \equiv 1$ (specific risk, $\sigma_1 = 0$),   (12.20b)

S_k gamma-distributed with variance σ_k^2, $k = 2, \ldots, N$.   (12.20c)

As mentioned in the introduction, the uniformly distributed factor S_0 enables one to incorporate negative correlations between upgradings and downgradings. This is achieved by letting upgradings and downgradings respond to S_0 in opposite ways. If one assigns positive factor loadings to upgradings, but negative factor loadings to downgradings, then realizations of S_0 larger (smaller) than 1 imply increasing (decreasing) upgrading rates and decreasing (increasing) downgrading rates, indicating an economic upturn (downturn). This justifies the interpretation of S_0 as an indicator of the business cycle. The factor S_0 might also prove useful in standard CreditRisk$^+$ when general correlation structures are to be modelled.

The constant *specific risk factor* S_1 plays an important role in model calibration. Varying the specific risk factor loadings while adjusting the systematic risk factor variances one can, e.g. keep default correlations fixed but change the probabilities of extremal events dramatically, cf. [3]. This degree of freedom is a unique feature of CreditRisk$^+$, absent in other credit risk models. We need S_1 in addition to compensate for negative factor loadings relative to the business cycle indicator S_0, cf. (12.25), (12.26) below.

The remaining *systematic risk factors* $S_2, \ldots, S_N$ are the drivers of correlations between obligors as usual. We propose to divide them into two disjoint subsets, one associated with positive value changes (upgradings), the other one with negative value changes (downgradings; see Remark 2 for an illustration). The choice of the variances σ_k^2 is a matter of model calibration and will not be discussed here.

Stochastic migration rates depend linearly on the risk factors. The *factor loadings* of obligor A are given by a matrix w^A,

$$w^A = \left(w_{rk}^A\right)_{\substack{r=0,\ldots,R \\ k=0,\ldots,N}}.$$

They determine the dependency of the transition probabilities $p_{Ar}(S)$ on S:

$$p_{Ar}(S) = p_{Ar} \sum_{k=0}^{N} w_{rk}^A S_k. \tag{12.21}$$

For fixed final rating category r, factor loadings have to satisfy

$$\sum_{k=0}^{N} w_{rk}^A = 1, \tag{12.22}$$

in order to preserve the unconditional transition probabilities

$$\mathbb{E}\big(p_{Ar}(S)\big) = p_{Ar}. \tag{12.23}$$

Covariances between transition probabilities are given by

$$\mathrm{cov}\big(p_{Ar}(S), p_{A'r'}(S)\big) = p_{Ar} p_{A'r'} \sum_{k=0}^{N} w_{rk}^A w_{r'k}^{A'} \sigma_k^2. \tag{12.24}$$

Choosing risk factors as in (12.20), the obligor-specific factor loadings have to satisfy the following constraints besides (12.22):

$$0 \leq w_{rk}^A \leq 1, \quad k = 2, \ldots, N \tag{12.25a}$$

$$-1 \leq w_{r0}^A \leq 1, \tag{12.25b}$$

$$0 \leq w_{r1}^A \leq 2, \tag{12.25c}$$

and

$$2w_{r0}^A + w_{r1}^A \geq 0. \tag{12.26}$$

By (12.25)–(12.26), the conditional migration rates given by (12.21) are always non-negative. In particular, condition (12.26) ensures that $w_{r0}^A S_0 + w_{r1}^A S_1 \geq 0$.

Remark 2. For interpretational issues it may be helpful to look at the factor S_0 from a slightly different perspective. Introduce another factor S_{-1}, defined as the "reflection" of S_0:

$$S_{-1} \equiv 2S_1 - S_0.$$

Then S_{-1} fulfils (12.19), is uniformly distributed in $[0, 2]$, and is countermonotonic relative to S_0, i.e. their correlation satisfies

$$\rho(S_0, S_{-1}) = -1.$$

One can consider a model where migration rates depend on the enlarged set $(S_k)_{k=-1,\ldots,N}$ of risk factors, but with all factor loadings being proper *weights*, i.e. (12.25a) applies to all k, including $k = -1, 0, 1$. Then all conditional migration rates are manifestly non-negative. Denote the corresponding weights by u_{rk}^A. Notice that one can introduce the constraint that, for all pairs (A, r), at least one of the weights $u_{r,-1}^A$, u_{r0}^A is zero. (This constraint just removes the ambiguity in the choice of the weights caused by the introduction of S_{-1}.) For $r < r(A)$ (upgrading) one would choose $u_{r,-1}^A = 0$, $u_{r0}^A > 0$, while for $r > r(A)$ (downgrading) one would choose $u_{r,-1}^A > 0$, $u_{r0}^A = 0$. In other words, upgradings are driven by S_0, downgradings by S_{-1}. This results in negative correlations between upgradings and downgradings.

Since S_{-1} is a linear combination of S_0 and S_1, this formulation of the model is equivalent to the one specified by (12.20). But the new formulation contains some redundancy thath is absent in (12.20). Another advantage of (12.20) is the independence of the $(S_k)_{k \geq 0}$, which will be used in Section 12.5.3. The transformation between weights and factor loadings in the two formulations is given by the equations

$$
\begin{aligned}
w_{r0}^A &= u_{r0}^A - u_{r,-1}^A \\
w_{r1}^A &= 2u_{r,-1}^A + u_{r1}^A
\end{aligned}
\qquad \text{resp.} \qquad
\begin{aligned}
u_{r0}^A &= (w_{r0}^A)^+ \\
u_{r,-1}^A &= (w_{r0}^A)^- \\
u_{r1}^A &= w_{r1}^A - 2(w_{r0}^A)^-,
\end{aligned}
$$

where the notation $w^{\pm} \equiv \frac{1}{2}(|w| \pm w) = \max\{\pm w, 0\}$ is used. The weight u_{r1}^A should be viewed as the "true" amount of specific risk of (A, r). It will in general be less than w_{r1}^A. The inequality (12.26) merely states that $u_{r1}^A \geq 0$.

Let us briefly illustrate the emergence of negative correlations. By (12.24), correlations between stochastic migration rates are given by the following expression:

$$\rho\big(p_{Ar}(S), p_{A'r'}(S)\big) = \frac{\sum_{k\neq 1} w_{rk}^A w_{r'k}^{A'} \sigma_k^2}{\left(\sum_{k\neq 1}(w_{rk}^A \sigma_k)^2\right)^{\frac{1}{2}} \left(\sum_{k'\neq 1}(w_{r'k'}^{A'} \sigma_{k'})^2\right)^{\frac{1}{2}}}. \qquad (12.27)$$

Consider the case $r < r(A)$ (upgrading) and $r' > r(A')$ (downgrading). We assume that upgradings and downgradings respectively are driven by disjoint sets of systematic risk factors. Then the numerator in (12.27) simplifies to $\frac{1}{3} w_{r0}^A w_{r'0}^{A'}$. Choosing the factor loadings of upgradings relative to S_0 as positive and the loadings of downgradings as negative, correlations between upgradings and downgradings become negative as desired. In the extreme case that $w_{r0}^A = 1$ and $w_{r'0}^{A'} = -1$ (which implies that $w_{rk}^A = 0$, $k \geq 1$, and $w_{r'1}^{A'} = 2$, $w_{r'k'}^{A'} = 0$, $k' \geq 2$), the value $\rho = -1$ is attained. (This case corresponds to weights $u_{r0}^A = 1$, $u_{r',-1}^{A'} = 1$, with all remaining weights vanishing.) $\qquad\square$

Next let $\mathcal{P}_j$ be a homogeneous subportfolio as introduced in Section 12.3.2. By (12.15), stochastic migration rates entail *stochastic Poisson parameters*:

$$\mu_{jr}(S) = \sum_{A\in\mathcal{P}_j} p_{Ar}(S) = \sum_{k=0}^{N} c_{rk}^j S_k \qquad (r \neq r_j), \qquad (12.28)$$

$$\text{with} \qquad c_{rk}^j \equiv \sum_{A\in\mathcal{P}_j} p_{Ar} w_{rk}^A. \qquad (12.29)$$

The random vector N_j representing the number of proper credit events in $\mathcal{P}_j$ now also depends on S. The conditional distribution of N_j given the hypothesis $S = s$ is a multivariate Poisson distribution as in (12.17):

$$\mathbb{P}[N_j = n \mid S = s] = e^{-|\mu_j(s)|} \frac{\mu_j(s)^n}{n!}, \qquad n \in \mathbb{N}^R.$$

The conditional PGF of N_j looks as follows (cf. (12.16)):

$$G_{N_j}(z_0, \ldots, \widehat{z_{r_j}}, \ldots, z_R \mid S) = \exp \sum_{r\neq r_j} \mu_{jr}(S)\,(z_r - 1). \qquad (12.30)$$

12.5 The Joint Distribution of Portfolio Profits and Losses

In the present section exact algorithms for the computation of the distribution of portfolio value changes are developed. These algorithms rest on two-dimensional recurrence schemes based on standard CreditRisk$^+$ [1] (Panjer's algorithm, Section 12.5.4) or, preferably, on Giese's approach [2] (Section 12.5.5).

Panjer's and Giese's recurrence algorithms are special cases of a general method to solve differential equations via power series expansion. The bivariate Panjer-like algorithm is obtained from the partial differential equation (PDE) (12.43) below by a comparison of coefficients. Likewise, Giese's algorithm is derived from the PDE (12.55). We provide additional algorithms to compute the inverse, resp. the square root of an analytic function, related to the PDEs (12.61) and (12.64).

12.5.1 Conditional value changes in homogeneous subportfolios

Let us start with the (conditional) distribution of value changes within a homogeneous subportfolio $\mathcal{P}_j$. Given S, we are interested in the distribution of the variable V_j defined in (12.18). Due to the Poisson approximation, this variable is typically unbounded from below *and* from above. We therefore follow Section 12.1.2 and introduce an auxiliary bivariate random vector $\tilde{V}_j$ with values in $\mathbb{N}^2$:

$$\tilde{V}_j \equiv \begin{pmatrix} V_j^+ \\ V_j^- \end{pmatrix}, \tag{12.31}$$

with components defined by

$$V_j^+ \equiv \sum_{r<r_j} v_{jr} N_{jr} \geq 0 \quad \text{(profits/upgradings)}, \tag{12.32}$$

$$V_j^- \equiv \sum_{r>r_j} |v_{jr}| N_{jr} \geq 0 \quad \text{(losses/downgradings)}, \tag{12.33}$$

such that the variable V_j is obtained as the difference of the components of $\tilde{V}_j$:

$$V_j = V_j^+ - V_j^-.$$

Using (12.30), the bivariate conditional probability-generating function of $\tilde{V}_j$ can be computed as follows:[10]

$$\begin{aligned} G_{\tilde{V}_j}(z_+, z_- \mid S) &= \mathbb{E}\left[z_+^{V_j^+} z_-^{V_j^-} \mid S \right] \\ &= G_{N_j}\left(z_+^{v_{j0}}, \dots, z_+^{v_{j,r_j-1}}, z_-^{|v_{j,r_j+1}|}, \dots, z_-^{|v_{jR}|} \mid S \right) \\ &= \exp\left(\sum_{r<r_j} \mu_{jr}(S)\left(z_+^{v_{jr}} - 1\right) + \sum_{r>r_j} \mu_{jr}(S)\left(z_-^{|v_{jr}|} - 1\right) \right) \\ &= \exp\left(\sum_{k=0}^N P_k^j(z_+, z_-) S_k \right), \end{aligned} \tag{12.34}$$

---

[10] The arguments of $G_{\tilde{V}_j}$ are denoted by $z_\pm$ to avoid confusion with the z_r used above.

where the $\mu_{jr}(S)$ are the stochastic Poisson parameters from (12.28), and the P_k^j are deterministic polynomials in $z_\pm$ having the following simple form:

$$P_k^j(z_+, z_-) = \sum_{r < r_j} c_{rk}^j z_+^{v_{jr}} + \sum_{r > r_j} c_{rk}^j z_-^{|v_{jr}|} - \sum_{r \neq r_j} c_{rk}^j. \qquad (12.35)$$

The coefficients c_{rk}^j have been defined in (12.29).

12.5.2 Aggregating over subportfolios

By assumption, the only source of correlation between credit events in the model is the joint dependence of migration rates on the risk vector S. Conditional on S, credit events are independent. This assumption allows us to aggregate over subportfolios (or obligors) in a very convenient way.

We assume that the total portfolio $\mathcal{P}_{\text{tot}}$ is partitioned into homogeneous subportfolios as in Section 12.3.2:

$$\mathcal{P}_{\text{tot}} = \mathcal{P}_1 \cup \cdots \cup \mathcal{P}_M.$$

Value changes in $\mathcal{P}_{\text{tot}}$ are described by the variable V, defined as the sum of the V_j:

$$V \equiv \sum_{j=1}^{M} V_j. \qquad (12.36)$$

Setting

$$V^\pm \equiv \sum_{j=1}^{M} V_j^\pm \geq 0, \qquad \tilde{V} \equiv \sum_{j=1}^{M} \tilde{V}_j = \binom{V^+}{V^-},$$

it follows that

$$V = V^+ - V^-, \qquad (12.37)$$

and the distribution of V can be obtained from the (bivariate) distribution of $\tilde{V}$ by the method described in Section 12.1.2, Equation (12.6).

Applying Proposition 1(g) and (12.34), we get the conditional PGF of $\tilde{V}$:

$$G_{\tilde{V}}(z_+, z_- \mid S) = \prod_{j=1}^{M} G_{\tilde{V}_j}(z_+, z_- \mid S) = \exp\left(\sum_{k=0}^{N} P_k(z_+, z_-) S_k\right), \qquad (12.38)$$

where the deterministic polynomials P_k are sums of the polynomials P_k^j from (12.35):

$$P_k = \sum_{j=1}^{M} P_k^j. \qquad (12.39)$$

12.5.3 Unconditional value changes within the total portfolio

By Proposition 1(f), the unconditional PGF of $\tilde{V}$ is obtained by averaging the conditional PGF

$$
G_{\tilde{V}}(z_+, z_-) = \mathbb{E}\big[G_{\tilde{V}}(z_+, z_- \mid S)\big] = \mathbb{E}\left[\exp \sum_{k=0}^{N} P_k(z_+, z_-)\, S_k\right]
$$

$$
= \prod_{k=0}^{N} \mathbb{E}\Big[\exp P_k(z_+, z_-)\, S_k\Big].
$$
(12.40)

Up to now the specification of the risk factors S_k has played no role in our computations. The only assumption we used above is the independence of the S_k.

   In order to proceed we have to specify the distributions of the S_k. We assume henceforth that the S_k are distributed according to (12.20). Then the moment-generating functions occurring in (12.40) can be computed by elementary integration. They are explicitly given by

$$
\mathbb{E}[\exp tS_k] = \begin{cases} \dfrac{\exp 2t - 1}{2t} & k = 0 \quad \text{(business cycle, } t \neq 0) \\[2mm] \exp t & k = 1 \quad \text{(specific risk)} \\[2mm] \left(1 - \sigma_k^2 t\right)^{-\sigma_k^{-2}} & k \geq 2 \quad \text{(systematic risk, } t < \sigma_k^{-2}). \end{cases}
$$
(12.41)

Inserting (12.41) into (12.40), we finally obtain the unconditional PGF of $\tilde{V}$:

$$
G_{\tilde{V}}(z_+, z_-) = \frac{\exp 2P_0(z_+, z_-) - 1}{2P_0(z_+, z_-)} \exp P_1(z_+, z_-)
$$

$$
\times \prod_{k=2}^{N} \left(1 - \sigma_k^2 P_k(z_+, z_-)\right)^{-\sigma_k^{-2}}.
$$
(12.42)

Thus we have found the PGF of $\tilde{V}$, which describes the joint distribution of profits and losses in $\mathcal{P}_{\text{tot}}$, in closed form. The remaining task is to derive from (12.42) the coefficients of the Taylor series expansion of $G_{\tilde{V}}$. Algorithms that achieve this goal are presented in Sections 12.5.4 and 12.5.5.

12.5.4 Generalizing Panjer's recurrence scheme

Although the generalization of Giese's approach [2] to the bivariate case is mathematically more appealing, it is instructive to look at the standard Panjer scheme [1] first.

   The standard CreditRisk$^+$ algorithm rests on the fact that the derivative of the logarithm of the probability-generating function of the loss variable is a rational function. This generalizes to the following ansatz. We assume that the gradient of $\log G_{\tilde{V}}$ is rational, i.e. can be written as

$$\nabla \log G_{\tilde{V}} = \frac{1}{B} \begin{pmatrix} A^+ \\ A^- \end{pmatrix} \tag{12.43}$$

with suitable polynomials $A^{\pm}, B$ in z_+, z_-. (One can always choose the same polynomial in the denominator, but this choice is also suggested by the explicit form of $G_{\tilde{V}}$, cf. Equation (12.52).)

Remark 3. Notice that the polynomials $A^{\pm}$ and B satisfy the following PDE:

$$A^+ \frac{\partial B}{\partial z_-} - B \frac{\partial A^+}{\partial z_-} = A^- \frac{\partial B}{\partial z_+} - B \frac{\partial A^-}{\partial z_+}. \tag{12.44}$$

Otherwise (12.43) would not admit a solution.

Note further that, for fixed $A^{\pm}$ and B, the logarithms of any two solutions $G, \tilde{G}$ of (12.43) differ only by a constant. This implies that G and $\tilde{G}$ must be proportional. A useful consequence is that the recurrence algorithm can be started with an arbitrary value g_{00}, if the resulting distribution is suitably rescaled afterwards. This freedom in the choice of g_{00} is of practical importance because in the case of large portfolios, the "true" value of g_{00} (i.e. the probability that no credit events occur, which equals $G_{\tilde{V}}(0,0)$) can be less than the smallest positive number a given compiler can distinguish from zero.[11] $\square$

The PDE (12.43) can be rewritten as follows:

$$B \frac{\partial G_{\tilde{V}}}{\partial z_{\pm}} = A^{\pm} G_{\tilde{V}}. \tag{12.45}$$

Let

$$A^{\pm}(z_+, z_-) = \sum_{m,n} a^{\pm}_{m,n} z^m_+ z^n_-, \quad B(z_+, z_-) = \sum_{m,n} b_{m,n} z^m_+ z^n_-, \tag{12.46}$$

and let the coefficients of $G_{\tilde{V}}$ be denoted by g_{mn}:

$$G_{\tilde{V}}(z_+, z_-) = \sum_{m,n \in \mathbb{N}^2} g_{mn} z^m_+ z^n_-. \tag{12.47}$$

Inserting (12.46) and (12.47) into (12.45), comparing coefficients, and solving for leading order terms, we obtain the following two recurrence relations (corresponding to the $\pm$-signs in (12.45)):

---

[11] For example, the smallest positive number ϵ my compiler can distinguish from zero is about $\epsilon = 2.2 \cdot 10^{-308}$. Assume that the true initial value is $G_{\tilde{V}}(0,0) = 10^{-400}$, which would be interpreted as zero (compare $\log G_{\tilde{V}}(0,0)$ with $\log \epsilon$ to check this). Then one can start the recurrence algorithm with e.g. $g_{00} = 10^{-300}$ and rescale the resulting distribution with a factor 10^{-100} at the end. Some coefficients of $G_{\tilde{V}}$ will then be indistinguishable from zero, but the bulk of the distribution will be obtained correctly.

$$g_{mn} = \frac{1}{b_{00}} \begin{cases} \dfrac{1}{m} \displaystyle\sum_{(k,l)<(m,n)} \left(a^+_{m-k-1,n-l} - kb_{m-k,n-l}\right)g_{kl}, & m > 0 \\[2ex] \dfrac{1}{n} \displaystyle\sum_{(k,l)<(m,n)} \left(a^-_{m-k,n-l-1} - lb_{m-k,n-l}\right)g_{kl}, & n > 0. \end{cases} \qquad (12.48)$$

In (12.48) we use the convention $a^{\pm}_{rs} = 0$ if $r < 0$ or $s < 0$, and we use the following partial ordering in $\mathbb{N}^2$:

$$(k,l) < (m,n) \iff k \leq m, \; l \leq n, \; (k,l) \neq (m,n). \qquad (12.49)$$

A coefficient g_{mn} is determined by its precursors g_{kl} with $(k,l) < (m,n)$. Note that the number of terms appearing in (12.48) is bounded by the degrees of the polynomials $A^{\pm}$ and B.

The recurrence scheme starts with the value

$$g_{00} \equiv G_{\tilde{V}}(0,0) \qquad (12.50)$$

which can be read off from (12.42). The coefficients g_{m0} ($m > 0$), resp. g_{0n} ($n > 0$) are obtained from g_{00} by applying the first, resp. second relation in (12.48). The remaining coefficients can be computed by using either relation; the result will anyway be the same. A possible choice is sketched in the following diagram, which displays coefficients up to degree 4:

$$\begin{array}{ccccccccc} g_{00} & \to & g_{01} & \to & g_{02} & \to & g_{03} & \to & g_{04} & \to \\ \downarrow & & \downarrow & & \downarrow & & \downarrow & & \downarrow & \\ g_{10} & & g_{11} & & g_{12} & & g_{13} & & & \\ \downarrow & & \downarrow & & \downarrow & & \downarrow & & & \\ g_{20} & & g_{21} & & g_{22} & & & & & \\ \downarrow & & \downarrow & & \downarrow & & & & & \\ g_{30} & & g_{31} & & & \ddots & & & & \\ \downarrow & & \downarrow & & & & & & & \\ g_{40} & & & & & & & & & \\ \downarrow & & & & & & & & & \end{array} \qquad (12.51)$$

Here vertical arrows correspond to the first relation in (12.48), and horizontal arrows to the second relation. The coefficients g_{mn} should be computed column-wise from left to right, up to the maximal degree one wishes to consider. Note that in this scheme, the coefficients of A^- are only used to compute the g_{0n} so that knowledge of the coefficients a^-_{rs} with $r > 0$ is dispensable.

Having discussed the two-dimensional recurrence scheme in general, let us return to the specific form of $G_{\tilde{V}}$ found above. By (12.42), the gradient of the logarithm of $G_{\tilde{V}}$ is given by

$$\nabla \log G_{\tilde{V}} = \left(\frac{2}{1 - \exp(-2P_0)} - \frac{1}{P_0}\right)\nabla P_0 + \nabla P_1 + \sum_{k=2}^{N} \frac{\nabla P_k}{1 - \sigma_k^2 P_k}. \qquad (12.52)$$

This is a rational function of $z_\pm$, except for the contribution of S_0, which has an exponential term in the denominator. In order to apply the recurrence scheme developed above, one has to introduce an appropriate cut-off of this exponential term.[12] The polynomials $A^\pm$ and B are then easily obtained from the polynomials P_k, just as in standard CreditRisk[+].

12.5.5 Enhancing Giese's recurrence scheme

The main benefits of Giese's approach to default-mode CreditRisk[+] are conceptual simplicity, greater flexibility, and improved numerical stability [2]. For the same reasons a generalization of Giese's approach to the present model of rating transitions appears to be attractive. In particular, it turns out that the business cycle indicator S_0 can be incorporated into a Giese-like framework without the need for further approximations (cf. the remark following Equation (12.52)).

To apply Giese's approach, we have to rewrite (12.42) in the following way:[13]

$$G_{\tilde{V}} = (\exp 2P_0 - 1)\frac{1}{2P_0}\exp\left(P_1 - \sum_{k=2}^{N}\sigma_k^{-2}\log(1 - \sigma_k^2 P_k)\right). \qquad (12.53)$$

Besides algorithms for the exp and log functions, (12.53) requires an algorithm to compute the inverse of a polynomial. A toolbox of bivariate recursive algorithms (all of which have obvious univariate analogues) is provided in the following sections. For later use, we have included an algorithm to compute the square root of a an analytic function that is positive at the origin.

Exponential Function

Let $Q(z_+, z_-)$ be analytic at the origin, with power series expansion

$$Q(z_+, z_-) = \sum_{kl} q_{kl} z_+^k z_-^l,$$

and let

$$P \equiv \exp Q. \qquad (12.54)$$

Then P satisfies the PDE

---

[12] As mentioned in the introduction, if one prefers to avoid the business cycle indicator S_0, one can treat upgradings as depending on the specific risk factor only and apply the above recurrence scheme without further approximations.

[13] One could try to incorporate the terms corresponding to S_0 into the exponential function by performing the substitution $\frac{\exp 2P_0 - 1}{2P_0} = \exp(\log(\exp 2P_0 - 1) - \log 2P_0)$. But because of (12.25b), and by the definition of the polynomial P_0, the value of P_0 at the origin can be negative, in which case the logarithms cannot be evaluated.

$$\nabla \log P = \nabla Q, \tag{12.55}$$

which can be rewritten as

$$\nabla P = P \nabla Q. \tag{12.56}$$

Inserting into (12.56) the power series expansions of Q and of P,

$$P(z_+, z_-) = \sum_{mn} p_{mn} z_+^m z_-^n,$$

and comparing coefficients, one finds the following relations for the coefficients of P:

$$p_{mn} = \begin{cases} \dfrac{1}{m} \displaystyle\sum_{k=1}^{m} \sum_{l=0}^{n} k q_{kl} p_{m-k, n-l}, & m > 0 \\[2ex] \dfrac{1}{n} \displaystyle\sum_{k=0}^{m} \sum_{l=1}^{n} l q_{kl} p_{m-k, n-l}, & n > 0. \end{cases} \tag{12.57}$$

These relations can be solved recursively by a scheme analogous to (12.51). That both relations in (12.57) yield the same coefficients p_{mn} can be viewed as a restatement of the PDE

$$\frac{\partial P}{\partial z_+} \frac{\partial Q}{\partial z_-} = \frac{\partial P}{\partial z_-} \frac{\partial Q}{\partial z_+}$$

which is implied by (12.56). Note that, as in the Panjer case (cf. Remark 3), any two solutions $P, \tilde{P}$ of (12.55) must be proportional. Therefore the remarks made in Section 12.5.4 on the choice of an "intermediate" initial value apply here as well. The "true" initial value is obtained by evaluating (12.54) at the origin:

$$p_{00} = \exp q_{00}. \tag{12.58}$$

Logarithm

Let a function $P(z_+, z_-)$ be given which is analytic and positive at the origin. As in the univariate case [2], the algorithm to compute the coefficients of the logarithm of P is obtained by inverting the relations of the exponential function, i.e. by solving (12.57) for leading coefficients of Q. For

$$Q \equiv \log P,$$

one immediately obtains from (12.57):

$$q_{mn} = \frac{1}{p_{00}} \begin{cases} \left(p_{mn} - \dfrac{1}{m} \displaystyle\sum_{(k,l)<(m,n)} k q_{kl} p_{m-k, n-l} \right), & m > 0 \\[3ex] \left(p_{mn} - \dfrac{1}{n} \displaystyle\sum_{(k,l)<(m,n)} l q_{kl} p_{m-k, n-l} \right), & n > 0. \end{cases} \tag{12.59}$$

Here we used the partial ordering defined in (12.49).

The relations (12.59) can again be solved recursively by a scheme analogous to (12.51), starting with the value

$$q_{00} = \log p_{00}. \tag{12.60}$$

For fixed P, any two solutions $Q, \tilde{Q}$ of (12.55) differ only by a constant. This means that one can start the recurrence algorithm with an arbitrary value, if the resulting distribution is appropriately shifted afterwards.

Inversion

Let $Q(z_+, z_-)$ be analytic at the origin, with $Q(0,0) \equiv q_{00} \neq 0$, and

$$P \equiv \frac{1}{Q}.$$

Differentiating the relation $PQ = 1$, one gets the PDE

$$\nabla(PQ) = 0. \tag{12.61}$$

Inserting the respective power series expansions, one finds the following recurrence relation,[14] valid for all indices $(m,n) > (0,0)$ (cf. (12.49)):

$$p_{mn} = -\frac{1}{q_{00}} \sum_{(k,l)<(m,n)} p_{kl} q_{m-k,n-l}. \tag{12.62}$$

Since P is determined by (12.61) up to a multiplicative constant, the remarks made in Section 12.5.4 on the choice of the initial value apply here once again. The "true" initial value is

$$p_{00} = \frac{1}{q_{00}}. \tag{12.63}$$

Square Root

Let $Q(z_+, z_-)$ be analytic and positive at the origin, and let P be its square root:
$$P \equiv \sqrt{Q}.$$

Then P satisfies the PDE
$$P\nabla P = \frac{1}{2}\nabla Q, \tag{12.64}$$

which entails the following two relations[15] for the coefficients of P:

---

[14] Relation (12.62) comprises two relations corresponding to the two components of (12.61).

[15] Note that, if $\tilde{P}$ is another solution of (12.64), then $\tilde{P}^2$ and P^2 differ at most by a constant.

$$p_{mn} = \frac{1}{p_{00}} \begin{cases} \left(\frac{q_{mn}}{2} - \frac{1}{m} \sum_{(k,l)<(m,n)} k p_{kl} p_{m-k,n-l}\right), & m > 0 \\ \left(\frac{q_{mn}}{2} - \frac{1}{n} \sum_{(k,l)<(m,n)} l p_{kl} p_{m-k,n-l}\right), & n > 0. \end{cases} \tag{12.65}$$

The initial value is

$$p_{00} = \sqrt{q_{00}}.$$

12.5.6 Calculating the distribution of profits and losses

We are finally in a position to calculate the total portfolio's distribution of value changes. Recall from (12.36)–(12.37) that value changes in $\mathcal{P}_{\text{tot}}$ are described by the random variable V, and that the latter is given by the difference of the components $V^{\pm}$ of the random vector $\tilde{V}$. Having computed the coefficients g_{mn} of the probability-generating function $G_{\tilde{V}}$ of $\tilde{V}$ by one of the methods explicated in Sections 12.5.4 or 12.5.5, it remains to apply (12.6) to obtain the distribution of V:

$$\begin{aligned} \mathbb{P}[V = n] &= \sum_{m \geq \max\{-n,0\}} \mathbb{P}[V^+ = m+n, \ V^- = m] \\ &= \sum_{m \geq \max\{-n,0\}} g_{m+n,m}, \qquad n \in \mathbb{Z}. \end{aligned} \tag{12.66}$$

If $G_{\tilde{V}}$ is expanded up to degree[16] D (the scheme (12.51) shows terms up to $D = 4$), then the distribution of V can be calculated in the range $V \in \{-D, \ldots, D\}$. In this case, the number of floating-point multiplications required by the algorithm from Section 12.5.5 is of order $\frac{D^4}{6}$ (see Appendix 2).

12.6 Gaussian Model Variants

In the present section we will demonstrate the (often denied) flexibility of the CreditRisk$^+$ framework. We will alter the model's distributional assumptions while keeping analytic tractability. Specifically, we will consider model variants with Gaussians (Section 12.6.1) resp. with squares of Gaussians (Section 12.6.2) as systematic risk factors. In particular, the latter model variant can serve to test the dependence of the model's output on distributional assumptions.

---

[16] In view of the apparent asymmetry of credit risk distributions, where losses dominate over profits, it might be more appropriate to choose distinct degrees $D^{\pm}$ for profits and losses, i.e. to restrict to coefficients g_{mn} with $m \leq D^+$ and $n \leq D^-$. The distribution of V would then be obtained in the range $V \in \{-D^-, \ldots, D^+\}$.

12.6.1 Gaussian background factors

We now assume instead of (12.20c) that the systematic risk factors S_k, $k \geq 2$, are normally distributed with mean one and variances σ_k^2. The positivity condition in (12.19) is then violated, which implies that conditional migration probabilities can be negative. For sufficiently small variances σ_k^2, this effect can, however, be neglected, as one obtains valid portfolio distributions in the end (cf. [3] for a similar observation in standard CreditRisk$^+$).

The formula (12.40), which expresses the probability-generating function of $\tilde{V}$ in terms of the moment-generating functions of the S_k is independent of distributional assumptions. In the Gaussian model the moment-generating functions of the S_k are given by

$$\mathbb{E}[\exp tS_k] = \exp\left(t\left(1 + \frac{\sigma_k^2 t}{2}\right)\right), \qquad k \geq 2. \tag{12.67}$$

Inserting (12.67) into (12.40), one gets the unconditional PGF of $\tilde{V}$ for the present model variant, replacing (12.42) and (12.53):

$$
\begin{aligned}
G_{\tilde{V}} &= \frac{\exp 2P_0 - 1}{2P_0} \exp P_1 \prod_{k=2}^{N} \exp\left(P_k\left(1 + \frac{\sigma_k^2 P_k}{2}\right)\right) \\
&= (\exp 2P_0 - 1)\frac{1}{2P_0} \exp\left(P_1 + \sum_{k=2}^{N} P_k\left(1 + \frac{\sigma_k^2 P_k}{2}\right)\right).
\end{aligned}
\tag{12.68}
$$

The coefficients of the systematic part ($k \geq 2$) of (12.68) can immediately be computed by the exponential algorithm from Section 12.5.5. Because of the absence of logarithmic terms, computation time is drastically reduced compared with the case of gamma-distributed systematic risk factors. The distribution of profits and losses in $\mathcal{P}_{\text{tot}}$ is finally obtained from the coefficients of $G_{\tilde{V}}$ as in (12.66).

12.6.2 Squares of Gaussians as background factors

Perhaps the simplest way to get rid of the negative conditional probabilities encountered in Section 12.6.1 is to consider squares of Gaussians instead of Gaussians. Let T_k, $k = 2, \ldots, N$, be normally distributed random variables with mean μ_k and variance τ_k^2, obeying the condition

$$\mu_k^2 + \tau_k^2 = 1. \tag{12.69}$$

Define the systematic risk factor S_k as the square of T_k:

$$S_k \equiv T_k^2. \tag{12.70}$$

The factors S_k then satisfy the condition (12.19), while the variance σ_k^2 of S_k is bounded by 2:

$$\sigma_k^2 = 2(1 - \mu_k^4) \in [0, 2]. \tag{12.71}$$

Notice that S_k is gamma distributed if and only if $\mu_k = 0$.

The moment-generating function of S_k is given by a Gaussian integral:

$$\mathbb{E}[\exp tS_k] = (1 - 2\tau_k^2 t)^{-\frac{1}{2}} \exp \frac{\mu_k^2 t}{1 - 2\tau_k^2 t}, \qquad k \geq 2 \tag{12.72}$$

for $t < \frac{1}{2\tau_k^2}$. The PGF of $\tilde{V}$ can therefore be written as

$$\begin{aligned}
G_{\tilde{V}} &= \frac{\exp 2P_0 - 1}{2P_0} \left(\prod_{k=2}^{N} (1 - 2\tau_k^2 P_k) \right)^{-\frac{1}{2}} \exp \left(P_1 + \sum_{k=2}^{N} \frac{\mu_k^2 P_k}{1 - 2\tau_k^2 P_k} \right) \\
&= \frac{\exp 2P_0 - 1}{2P_0} \exp \left(P_1 + \sum_{k=2}^{N} \left(-\frac{1}{2} \log(1 - 2\tau_k^2 P_k) + \frac{\mu_k^2 P_k}{1 - 2\tau_k^2 P_k} \right) \right).
\end{aligned} \tag{12.73}$$

In view of the second line in (12.73), the coefficients of $G_{\tilde{V}}$ can be computed via the algorithms for exp, log, and inversion developed in Section 12.5.5. Alternatively, one can apply the algorithms for exp, inversion, and square root to the first line in (12.73). Note that the gradient of $\log G_{\tilde{V}}$ is "almost" rational in the sense explained below (12.52), so that the Panjer-like algorithm from Section 12.5.4 also applies. The computation of the distribution of profits and losses proceeds from here as usual.

Appendix 1: Conditional Independence

The notion of "conditional independence" is omnipresent in the literature on portfolio credit risk models. We have, however, never met a formal definition of conditional independence. Such a definition may well be dispensable because the meaning might be obvious.

Nevertheless we would like to provide a definition here, because conditional independence plays a central role in most credit risk models, and because CreditRisk[+] relies on Proposition 1(g) from Section 12.1.1, which in turn follows from Proposition 2 below.

Definition 2. *Let $(\Omega, \mathcal{A}, \mathbb{P})$ be a probability space, and let $\mathcal{B} \subset \mathcal{A}$ be a sub-σ-algebra. A collection of events $A_1, \ldots, A_M \in \mathcal{A}$ ($M \leq \infty$) is called conditionally independent under the hypothesis $\mathcal{B}$ if the following equality holds almost surely (a. s.) for any finite subset $\{i_1, \ldots, i_m\} \subset \{1, \ldots, M\}$:*

$$\mathbb{P}[A_{i_1} \cap \cdots \cap A_{i_m} \mid \mathcal{B}] = \mathbb{P}[A_{i_1} \mid \mathcal{B}] \cdots \mathbb{P}[A_{i_m} \mid \mathcal{B}]. \tag{12.74}$$

The conditional probability of an event A given $\mathcal{B}$ appearing above is defined as the conditional expectation of the corresponding indicator variable:

$$\mathbb{P}[A \mid \mathcal{B}] \equiv \mathbb{E}[\mathbf{1}_A \mid \mathcal{B}].$$

(12.74) is therefore equivalent to

$$\mathbb{E}\big[\mathbf{1}_{A_{i_1}} \cdots \mathbf{1}_{A_{i_m}} \,\big|\, \mathcal{B}\big] = \mathbb{E}\big[\mathbf{1}_{A_{i_1}} \,\big|\, \mathcal{B}\big] \cdots \mathbb{E}\big[\mathbf{1}_{A_{i_m}} \,\big|\, \mathcal{B}\big] \quad \text{a.s.}$$

Definition 2 extends in the usual way from events to sub-σ-algebras and random variables:

- A collection of *sub-σ-algebras* $\mathcal{A}_1, \ldots, \mathcal{A}_M \subset \mathcal{A}$ is *conditionally independent* under the hypothesis $\mathcal{B}$ if and only if every collection $\{A_1, \ldots, A_M\}$ of events $A_i \in \mathcal{A}_i$ is conditionally independent.
- A collection of *random variables* $X_1, \ldots, X_M$ on $(\Omega, \mathcal{A}, \mathbb{P})$ is *conditionally independent* under the hypothesis $\mathcal{B}$ if and only if the sub-σ-algebras generated by $X_1, \ldots, X_M$ in $\mathcal{A}$ are conditionally independent.

The latter condition can equivalently be stated as follows. For all finite subsets $\{i_1, \ldots, i_m\} \subset \{1, \ldots, M\}$ one has

$$\mathbb{P}\big[X_{i_1} \in B_{i_1}, \ldots, X_{i_m} \in B_{i_m} \,\big|\, \mathcal{B}\big]$$
$$= \mathbb{P}\big[X_{i_1} \in B_{i_1} \,\big|\, \mathcal{B}\big] \cdots \mathbb{P}\big[X_{i_m} \in B_{i_m} \,\big|\, \mathcal{B}\big], \quad \text{a.s.}$$

where B_i is an arbitrary measurable set in the target space of X_i.

Finally, if Y is a random variable on $(\Omega, \mathcal{A}, \mathbb{P})$, then the phrase "conditionally independent under the hypothesis Y" should be interpreted as "conditionally independent under the hypothesis (sub-σ-algebra generated by Y)".

The factorization property of the conditional probability-generating function of the sum of conditionally independent random vectors stated in Proposition 1(g) in Section 12.1.1 is obtained by applying the following proposition to random vectors $X_1, \ldots, X_M$ with values in $\mathbb{N}^N$, and to the bounded functions

$$f_1(n) \equiv \cdots \equiv f_M(n) \equiv z^n, \quad n \in \mathbb{N}^N,$$

where $z \in [-1, 1]^N$ is fixed.

Proposition 2. *Let* $(X_j)_{j=1,\ldots,M}$ *be a collection of random variables on* $(\Omega, \mathcal{A}, \mathbb{P})$ *with values in measurable spaces* $(\Omega_j, \mathcal{A}_j)$, *and let* $(f_j)_{j=1,\ldots,M}$ *be measurable real functions on* Ω_j *which are essentially bounded w.r.t. the distribution of* X_j. *If* $X_1, \ldots, X_M$ *are conditionally independent given a hypothesis* $\mathcal{B}$, *then*

$$\mathbb{E}\big[f_{i_1}(X_{i_1}) \cdots f_{i_m}(X_{i_m}) \,\big|\, \mathcal{B}\big] = \mathbb{E}\big[f_{i_1}(X_{i_1}) \,\big|\, \mathcal{B}\big] \cdots \mathbb{E}\big[f_{i_m}(X_{i_m}) \,\big|\, \mathcal{B}\big] \quad \text{a.s.}$$

for all finite subsets $\{i_1, \ldots, i_m\} \subset \{1, \ldots, M\}$.

Proof. Let $\mathcal{S}_j$ be the real linear span of the characteristic functions $1_{[X_j \in B_j]}$ with $B_j \in \mathcal{A}_j$. Using conditional independence and linearity, we get for $Y_j \in \mathcal{S}_j$:

$$\mathbb{E}\big[Y_{i_1} \cdots Y_{i_m} \,\big|\, \mathcal{B}\big] = \mathbb{E}\big[Y_{i_1} \,\big|\, \mathcal{B}\big] \cdots \mathbb{E}\big[Y_{i_m} \,\big|\, \mathcal{B}\big] \quad \text{a.s.}$$

Since $f_j(X_j)$ is essentially bounded and $\mathcal{S}_j$ is dense in $L_\mathbb{R}^\infty(\Omega, X_j^{-1}(\mathcal{A}_j), \mathbb{P})$, we can approximate $f_j(X_j)$ by a (bounded) sequence $Y_j^{(n)}$ in $\mathcal{S}_j$ relative to the L^∞-norm. The claim then follows from the continuity of conditional expectations.   □

Appendix 2: Computation Time

We would like to give a rough estimate of the number of floating-point multiplications required by the algorithm proposed in Section 12.5.5. We consider the case that all relevant functions are expanded up to some maximal degree D.

We start with the observation that the computation time involved in the following steps can be neglected:

- Calculation of the polynomials P_k.
- Calculation of $\exp 2P_0$, of $\frac{1}{2P_0}$, and of the log-terms in (12.53).[17]
- Calculation of the distribution of V via (12.66).

We may then concentrate on the remaining steps in (12.53):

- Calculation of the product of two polynomials of degree D up to degree D.
- Calculation of the exponential of a polynomial of degree D up to degree D.

Product of Polynomials

Let P and Q be polynomials of degree $\leq D$ in two variables, with coefficients p_{mn}, q_{mn} $(m + n \leq D)$. The product PQ has coefficients

$$(PQ)_{mn} = \sum_{k=0}^{m} \sum_{l=0}^{n} p_{kl} q_{m-k, n-l}.$$

One needs $(m+1)(n+1) \approx mn$ multiplications to compute $(PQ)_{mn}$. In leading order, the number of multiplications needed to compute PQ up to degree D is therefore given by

$$\sum_{m=1}^{D} \sum_{n=1}^{D-m} mn \approx \frac{1}{2} \sum_{m=1}^{D} m(D-m)^2 \tag{12.75}$$

$$\approx \frac{1}{2} \int_0^D m(D-m)^2 \, dm = \frac{D^4}{24}.$$

---

[17] The contribution of these terms can be neglected because the number of nonvanishing coefficients of the P_k is much smaller than D^2.

Exponential of a Polynomial

Let Q be as above, and set $P \equiv \exp Q$. The calculation of a coefficient p_{mn} of P by the recurrence scheme (12.57) requires approximately $2mn$ multiplications. In leading order, the number of multiplications needed to compute P up to degree D is therefore equal to (cf. (12.75))

$$2 \sum_{m=1}^{D} \sum_{n=1}^{D-m} mn \approx \frac{D^4}{12}.$$

Overall Computation Time

Equation (12.53) involves two multiplications of polynomials and one evaluation of an exponential of a polynomial, where the respective polynomials typically have no vanishing coefficients. The number of floating-point multiplications required by the model is therefore (in leading order) given by $2 \cdot \frac{D^4}{24} + \frac{D^4}{12} = \frac{D^4}{6}$. This should be contrasted with default-mode CreditRisk$^+$, where the number of floating-point multiplications is of order D^2 (using either Panjer's or Giese's algorithm, with D being equal to the number of recurrence steps).

Acknowledgements. The author is indebted to Jürgen Allinger and Thomas Jacobi for stimulating discussions. Further thanks are due to Dirk Tasche for useful comments on an early version of the manuscript.

References

1. Credit Suisse Financial Products. CreditRisk$^+$: A credit risk management framework. London, 1997. Available at http://www.csfb.com/creditrisk.
2. G. Giese. Enhancing CreditRisk$^+$. *Risk*, 16(4):73–77, April 2003.
3. M. B. Gordy. A comparative anatomy of credit risk models. *J. Banking & Finance*, 24:119–149, 2000.
4. B. Rolfes and F. Bröker. Good migrations. *Risk*, 11(11):72–73, 1998.

13

Dependent Sectors and an Extension to Incorporate Market Risk

Oliver Reiß*

Summary. In standard CreditRisk$^+$ the risk factors are assumed to be independently gamma distributed and as a consequence the model can be computed analytically. If one extends the model such that the risk factors are dependently distributed with quite arbitrary distributions, one has to give up the existence of a closed-form solution. The advantage of this approach is that one gains interesting generalizations and the computational effort to determine the loss distribution still remains quite small.

In the first step the model will be generalized such that the risk factors are dependently distributed with quite arbitrary distributions and as a suitable choice in practice, the dependent lognormal distribution is suggested. This model can then be transferred to a time continuous model and the risk factors become processes, more precisely geometric Brownian motions. Having a time continuous credit risk model is an important step to combining this model with market risk. Additionally a portfolio model will be presented where the changes of the spreads are driven by the risk factors. Using a linear expansion of the market risk, the distribution of this portfolio can be determined. In the special case that there is no credit risk, this model yields the well-known delta normal approach for market risk, hence a link between credit risk and market risk has been established.

Introduction

In standard CreditRisk$^+$, see Chapter 2, the risk factors are assumed to be independently gamma distributed. From a practitioner's point of view, this assumption is too rigorous, since one would like to introduce the sectors due to general classes of business, and typical sectors could be construction, banking, utility industry, transportation, etc. Also economic sectors like gross domestic product or the business activity of certain countries or currency areas can be taken into account. Using such sectors, it is quite easy to estimate the

---

* Supported by the DFG Research Center "Mathematics for key technologies" (FZT 86) in Berlin.

sector affiliations of each obligor, but these sectors are obviously not independent. One way to abstain from the independence assumption is presented in [1] and this study showed that there is an effect of dependent risk factors on the variance of the credit loss. A more general approach to handling dependent risk factors will be presented in Section 13.1. Since it is difficult in general to describe the dependency of random variables completely, it is suggested for practical application that dependent lognormal risk factors are introduced, because the dependency of lognormal random variables can easily be described by a covariance matrix. The resulting model is more complex and the valuation of the characteristic function of the credit loss can be done by Monte Carlo only, but the computational effort is tolerable. Since the distribution of any real-valued random variable is completely described by its characteristic function, one can also obtain the loss distribution, e.g. by fast Fourier transformation.

Standard CreditRisk$^+$ or the model presented in Section 13.2 give an answer to the question of the size of the losses at the end of a fixed period. One would also like to analyse the loss process during a certain period. In this context, the loss and the factors become processes. In Section 13.3 such a model with lognormal factor processes is introduced. The description of the default events in this model arise quite naturally according to standard CreditRisk$^+$.

With a model that allows a time continuous description of credit risk, one has made a large step to combining market and credit risk. The next step is done in Section 13.3, where the portfolio valuation in a combined model is introduced. In this model, credit spreads are modelled by the sector processes. The profit and loss distribution of this portfolio can be obtained with the techniques presented in Section 13.2. One can consider two special cases of this model. If there is no market risk one obtains the model presented in Section 13.2; if there is no credit risk, one gets the well-known delta normal approach to determining market risk.

13.1 Lognormal or Other Risk Factors

In standard CreditRisk$^+$ the risk factors S_k are assumed to be independently gamma distributed. The reason for this assumption is based on the simple calculations resulting from this supposition and not on statistical evidence. Hence one may suggest another distribution for the risk factors and one possibility is to use dependent lognormal sectors. The advantage of this choice is that the dependence of the sectors can be described by a correlation matrix. To estimate these correlations one can use as a first approximation correlations between the corresponding asset indices. On the other hand, the characteristic function of the credit loss can no longer be obtained by a closed-form solution, but it can be computed numerically by a Monte Carlo method.

To illustrate that the changeover from the gamma distribution to the lognormal distribution has no dramatic impact on the distribution, the gamma distribution and the lognormal distribution each with mean 1 and variance σ^2 are plotted. The gamma distribution and the lognormal distribution each with the same first two moments look quite similar (Figure 13.1).

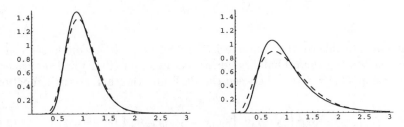

Fig. 13.1. The lognormal density (solid) and the density of the gamma distribution (dashed), both with mean 1 and $\sigma = 0.3$ (left), $\sigma = 0.5$ (right)

As shown in Subsection 13.1.2, the expectation and the variance of the credit loss depend only on the mean (which is fixed to be 1) and the covariances of the risk factors. Hence, if the risk factors are independent, one obtains the same first two moments for the loss as in the standard CreditRisk$^+$ model.

However, the computations in this section are valid for any jointly distributed risk factors – as long as all pairwise covariances exist.

13.1.1 Introduction of the model

The model analysed in this section is a modification of standard CreditRisk$^+$: the risk factors S_k are not independently gamma distributed but are somehow dependently distributed such that $\text{cov}[S_k, S_l]$ exist and $\mathbb{E}[S_k] = 1$ for all $k, l = 1, \ldots N$. The other ingredients of the model are unchanged. Recall that for each of the K obligors A the factor loadings $w_{Ak} \geq 0$ are given with $\sum_{k=1}^{N} w_{Ak} \leq 1$ and the idiosyncratic risk of each obligor is given by $w_{A0} = 1 - \sum_{k=1}^{N} w_{Ak}$. For each obligor there is a the probability of default p_A within one year. The default intensity of obligor A is $p_A^S = p_A \left(w_{A0} + \sum_{k=1}^{N} w_{Ak} S_k \right)$. The time of default T_A of obligor A is an exponentially distributed random variable with intensity p_A^S. Usually one is interested in the credit loss within a certain time horizon T and one introduces the default indicator variable

$$\mathbf{1}_A := \begin{cases} 0 & \text{if } T < T_A \\ 1 & \text{if } T \geq T_A \end{cases}.$$

If $\mathbf{1}_A = 0$, the obligor A survives up to T and if $\mathbf{1}_A = 1$, the obligor A defaults before T. If the obligor A defaults, the loss due to this default event is given by ν_A, the so-called exposure. In standard CreditRisk$^+$ it is assumed that all ν_A are integer multiples of a basic loss unit. Since the computations in this chapter are based on characteristic functions instead of on probability-generating functions, here is no need to introduce such a basic loss unit. The loss X of a loan portfolio up to time T is hence given by

$$X := \sum_A \mathbf{1}_A \nu_A.$$

Thus one can inherit the characteristic function of the credit losses conditional on S using the probability-generating function (PGF) and the general relationship $\Phi(z) = G(e^{iz})$ between the PGF and the characteristic function:

$$\Phi_{X|S}(z) = \exp\left(\sum_A p_A T \left(w_{A0} + \sum_{k=1}^{N} w_{Ak} S_k\right)(e^{i\nu_A z} - 1)\right). \quad (13.1)$$

Based on this expression for the characteristic function of X conditional on S one can determine the expectation and the variance of X. It turns out that these moments are independent of the specific distribution of the risk factors. The first two moments of X depend only on the first two moments of S, hence all pairwise covariances must be well-defined.

13.1.2 The first moments of X

Theorem 1. *The first two moments of X are given by*

$$\mathbb{E}[X] = T \sum_A p_A \nu_A,$$

$$\mathbb{E}[X^2] = T \sum_A p_A (\nu_A)^2 + T^2 \left[\sum_A p_A \nu_A\right]^2$$

$$+ T^2 \sum_{A,\tilde{A}} p_A \nu_A p_{\tilde{A}} \nu_{\tilde{A}} \sum_{k,l=1}^{N} w_{Ak} w_{\tilde{A}l} \mathrm{cov}[S_k, S_l],$$

$$\mathrm{var}[X] = T \sum_A p_A (\nu_A)^2 + T^2 \sum_{A,\tilde{A}} p_A \nu_A p_{\tilde{A}} \nu_{\tilde{A}} \sum_{k,l=1}^{N} w_{Ak} w_{\tilde{A}l} \mathrm{cov}[S_k, S_l].$$

Proof. To determine the nth moment one needs to compute[1]

$$\mathbb{E}[X^n] = \lim_{z \to 0} \frac{1}{i^n} \frac{d^n}{dz^n} \mathbb{E}[\Phi_{X|S}(z)].$$

In fact, there are three limits in this expression: the expectation (integration), a differentiation and the valuation at one point. From (13.1) it is clear that

---

[1] There are no difficulties with the limit in the equation; the notation has only been chosen to indicate clearly when the valuation at $z = 0$ takes place.

$\Phi_{X|S}(z)$ is analytical and may be written as a power series in z. Since power series converge uniformly on a compact interval, the exchange of the limits is allowed:

$$\mathbb{E}[X^n] = \mathbb{E}[\lim_{z \to 0} \frac{1}{i^n} \frac{d^n}{dz^n} \Phi_{X|S}(z)].$$

The first two derivatives of $\Phi_{X|S}(z)$ are

$$\Phi'_{X|S}(z) = \Phi_{X|S}(z) \sum_A p_A T \left(w_{A0} + \sum_{k=1}^{N} w_{Ak} S_k \right) e^{i\nu_A z} (i\nu_A),$$

$$\Phi''_{X|S}(z) = \Phi_{X|S}(z) \sum_A p_A T \left(w_{A0} + \sum_{k=1}^{N} w_{Ak} S_k \right) e^{i\nu_A z} (i\nu_A)^2 \qquad (13.2)$$

$$+ \Phi_{X|S}(z) \left[\sum_A p_A T \left(w_{A0} + \sum_{k=1}^{N} w_{Ak} S_k \right) e^{i\nu_A z} (i\nu_A) \right]^2.$$

Valuation at $z = 0$ yields

$$\frac{1}{i} \Phi'_{X|S}(0) = T \sum_A p_A \nu_A \left(w_{A0} + \sum_{k=1}^{N} w_{Ak} S_k \right),$$

$$\frac{1}{i^2} \Phi''_{X|S}(0) = T \sum_A p_A (\nu_A)^2 \left(w_{A0} + \sum_{k=1}^{N} w_{Ak} S_k \right)$$

$$+ \left[T \sum_A p_A \nu_A \left(w_{A0} + \sum_{k=1}^{N} w_{Ak} S_k \right) \right]^2.$$

Take the expectation over S and recall that $\mathbb{E}[S_k] = 1$, $w_{A0} + \sum_{k=1}^{N} w_{Ak} = 1$ to obtain the final result. $\qquad\qquad\qquad\qquad\qquad\qquad\qquad\qquad\qquad\qquad\quad \square$

Remark 1. Note that in the case of uncorrelated risk factors and $T = 1$ the first two moments are identical to (2.13) and (2.15) of Chapter 2, since the variance of the random variables S_k is given by $\sigma_k^2 = \text{cov}[S_k, S_k]$.

13.1.3 The characteristic function by a Monte Carlo approach

From (13.1) the characteristic function of X conditional on S is known and the characteristic function of X can be obtained by taking the expectation. As the risk factors are dependent, there is no closed solution for the characteristic function Φ_X. In order to determine Φ_X numerically one can evaluate the expectation by Monte Carlo:

$$\Phi_X(z) = \mathbb{E}[\Phi_{X|S}(z)]$$

$$= e^{\sum_A p_A T w_{A0}(e^{i\nu_A z}-1)} \mathbb{E}\left[\exp\left(\sum_{k=1}^{N} S_k \sum_A T p_A w_{Ak}(e^{i\nu_A z}-1)\right)\right]$$

$$= \exp\left(\xi_0(z)\right) \mathbb{E}\left[\exp\left(\sum_{k=1}^{N} S_k \xi_k(z)\right)\right]$$

where $\xi_k(z) := \sum_A T p_A w_{Ak}(e^{i\nu_A z} - 1)$ for $k = 0, \dots N$.

First one has to decide for which z one wants to compute the characteristic function $\Phi_X(z)$. Let Z denote the number of sample points. Since the functions $\xi_k(z)$ are deterministic functions, these functions can be valuated before starting the Monte Carlo loop for each z considered. The advantage of this organization is that each diced tuple of S can be used for each z to compute the expectation. Hence the computation time for the generation of the random numbers is independent of S. In the case that the S_k are dependently lognormal distributed, the effort to dice one sample is given by $\mathcal{O}(N^2)$, however, the same effort will mostly hold for other multivariate distributions. If M is the number of Monte Carlo iterations and as K is the number of obligors, the computational effort of this method is given by

$$\underbrace{\mathcal{O}(KNZ)}_{\text{Computing } \xi_k} + \underbrace{\mathcal{O}(MN^2)}_{\text{Dicing } s^k} + \underbrace{\mathcal{O}(MZN)}_{\text{Valuation of the Monte Carlo sum}} \qquad (13.3)$$

Note that in practice K and M are larger than N, Z, and that the computational effort does not contain a term KM. So the effort of this Monte Carlo method is tolerable, in contrast to a direct simulation. A direct Monte Carlo approach to the model for X would mean the effort for each Monte Carlo simulation first to sample the risk factors S, then to compute the default intensities and sample the time of default of each obligor. Hence the direct simulation approach is not feasible due to a computational effort of $\mathcal{O}(KMN)$.

The proposed algorithm admits two possibilities for parallel computing. The first possibility is the usual Monte Carlo parallelism, i.e. evaluating the expression parallel for several diced random numbers. Secondly, one can also compute independently the expression for each z and fixed dice of random numbers S. Since one is interested in the valuation of an expectation and not in the simulation of a stochastic process, it may be favourable to use quasi-Monte Carlo methods, due to a potentially much better convergence.

13.1.4 The credit loss distribution

Recall from Chapter 8 that the distribution of a real-valued random variable X is uniquely determined by its characteristic function Φ_X. If Φ_X is integrable, then X has a density f that can be computed by Fourier inversion:

$$f(x) = \frac{1}{2\pi} \int\limits_{-\infty}^{\infty} e^{-izx} \Phi_X(z)\, dz.$$

Even if the characteristic function is not integrable, one can obtain the distribution of X by Fourier inversion. As shown in Chapter 8 or [8], the Fourier inversion by the fast Fourier transformation (FFT), see e.g. [7], works fine also in this situation. Hence, if one performs a Fourier inversion by FFT one needs not care whether Φ_X is integrable or not.

For the application in credit risk one knows that the density f resides on the interval $[0, \sum_A \nu_A]$, since the credit loss is greater or equal to 0 (the best case that no obligor defaults) and the credit loss is less than or equal to $\sum_A \nu_A$, which corresponds to the worst case that all obligors default. The last parameter required for the FFT algorithm is the number of sample points Z. Usually Z is a power of two, $Z = 2^n$, where n takes values round about 10.

13.1.5 Numerical example: the impact of sector correlations

In this example a portfolio with 1000 loans, each of size 1, is studied: 500 obligors belong to the first sector and the other 500 obligors belong to the second sector; there is no idiosyncratic risk. The volatility of both sectors is assumed to be 30% and the time horizon of the analysis is one year. It is assumed that all obligors have an annual default probability of 2.5%. In Figure 13.2 the densities of the credit loss are plotted for the sector correlation $\rho = -1, 0, 1$.

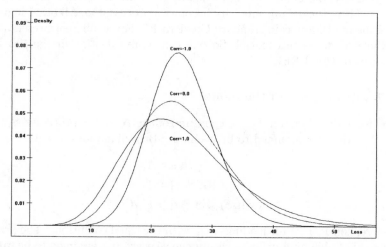

Fig. 13.2. Densities of credit loss for sector correlations $\rho = -1, 0, 1$

Note that the case $\rho = -1$ corresponds to the case that all obligors only have idiosyncratic risk; the case $\rho = 0$ corresponds to the usual CreditRisk$^+$

approach with independent sectors and the case $\rho = 1$ corresponds to the case that all obligors belong to the same sector.

Knowing the density of the credit loss distribution one can easily determine the usual risk management ratios. Table 13.1 shows that the factor correlation has a large impact on the standard deviation, the value-at-risk and the expected shortfall. Hence for applications in practice it is rather important to have a model that allows computations for any sector correlation and thus provides more precise risk valuations.

	$\rho = -1$	$\rho = 0$	$\rho = 1$
Standard deviation	5.3	7.4	9.2
Value-at-risk (95%)	34.0	38.0	41.8
Expected shortfall (95 %)	36.6	42.7	47.9

Table 13.1. Impact of sector correlations on risk measures

13.2 A Time Continuous Model

In standard CreditRisk$^+$ as well as in the generalization presented in the previous section only the loss at a certain time horizon is studied. One may also be interested in a model that allows a description of credit risk in continuous time. In this section such a model is presented. The risk factors are now described by processes in continuous time instead of discrete random variables, more precisely, the factors are modelled by (dependent) geometric Brownian motions. Then, the default intensity becomes a process, too. All other ingredients of the model are as in standard CreditRisk$^+$. For a coherent presentation the whole model is introduced. So one can easily identify similarities and differences to CreditRisk$^+$.

13.2.1 Introduction of the model

Let us fix a time horizon $T_\infty > 1$ and introduce N factor processes $S_{k,t}$ on $t \in [0, T_\infty]$ that are assumed to be geometric Brownian motions with

$$S_{k,0} = 1,$$
$$\mathbb{E}[S_{k,t}] = 1,$$
$$\text{cov}[\ln S_{k,t}, \ln S_{l,t}] = C_{kl}t.$$

As before, consider factor loadings w_{Ak} with the usual conditions. The probability p_A that obligor A defaults within one year again is assumed to be rather small. The default intensity now becomes a stochastic process:

$$p_{A,t}^S := p_A \left(w_{A0} + \sum_{k=1}^{N} w_{Ak} S_{k,t} \right).$$

The time of default of obligor A is denoted by T_A and is exponentially distributed with the intensity process $p_{A,t}^S$. Hence one may define default indicator processes $\mathbf{1}_{A,t}$ by

$$\mathbf{1}_{A,t} := \begin{cases} 0 & \text{if } t < T_A \\ 1 & \text{if } t \geq T_A \end{cases}.$$

Conditional on S the indicator processes are independent and

$$\mathbb{P}[T_A < t] = \mathbb{P}[\mathbf{1}_{A,t} = 1] = 1 - \exp\left(-\int_0^t p_{A,s}^S ds\right).$$

If obligor A defaults, the loss due to this default is given by ν_A and hence the process of cumulated defaults is given by

$$X_t := \sum_A \nu_A \mathbf{1}_{A,t}.$$

Remark 2 (Default indicator and Poisson process). Since $\mathbf{1}_{A,t}$ is a jump process, one may also be interested in the relationship between $\mathbf{1}_{A,t}$ and the Poisson process. Let Y_t be a Poisson process with time-dependent default intensity $p_{A,t}^S \geq 0$, which has right continuous paths and existing left limits such that for all $t > s \geq 0$:

$$Y_0 = 0, \qquad Y_t - Y_s \in \mathbb{N},$$

$$\mathbb{P}[Y_t - Y_s = n] = \frac{1}{n!}\left(\int_s^t p_{A,u}^S du\right)^n \exp\left(-\int_s^t p_{A,u}^S du\right),$$

$$Y_t - Y_s \quad \text{is independent of } Y_{[0,s]}.$$

Then the following relation holds for the default indicator process:

$$\mathbf{1}_t = \begin{cases} 0 & \text{if } Y_t = 0 \\ 1 & \text{if } Y_t > 0 \end{cases}.$$

13.2.2 The characteristic function by Monte Carlo methods

Let $\mathcal{F}_t$ be the natural filtration induced by the processes S_k. If one defines the risk factors in Section 13.1 to be $\frac{1}{t}\int_0^t S_{k,\tau}\,d\tau$, one obtains the characteristic function of X_t conditional on $\mathcal{F}_t$:

$$\Phi_{X_t|\mathcal{F}_t}(z) = \exp\left(\sum_A (e^{i\nu_A z} - 1)p_A\left(w_{A0}t + \sum_{k=1}^N w_{Ak}\int_0^t S_{k,\tau}\,d\tau\right)\right).$$

Since the factor processes $S_{k,t}$ are not independent, the expectation to obtain the characteristic function $\Phi_{X_t}(z)$ can be evaluated by Monte Carlo methods only:[2]

$$\Phi_{X_t}(z) = \mathbb{E}[\Phi_{X_t}|_{\mathcal{F}}(z)] = \exp\left(\sum_A (e^{i\nu_A z} - 1)p_A w_{A0} t\right) \times$$

$$\mathbb{E}\left[\exp\left(\sum_A \sum_{k=1}^N (e^{i\nu_A z} - 1)p_A w_{Ak} \int_0^t S_{k,\tau} d\tau\right)\right]$$

$$= \exp(\xi_0(z)t) \, \mathbb{E}\left[\exp\left(\sum_{k=1}^N \xi_k(z) \int_0^t S_{k,\tau} d\tau\right)\right]$$

$$\text{where} \quad \xi_k(z) := \sum_A (e^{i\nu_A z} - 1)p_A w_{Ak} \qquad \text{for } k = 0, \ldots N.$$

Let $\mathcal{Z}$ be the set of values of z for which one wants to evaluate the characteristic function. It is then possible to compute the expressions $\xi_k(z)$ for all $z \in \mathcal{Z}$ before starting the Monte Carlo procedure. For each Monte Carlo iteration one has to sample the random numbers

$$I^k \sim \int_0^t S_{k,\tau} d\tau$$

with dependent geometric Brownian motions $S_{k,\tau}$. Using one draw of I^k one computes $\sum_{k=1}^N \exp(\xi_k(z)I^k)$ for each $z \in \mathcal{Z}$ and the total cost of this Monte Carlo method is comparable to the computation time to handle the model presented in Section 13.1. There is only a larger effort in dicing I^k, since one has to sample paths and to integrate them instead of dicing one lognormal random number. The asymptotic costs are again given by (13.3).

Remark 3. To sample I^k one can proceed as follows. Let B be the Cholesky decomposition of the log–covariance matrix C and let W_t^l be independent Brownian motions. Then $S_{k,t}$ is given by

$$S_{k,t} = \exp\left(\sum_{l=1}^N B_{kl} W_t^l - \frac{1}{2} C_{kk} t\right).$$

To compute the integral one divides the interval $[0, t]$ into D intervals of equal length and computes the integral by the approximation

$$I^k \approx \frac{t}{2D} \sum_{i=1}^D (S_{k, \frac{(i-1)t}{D}} + S_{k, \frac{it}{D}}).$$

---

[2] To the author's knowledge, there is no closed-form solution for the characteristic function even in the case of independent lognormal processes $S_{k,t}$.

There is much literature on properties of the integral of the geometric Brownian motion, see for example [4, 5, 6, 9].

Since there is a numerical algorithm to compute the characteristic function of the credit loss X_t, one can determine an approximate density of X_t by Fourier inversion as presented in Subsection 13.1.4. To round off the discussion, the first two moments of X_t for a fixed t will be computed now.

13.2.3 Moments of X_t

Lemma 1. *Let $S_{k,t}$ and $S_{l,t}$ be two geometric Brownian motions where $(\ln S_{k,t} + \frac{1}{2}C_{kk}t)$ and $(\ln S_{l,t} + \frac{1}{2}C_{ll}t)$ are normally distributed with mean 0 and covariance $C_{kl}t$. Then*

$$\mathbb{E}\left[\int_0^t S_{k,\tau}\,d\tau\right] = t,$$

$$\mathbb{E}\left[\int_0^t S_{k,\tau}\,d\tau \int_0^t S_{l,\theta}\,d\theta\right] = \frac{2}{(C_{kl})^2}\left(e^{C_{kl}t} - 1 - C_{kl}t\right).$$

Proof. To prove the first equation, one only has to use Fubini's theorem. For the second statement, let $\tau \le \theta$:

$$\mathbb{E}[S_{k,\tau}S_{l,\theta}] = \mathbb{E}[S_{k,\tau}S_{l,\tau}\frac{S_{l,\theta}}{S_{l,\tau}}] = \mathbb{E}[S_{k,\tau}S_{l,\tau}] = \exp(C_{kl}\tau).$$

Hence,

$$\mathbb{E}\left[\int_0^t S_{k,\tau}\,d\tau \int_0^t S_{l,\theta}\,d\theta\right] = \int_0^t d\tau \int_0^t d\theta \,\mathbb{E}[S_{k,\tau}S_{l,\theta}]$$

$$= 2\int_0^t d\tau \int_\tau^t d\theta\, e^{C_{kl}\tau} = 2\int_0^t d\tau\, e^{C_{kl}\tau}(t - \tau)$$

$$= \frac{2}{(C_{kl})^2}\left(e^{C_{kl}t} - 1 - C_{kl}t\right).$$

$\square$

This lemma and Theorem 1 prove the following.

Theorem 2. *The expectation and variance of X_t are given by*

$$\mathbb{E}[X_t] = t\sum_A p_A \nu_A,$$

$$\mathrm{var}[X_t] = t\sum_A p_A(\nu_A)^2 + \sum_{A,\tilde{A}} p_A \nu_A p_{\tilde{A}}\nu_{\tilde{A}}\sum_{k,l=1}^N w_{Ak}w_{\tilde{A}l}\left(\frac{2(e^{C_{kl}t} - 1 - C_{kl}t)}{(C_{kl})^2} - t^2\right).$$

13.3 Combining Market Risk and Credit Risk

Up to now, only models with fixed exposure have been studied in this chapter. In practice, the loss that occurs if an obligor defaults may depend on market fluctuations. An example would be a future or option contract that is traded over the counter and so the value of such deal at maturity is fraught with counterparty risk, which is a credit risk in fact.

In this section our model will be extended in such a way that it incorporates market risk. If there is no credit risk, this approach yields the well-known delta normal approach to assessing market risk. The idea presented here differs from other models that combine market and credit risk (see e.g. [3]).

Whenever one talks about market risk, there is a portfolio depending on market risk factors and one has the possibility to evaluate a portfolio for a given state of these factors. Here a model of a portfolio that combines market risk and credit risk is introduced. Finally, the computation of the profit and loss distribution (P&L) of this portfolio is achieved and an approximative solution of this task can be given.

In addition to the assumption that default probabilities are small, it is assumed – according to the delta normal model – that the effect of the (normally distributed) market fluctuations may be linearized. An algorithm to compute the P&L is presented that is based on these assumptions. The computational effort of this algorithm is tolerable for a medium number of obligors and is independent of the number of market risk factors.

13.3.1 A portfolio with market risk and credit risk

The credit risk driving factors are the default indicators $\mathbf{1}_{A,t}$ and the factor processes $S_{k,t}$ of the credit risk model presented in Section 13.2. Due to changes of the factor processes $S_{k,t}$ the obligors, default intensities alter and this affects the probability of default of each obligor. In fact, this is the so-called spread risk, which is modelled by the factor processes in this way.

Additional to the (credit) risk factors there are market risk factors M_t^l, $l = 1, \ldots, \tilde{N}$. Examples of market risk factors are returns of stock prices, foreign exchange rates or interest rates. Let M_t^l be dependent Brownian motions with drift 0 and different volatilities. Then the dependencies between $S_{k,t}$ and M_t^l are described by a covariance matrix C, which has the following structure:

$$C = \begin{pmatrix} C^1 & C^{2\top} \\ C^2 & C^3 \end{pmatrix}.$$

C^1 is an $N \times N$ matrix, C^3 is an $\tilde{N} \times \tilde{N}$ matrix, and C^2 is an $\tilde{N} \times N$ matrix; the entries of these matrices are given by

$$C_{kl}^1 = \frac{1}{t}\text{cov}(\ln S_{k,t}, \ln S_{l,t}),$$

$$C_{kl}^2 = \frac{1}{t}\text{cov}(M_t^k, \ln S_{l,t}),$$

$$C_{kl}^3 = \frac{1}{t}\text{cov}(M_t^k, M_t^l).$$

Let $\nu_A(M_t, t)$ denote the nominal amount of obligor A at time t, discounted by the risk-free interest rate. For simplicity, let us assume that there is only one and fixed settlement date T_A for each obligor. The probability that obligor A survives up to T_A under the condition that no default has occurred up to time t is given by $P[\mathbf{1}_{A,T_A} = 0|\mathbf{1}_{A,t} = 0]$. These are functions of the state of the factor processes:

$$q_A(s,t) := \mathbb{P}[\mathbf{1}_{A,T_A} = 0|S_t = s \wedge \mathbf{1}_{A,t} = 0]$$

$$= \mathbb{E}\left[\exp\left(-p_A \int_t^{T_A}(w_{A0} + \sum_{k=1}^N w_{Ak}S_{k,\tau})d\tau\right)\Bigg| S_{k,t} = s_k\right]. \quad (13.4)$$

Since any claim fraught with credit risk is discounted by this probability, the value process of the portfolio can be written

$$V_t = \sum_A \nu_A(M_t, t)q_A(S_t, t)(1 - \mathbf{1}_{A,t}).$$

13.3.2 Obtaining an approximative P&L

Besides the assumption that p_A is small, the following linearization is also used. Let us assume that the nominal functions may be approximated by linear functions:

$$\nu_A(M_t, t) \approx \nu_A(M_0, 0) + \frac{\partial \nu_A}{\partial t}(M_0, 0)t + \frac{\partial \nu_A}{\partial M_t}(M_0, 0) \cdot M_t$$

$$= \overset{0}{\nu_A}(t) + \overset{\Delta}{\nu_A} \cdot M_t$$

where $\overset{0}{\nu_A}(t) := \nu_A(M_0, 0) + \frac{\partial \nu_A}{\partial t}(M_0, 0)t,$

$$\overset{\Delta}{\nu_A} := \frac{\partial \nu_A}{\partial M_t}(M_0, 0).$$

By this assumption the value process is given by

$$V_t = \sum_A(\overset{0}{\nu_A}(t) + \overset{\Delta}{\nu_A} \cdot M_t)\, q_A(S_t, t)\, (1 - \mathbf{1}_{A,t}).$$

In the next step the dependence between M_t and S_t will be analysed. Let B be the Cholesky decomposition of the covariance matrix C, with the structure

$$C = \begin{pmatrix} C^1 & C^{2\mathsf{T}} \\ C^2 & C^3 \end{pmatrix} = \begin{pmatrix} B^1 & 0 \\ B^2 & B^3 \end{pmatrix} \begin{pmatrix} B^1 & 0 \\ B^2 & B^3 \end{pmatrix}^{\mathsf{T}} = BB^{\mathsf{T}}.$$

Then the usual method to sample $\ln S_t$ and M_t is to choose two vectors W_t^1, W_t^2 of independent Brownian motions and to use the transformation

$$\begin{pmatrix} \ln S_{k,t} + \frac{1}{2}C_{kk}t \\ M_t \end{pmatrix} = \begin{pmatrix} B^1 & 0 \\ B^2 & B^3 \end{pmatrix} \begin{pmatrix} W_t^1 \\ W_t^2 \end{pmatrix}.$$

Hence the state of the factor process is given by

$$S_{k,t} = \exp\left((B^1 W_t^1)_k - \frac{1}{2}C_{kk}^1 t \right)$$

and M_t can be expressed by

$$M_t = B^2 W_t^1 + B^3 W_t^2.$$

Hence, the portfolio has the representation

$$V_t = \sum_A q_A(S_t, t)(1 - \mathbf{1}_{A,t})\left(\overset{0}{\nu_A}(t) + B^{2\mathsf{T}}\overset{\Delta}{\nu_A} W_t^1 + B^{3\mathsf{T}}\overset{\Delta}{\nu_A} W_t^2 \right). \qquad (13.5)$$

The P&L may be determined by Fourier inversion of the characteristic function of V_t. The first step to computing the latter is to determine the characteristic function conditional on the state of S_t and $\mathbf{1}_t$:

$$\Phi_{V_t|(S_t,\mathbf{1}_t)}(z) = \mathbb{E}[e^{izV_t}|S_t, \mathbf{1}_t]$$

$$= \exp\left(iz\sum_A q_A(S_t,t)(1-\mathbf{1}_{A,t})(\overset{0}{\nu_A}(t) + \overset{\Delta}{\nu_A} B^2 W_t^1) \right) \times$$

$$\exp\left(-\frac{1}{2}z^2 \left\| \sum_A q_A(S_t,t)(1 - \mathbf{1}_{A,t})B^{3\mathsf{T}}\overset{\Delta}{\nu_A} \right\|^2 t \right).$$

As before, the default probability for each obligor is assumed to be quite small. Hence one may approximate (13.4) by

$$q_A(s,t) \approx \exp\left(-p_A(T_A - t)(w_{A0} + \sum_{k=1}^N w_{Ak}s_k) \right). \qquad (13.6)$$

The characteristic function of V_t can now be evaluated by a Monte Carlo method using the representation

$$\Phi_{V_t}(z) = \mathbb{E}\left[\Phi_{V_t|(S_t,\mathbf{1}_t)}(z) \right].$$

For one Monte Carlo sample one has to sample the paths W_t^1 of independent Brownian motions. By (13.5) one can determine $S_{k,t}$ and then by (13.6) one obtains $q_A(S_t, t)$. Then for each obligor, one has to sample the binary random variable

$$
\mathbf{1}_{A,t} = \begin{cases} 0 & \text{with probability } 1 - \exp\left(-p_A \int_0^t (w_{A0} + \sum_{k=1}^N w_{Ak} S_{k,\tau}) d\tau\right) \\ 1 & \text{with probability } \exp\left(-p_A \int_0^t (w_{A0} + \sum_{k=1}^N w_{Ak} S_{k,\tau}) d\tau\right) \end{cases}.
$$

Of course, with one sample of $\mathbf{1}_t$ and S_t one can evaluate the expression under the expectation for several s. If one proceeds in this way, the computational effort of this method is obviously of order $\mathcal{O}(M \cdot (K + N + N \cdot K + Z))$ where M denotes the number of Monte Carlo valuations and Z is the number of sample points of $\Phi_{V_t}(z)$. Hence this method can be used in practice if the number of obligors is not too big. Since the effort is independent of the number of market risk factors $\tilde{N}$, this approach can be applied even for large $\tilde{N}$. In order to obtain the density of the portfolio's profit and loss distribution one can use the Fourier inversion technique based on the fast Fourier transformation described in Subsection 13.1.4.

13.4 Conclusion

A model that combines market risk and credit risk has been presented. In the special case that there is only one obligor who never defaults, the previous analysis of the P&L distribution gives the delta normal approach, which is the most simple and well-known idea to deal with market risk alone. In the case that there is no market risk, the model conforms with the credit risk model presented in Section 13.2. This underlines that the presented model is a natural generalization to describe market risk and credit risk.

In order to find other and more general models that combine market risk and credit risk it is necessary to find other credit risk models that allow the handling of a stochastic exposure. One approach in this direction can be found in Chapter 9 or in [2], where a stochastic independence between the risk factors and the processes, which affect the size of the exposure, is assumed. They also remark on how to overcome this assumption, but a deeper analysis remains to be done. In [3] another approach is presented to modelling market and credit risk. Duffie and Pan split the portfolio into a *value component* and a *default component* and they treat each component separately.

For applications in risk management it seems to be necessary to model dependencies between the sectors and the exposure. For example, assume that you hold put options on the stock of a bank that are sold by this bank. In the case of default of this bank the stock will fall, but your (theoretical) win on the puts is worthless since the bank is unable to pay them out. The model presented in Section 13.3 provides a first approach to dealing with such situations.

Based on the density of the portfolio's P&L one can perform many more computations. Most risk measures used in practice are functionals of the P&L that can be determined, since the Fourier inversion yields the whole distribution of a portfolio. The same argument also holds for the pricing of derivatives on such portfolios. For example, the price of a European option can be represented as an expectation of a function of the portfolio return. There are plain credit risk portfolios traded, e.g. asset-backed securities, but it seems very likely that options on more complex portfolios will be sold in future and hence there is a further demand on models which unify market and credit risk.

Acknowledgements. The author thanks Hermann Haaf (Commerzbank AG) for raising interesting questions regarding the CreditRisk$^+$ model and John Schoenmakers (Weierstrass Institute) for many valuable comments and suggestions. The research was done in cooperation with the bmb+f project "Effiziente Methoden zur Bestimmung von Risikomaßen", which is also supported by Bankgesellschaft Berlin AG.

References

1. P. Bürgisser, A. Kurth, A. Wagner, and M. Wolf. Integrating correlations. *Risk*, 12(7):57–60, 1999.
2. P. Bürgisser, A. Kurth, and A. Wagner. Incorporating severity variations into credit risk. *Journal of Risk*, 3(4):5–31, 2001.
3. D. Duffie and J. Pan. Analytical value-at-risk with jumps and credit risk. *Finance and Stochastics*, 5(2):155–180, 2001.
4. D. Dufresne. The distribution of a perpetuity, with applications to risk theory and pension funding. *Scandinavian Actuarial Journal*, 39–79, 1990.
5. D. Dufresne. The integral of geometric Brownian motion. *Advances in Applied Probability*, 33:223–241, 2001.
6. H. Geman and M. Yor. Bessel processes, Asian options, and perpetuities. *Mathematical Finance*, 3(4):349–375, 1993.
7. W. H. Press, S. A. Teukolsky, W. T. Vetterling and B. P. Flannery. *Numerical Recipes in C: The Art of Scientific Computing.* Second Edition, Cambridge University Press, Cambridge, 1992.
8. O. Reiß. *Fourier inversion algorithms for generalized CreditRisk$^+$ models and an extension to incorporate market risk.* WIAS–Preprint No. 817, 2003.
9. M. Yor. On some exponential functionals of Brownian motion. *Advances in Applied Probability* 24(3):509–531, 1992.

Econometric Methods for Sector Analysis

Leif Boegelein, Alfred Hamerle, Michael Knapp and Daniel Rösch

Summary. Default event correlation and dependency plays a key role in determining portfolio credit risk. We present two econometric methods that can be employed to forecast default rates and describe correlations of the forecasts. The first approach is company-specific, based on a Merton-style threshold model. The second focuses on the systematic risk factors influencing the obligors of a specific industry or country. Both approaches are particularly well suited to estimating default rate development and correlation in a conditional independence framework such as CreditRisk$^+$. Based on panel data covering a variety of countries and industries we present an implementation example for the second method of seemingly unrelated regression and illustrate the implications of dynamic and correlated default rates on the risk of a large portfolio of credits.

Introduction

In CreditRisk$^+$ each obligor of a given portfolio of credits can be allocated to a set of sectors, independent risk factors represented by a continuous random variable of known distribution. The default frequencies of obligors sharing a sector are dependent on a common random factor and therefore allow for the modelling of correlated default events. The definition and parameterization of sectors plays a key role in model implementation. Not only is the distribution of losses heavily influenced by the chosen sector composition and its parameters, the relative risk contributions of single transactions and subportfolios depend directly on the parameters and model assumptions about sector characteristics. In the following we will present two methods to implement sector analysis in CreditRisk$^+$ based on statistical tests. Both methods expand the original model where default frequencies are static model input to include the influence of time-varying risk factors on default frequencies. It is shown that sector analysis can be implemented from the econometric results.

Under the individual approach, single obligors' default frequencies are first estimated with an econometric random factor model. It can be shown that the influence of the random factor has a direct counterpart in CreditRisk$^+$ and is responsible for the implicit pairwise correlation of credit events. In the second step, the validity of a given sector definition is tested based on a statistical test of significance and correlation of the random factors.

The second method is fitted to the usual data limitations of banks' SME (small and medium-size enterprise) portfolios. It estimates the dependency of default rates aggregated on industry and country level on systematic risk factors with seemingly unrelated regression (SUR). The validity of a given definition of sectors is then examined based on a statistical test of the correlation between regression residuals. Both methods achieve three goals:

- Derive a sector composition in line with the basic model assumptions.
- Provide the model with dynamic forecasts for default frequencies rather than static historical averages.
- Provide estimates for sector parameters.

We present an example of the implementation of sector analysis using the second approach where systematic risk factors are identified as macroeconomic variables.

14.1 Basics for Sector Analysis in CreditRisk$^+$

In the following we recall the relevant basic ideas from CreditRisk$^+$. We consider a portfolio of K different loans, written to different obligors A with unconditional default frequencies p_A. Obligors are allocated to N sectors $S_1, \ldots, S_N$ represented by the random variables $S_1, \ldots, S_N$ with expectations $\mu_1, \ldots, \mu_N$ and standard deviations $\sigma_1, \ldots, \sigma_N$. The relative influence of a certain sector S_k on the unconditional default frequency is expressed by the factor loading w_{Ak}, where

$$\sum_{k=1}^{N} w_{Ak} = 1 \quad \text{and} \quad 0 \leq w_{Ak} \leq 1.$$

The default frequency of obligor A conditional on the random variables is

$$p_A^S = p_A \sum_{k=1}^{N} w_{Ak} \frac{S_k}{\mu_k}. \tag{14.1}$$

Recall that p_A^S is a random variable with $\mathbb{E}[p_A^S] = p_A$ and $\text{var}[p_A^S] = \sigma_A^2 = p_A^2 \sum_{k=1}^{N} w_{Ak}^2 \frac{\sigma_k^2}{\mu_k^2}$. As is shown in [5] or Chapters 2 and 10, in particular (2.12), (2.23) and (10.1), the pairwise correlation of default events between two obligors A and B can be approximated by

$$\rho_{AB} := \sqrt{p_A p_B} \sum_{k=1}^{N} w_{Ak} w_{Bk} \left(\frac{\sigma_k}{\mu_k}\right)^2. \tag{14.2}$$

From (14.2) it becomes apparent that default events of obligors that do not share a sector or are allocated fully to a sector where $\sigma_k = 0$ are uncorrelated. Therefore, any practical implementation of sector analysis has to provide an estimation for the parameters μ_k and σ_k as well as factor loadings that reflect actual correlation patterns of default events. The model poses two restrictions on the definition of sectors for any implementation. The probability-generating function (PGF) of the portfolio loss for a sector S_k is obtained by multiplying the PGF of individual obligors' losses G_A conditional on S_k:

$$G_k(z|S_k) = \prod_A G_A(z|S_k).$$

This implies the restriction of independence between individual loss events conditional on the realization of S_k. The PGF of unconditional losses $G_k(z)$ for sector S_k given the density function $f_k(s)$ of the sector defining random variable S_k is given by

$$G_k(z) = \int G_k(z|S_k = s) f_k(s) ds,$$

and the PGF of unconditional portfolio loss $G(z)$ is

$$G(z) = \prod_{k=1}^{N} G_k(z). \tag{14.3}$$

(14.3) illustrates the second assumption of the model, namely the independence of loss events between different sectors. In the following we present two different approaches to estimating parameters and test whether a given sector composition satisfies these restrictions. The suggested approaches are both econometric models, and thus model the dynamics of default frequencies over time. That is, an extension of the notation to a dynamic view is required, denoting the index for the time series observations by t, where $t = 1, \ldots, T$.

14.2 An Individual Approach: The Random Effects Hazard Rate Model

In this section we consider an individual Merton-type threshold-value model for the default probability. The event that an obligor is unable to fulfil his payment obligations is defined as default. The default event for obligor A in the time period t is random and modelled using an indicator variable $\mathbf{1}_{At}$, i.e.

$$\mathbf{1}_{At} = \begin{cases} 1 & \text{borrower } A \text{ defaults in } t \\ 0 & \text{otherwise} \end{cases} \quad \text{for } A = 1, \ldots, K_t, \ t = 1, \ldots, T.$$

The default event is assumed to be observable. In addition, a continuous-state variable r_{At} is defined that may be interpreted as the logarithmic return of a firm's assets. r_{At} is not observable. For the relationship between r_{At} and the default event $\mathbf{1}_{At}$ a threshold model is assumed. Default is equivalent to the return of a firm's assets falling below a threshold c_{At}, i.e.

$$r_{At} \leq c_{At} \quad \Leftrightarrow \quad \mathbf{1}_{At} = 1 \text{ for } A = 1, \ldots, K_t, \ t = 1, \ldots, T. \tag{14.4}$$

Implicitly, (14.4) makes the assumption that default has not occurred in previous periods of time. Therefore, the unconditional default frequency

$$p_{At} = \mathbb{P}[\mathbf{1}_{At} = 1] = \mathbb{P}[r_{At} \leq c_{At}]$$

can be interpreted as discrete-time hazard rate. For the latent variable r_{At} we assume a linear random effect panel model that may include fundamental, macroeconomic and statistical risk factors. Denoting vectors by boldface letters, the model can be written as

$$r_{At} = \beta_0 + \boldsymbol{\beta}^\top \mathbf{x}_{At-1} + \boldsymbol{\gamma}^\top \mathbf{z}_{t-1} + \sigma_k S_{kt} + \nu u_{At}$$

for $A = 1, \ldots, K_t$, $t = 1, \ldots, T$. $\mathbf{x}_{At-1}$ denotes a vector of time-lagged firm-specific risk factors such as the return on equity of the firm's last period financial statement or the number of employees one or more periods ago. Alternatively, $\mathbf{x}_{At-1}$ may be a risk score representing the result of a rating process in period $t-1$. The parameter vector $\boldsymbol{\beta}$ contains the exposures to these individual risk factors. Correspondingly, $\mathbf{z}_{t-1}$ denotes a vector of systematic risk factors, such as the unemployment rate of the previous period or the money market rate two years ago and $\boldsymbol{\gamma}$ is the vector of exposures. S_{kt} is a standard normally distributed random variable with loading σ_k, u_{At} is an error term the distribution of which will be specified below. The idiosyncratic error terms u_{At} are assumed to be independent from the S_{kt} and independent for different borrowers. All random variables are serially independent. ν and β_0 are constants. The idiosyncratic risk factors $\mathbf{z}_{t-1}$ play a crucial role in sector analysis. It is assumed that the firms are homogenous within a sector regarding the relevant risk factors and the factor exposures. Parameters and risk factors are allowed to differ between sectors. Note that the econometric approach assumes that a borrower is influenced by exactly one sector specific random variable S_{kt}, i.e. $w_{Ak} = 1$ for observed defaults. This assumption, however, is only made for the estimation and the identification of independent sectors. For the implementation in CreditRisk$^+$ a borrower can be attributed to more than one sector once the sectors are identified.

Furthermore, in the econometric approach different distribution assumptions are used for the random factors and a different threshold model for the latent variable is assumed. While our approach assumes a normal distribution for the factors, CreditRisk$^+$ uses a gamma distribution. On the other hand, the CreditRisk$^+$ sector random variable is applied in multiplicative form as specified in (14.1), whereas we use a linear model. For a mapping of the two models, see [6]. Nevertheless we use the same notation in order to maintain the analogy. As will be shown later on, our assumptions lead to familiar econometric models and estimation procedures and thus are easy to implement. In the latent model the asset returns are not observable. Only the risk factors and the realization of the default indicator $\mathbf{1}_{At}$ can be observed. The link between the risk factors and the default frequency is described by the threshold model (14.4). Given that default has not happened before t, one obtains for the default frequency of the time period t, given information about the observable factors and factor exposures until time period $t-1$

$$p_{At}^S(\mathbf{x}_{At-1}, \mathbf{z}_{t-1}) = \mathbb{P}[u_{At} < c_{At} - \beta_0 - \boldsymbol{\beta}^\mathsf{T}\mathbf{x}_{At-1} - \boldsymbol{\gamma}^\mathsf{T}\mathbf{z}_{t-1} - \sigma_k S_{kt} | \mathbf{x}_{At-1}, \mathbf{z}_{t-1}]$$
$$= F(\tilde{\beta}_{0At} - \boldsymbol{\beta}^\mathsf{T}\mathbf{x}_{At-1} - \boldsymbol{\gamma}^\mathsf{T}\mathbf{z}_{t-1} - \sigma_k S_{kt})$$

where $\tilde{\beta}_{0At} = c_{At} - \beta_0$, and $F(.)$ denotes the distribution function of the error term u_{At}. Since the thresholds c_{At} cannot be observed, $\tilde{\beta}_{0At}$ is not estimable and we restrict the intercept to $\tilde{\beta}_0$. Different assumptions about the error distribution function $F(.)$ lead to different models. The random effect probit specification is given by

$$p_{At}^S(\mathbf{x}_{At-1}, \mathbf{z}_{t-1}) = \Phi(\beta_0 + \boldsymbol{\beta}^\mathsf{T}\mathbf{x}_{At-1} + \boldsymbol{\gamma}^\mathsf{T}\mathbf{z}_{t-1} + \sigma_k S_{kt}) \tag{14.5}$$

where $\Phi(.)$ denotes the the standard normal distribution function. It can be shown that (14.5) is an extension of the Basel II model, and explains default frequencies with time-varying risk factors. Here we consider in more detail the logistic specification, which can be written as

$$p_{At}^S(\mathbf{x}_{At-1}, \mathbf{z}_{t-1}) = L(\beta_0 + \boldsymbol{\beta}^\mathsf{T}\mathbf{x}_{At-1} + \boldsymbol{\gamma}^\mathsf{T}\mathbf{z}_{t-1} + \sigma_k S_{kt}) \tag{14.6}$$

where $L(g) = \frac{\exp(g)}{(1+\exp(g))}$ is the distribution function of the logistic distribution. Assuming a standard normal distribution for the random effect S_{kt} the "unconditional" default frequency (in Basel II terminology meaning that we have only information about $\mathbf{x}_{At-1}$ and $\mathbf{z}_{t-1}$ but not the value of S_{kt}) is given by the expectation of (14.6) as

$$p_{At}(\mathbf{x}_{At-1}, \mathbf{z}_{t-1}) = \int_{-\infty}^{+\infty} L(\beta_0 + \boldsymbol{\beta}^\mathsf{T}\mathbf{x}_{At-1} + \boldsymbol{\gamma}^\mathsf{T}\mathbf{z}_{t-1} + \sigma_k S_{kt})\varphi(S_{kt})dS_{kt} \tag{14.7}$$

where $\varphi(.)$ denotes the density function of the standard normal distribution. Suppose we have observed a time series of defaults $\mathbf{1}_{At}$, numbers K_t of borrowers and risk factors $\mathbf{x}_{At-1}$ and $\mathbf{z}_{t-1}$ ($A = 1, \ldots, K_t, t = 1, \ldots, T$). Then the marginal loglikelihood function for the observed defaults is

$$l(\beta_0, \boldsymbol{\beta}, \boldsymbol{\gamma}, \sigma_k) =$$
$$\sum_{t=1}^T \ln \int_{-\infty}^{+\infty} \prod_{A=1}^{K_t} p_{At}^S(\mathbf{x}_{At-1}, \mathbf{z}_{t-1})^{1_{At}} [1 - p_{At}^S(\mathbf{x}_{At-1}, \mathbf{z}_{t-1})]^{1 - 1_{At}} \varphi(S_{kt})dS_{kt}$$

where $p_{At}^S(\mathbf{x}_{At-1}, \mathbf{z}_{t-1})$ is given by (14.5) or (14.6) respectively. As an extension of common logit or probit models an important part of the method is the integral over the random effect. The integral approximation can for example be conducted by the adaptive Gaussian quadrature as described in [2]. Usually the logarithm of the likelihood function is minimized numerically with respect to the unknown parameters for which several algorithms, such as the Newton–Raphson method, exist and are implemented in statistical software packages. Under usual regulatory conditions the resulting estimators $\hat{\beta}_0, \hat{\beta}, \hat{\gamma}$ and $\hat{\sigma}_k$ asymptotically exist, are consistent and converge against normality. Thus common statistical tests for significance can be conducted.

The default correlation is the key parameter in sector analysis. In the following, we first consider default correlations within a sector, and in the second step we model default correlation between sectors. First we note that the default indicators $\mathbf{1}_{At}$ and $\mathbf{1}_{Bt}$ for different obligors A and B are binary random variables taking only the values 0 and 1. For binary variables the correlation coefficient ρ_{AB} for period t and for given values of the risk factors $\mathbf{x}_{At-1}, \mathbf{x}_{Bt-1}$ and $\mathbf{z}_{t-1}$ can be written as

$$\rho_{AB} = \frac{p_{ABt} - p_{At}p_{Bt}}{\sqrt{p_{At}(1 - p_{At})}\sqrt{p_{Bt}(1 - p_{Bt})}} \tag{14.8}$$

where for convenience we have left out the arguments of $p_{At}(\mathbf{x}_{At-1}, \mathbf{z}_{t-1})$, $p_{Bt}(\mathbf{x}_{Bt-1}, \mathbf{z}_{t-1})$ and $p_{ABt}(\mathbf{x}_{At-1}, \mathbf{x}_{Bt-1}, \mathbf{z}_{t-1})$. Note that the probabilities $p_{At}(\mathbf{x}_{At-1}, \mathbf{z}_{t-1})$ and $p_{Bt}(\mathbf{x}_{Bt-1}, \mathbf{z}_{t-1})$ can be calculated according to (14.7). Finally, the quantity $p_{ABt}(\mathbf{x}_{At-1}, \mathbf{x}_{Bt-1}, \mathbf{z}_{t-1})$ is the joint probability that in period t both firm A and B will default, given that neither firm has defaulted before. In order to determine this joint default frequency we use the fact that for given S_{kt} the defaults of firms A and B are independent, i.e.

$$p^S_{ABt}(\mathbf{x}_{At-1}, \mathbf{x}_{Bt-1}, \mathbf{z}_{t-1}) = L(\beta_0 + \beta^\top \mathbf{x}_{At-1} + \gamma^\top \mathbf{z}_{t-1} + \sigma_k S_{kt})$$
$$\times L(\beta_0 + \beta^\top \mathbf{x}_{Bt-1} + \gamma^\top \mathbf{z}_{t-1} + \sigma_k S_{kt}) \tag{14.9}$$

However, since S_{kt} is not observable, (14.9) cannot be calculated. For the unconditional joint default frequency $p_{ABt}(\mathbf{x}_{At-1}, \mathbf{x}_{Bt-1}, \mathbf{z}_{t-1})$ of the observable default events $\{\mathbf{1}_{At} = 1\}$ and $\{\mathbf{1}_{Bt} = 1\}$ for given values of $\mathbf{x}_{At-1}, \mathbf{x}_{Bt-1}$ and $\mathbf{z}_{t-1}$ and under the condition that none of the firms has defaulted before, we obtain

$$p_{ABt}(\mathbf{x}_{At-1}, \mathbf{x}_{Bt-1}, \mathbf{z}_{t-1}) = \int_{-\infty}^{+\infty} \tilde{L}_{At}(\sigma_k S_{kt})\tilde{L}_{Bt}(\sigma_k S_{kt})\varphi(S_{kt})dS_{kt}$$

where $\tilde{L}_{At}(\sigma_k S_{kt})$ stands for $L(\beta_0 + \beta^\top \mathbf{x}_{At-1} + \gamma^\top \mathbf{z}_{t-1} + \sigma_k S_{kt})$ and the abbreviation $\tilde{L}_{Bt}(\sigma_k S_{kt})$ represents $L(\beta_0 + \beta^\top \mathbf{x}_{Bt-1} + \gamma^\top \mathbf{z}_{t-1} + \sigma_k S_{kt})$. From the derivation above it can be seen that the parameter σ_k of the random effect plays the major role for the default correlation. If $\sigma_k = 0$ then $p_{ABt}(\mathbf{x}_{At-1}, \mathbf{x}_{Bt-1}, \mathbf{z}_{t-1}) = p_{At}(\mathbf{x}_{At-1}, \mathbf{z}_{t-1})p_{Bt}(\mathbf{x}_{Bt-1}, \mathbf{z}_{t-1})$ and from (14.8) we obtain $\rho_{AB} = 0$. Thus the test of the null hypothesis

$$H_0 : \rho_{AB} = 0$$

for all A,B in a sector, can be conducted by testing

$$H_0 : \sigma_k = 0.$$

If the null hypothesis cannot be rejected, the random time effect may be neglected. Then, the default events within the sector may be considered to be independent, supporting a sector implementation of one sector with $\sigma_k = 0$.

In the next step two different sectors are considered. Then for an obligor from Sector 1 the default frequency is given by

$$p^S_{At}(\mathbf{x}^{(1)}_{At-1}, \mathbf{z}^{(1)}_{t-1}) = L(\beta_{01} + \boldsymbol{\beta}^{\mathsf{T}}_1 \mathbf{x}^{(1)}_{At-1} + \boldsymbol{\gamma}^{\mathsf{T}}_1 \mathbf{z}^{(1)}_{t-1} + \sigma_1 S_{1t})$$

and for an obligor from Sector 2 the probability of default can be written as

$$p^S_{Bt}(\mathbf{x}^{(2)}_{Bt-1}, \mathbf{z}^{(2)}_{t-1}) = L(\beta_{02} + \boldsymbol{\beta}^{\mathsf{T}}_2 \mathbf{x}^{(2)}_{Bt-1} + \boldsymbol{\gamma}^{\mathsf{T}}_2 \mathbf{z}^{(2)}_{t-1} + \sigma_2 S_{2t}).$$

In general, not only $\mathbf{x}^{(1)}_{At-1}$ and $\mathbf{x}^{(2)}_{Bt-1}$, but also $\mathbf{z}^{(1)}_{t-1}$ and $\mathbf{z}^{(2)}_{t-1}$ may differ with respect to the contained risk factors as well as with respect to the numbers of factors. Moreover, the random effects S_{1t} and S_{2t} may be different since they may exhibit different realizations in different sectors. Given the observable risk factors $\mathbf{x}^{(1)}_{At-1}, \mathbf{x}^{(2)}_{Bt-1}, \mathbf{z}^{(1)}_{t-1}$ and $\mathbf{z}^{(2)}_{t-1}$ in each segment the default correlation depends on the joint distribution of the random effects and the joint default frequency can be written as

$$p_{ABt}(\mathbf{x}^{(1)}_{At-1}, \mathbf{x}^{(2)}_{Bt-1}, \mathbf{z}^{(1)}_{t-1}, \mathbf{z}^{(2)}_{t-1}) =$$
$$\int_{-\infty}^{+\infty} \int_{-\infty}^{+\infty} \tilde{L}_{At}(\sigma_1 S_{1t}) \tilde{L}_{Bt}(\sigma_2 S_{2t}) \varphi(S_{1t}, S_{2t}) dS_{1t} dS_{2t}$$

where $\varphi(S_{1t}, S_{2t})$ denotes the joint density function of the random effects, $\tilde{L}_{At}(\sigma_1 S_{1t})$ stands for $L(\beta_{01} + \boldsymbol{\beta}^{\mathsf{T}}_1 \mathbf{x}^{(1)}_{At-1} + \boldsymbol{\gamma}^{\mathsf{T}}_1 \mathbf{z}^{(1)}_{t-1} + \sigma_1 S_{1t})$, while $\tilde{L}_{Bt}(\sigma_2 S_{2t})$ represents $L(\beta_{02} + \boldsymbol{\beta}^{\mathsf{T}}_2 \mathbf{x}^{(2)}_{Bt-1} + \boldsymbol{\gamma}^{\mathsf{T}}_2 \mathbf{z}^{(2)}_{t-1} + \sigma_2 S_{2t})$.

If the random effects of the two different sectors are uncorrelated, i.e. if their correlation ρ_{12} for all t equals zero, then this joint default frequency for two borrowers in different segments reduces to the product of the marginal default frequencies, thus rendering their default events uncorrelated. The same holds if at least one of two sectors is not exposed to a random effect at all, that is, if its σ due to the factor equals zero. Thus, for sector analysis, it firstly can be tested if σ_1 and/or σ_2 equal zero as described above. If both hypotheses are rejected the empirical Bayes estimates for the realizations of the random effects can be computed for each year. Using the empirical correlation coefficient of the two time series a standard statistical test of the hypothesis

$$H_0 : \rho_{12} = 0$$

can be conducted. If the null hypothesis cannot be rejected, the random effects, and thus the default events of different sectors may be uncorrelated, thus supporting a sector implementation at the level of confidence used for the test. Thus, for pre-specified sectors the condition of uncorrelated random effects can be tested as described.

14.3 A Sector-Specific Approach: The Seemingly Unrelated Regressions (SUR) Model

In this section we describe a sector-specific approach using sector-specific default rates. For the sector analysis considered in this section we do not use firm-specific information. So in the following the term $\beta^\top \mathbf{x}_{At-1}$ in model (14.9) is not relevant. However, it should be noted that in general, rating information is important and should be used to calculate forecasts of the default probabilities. For sector analysis the focus is on investigating the dependencies between different sectors. For these dependencies common risk factors are responsible. Identifying the common risk factors and taking them into account, the remaining correlation should be negligible. From (14.6) we have

$$p_{At}^S(\mathbf{z}_{t-1}) = L(\beta_0 + \gamma^\top \mathbf{z}_{t-1} + \sigma_k S_{kt}). \tag{14.10}$$

Since we do not take into account firm-specific risk factors, all obligors in a sector are treated as homogenous with the same time-varying average default frequency $p_{At}^S(\mathbf{z}_{t-1})$. (14.10) can be rewritten as

$$ln\frac{p_{At}^S(\mathbf{z}_{t-1})}{1 - p_{At}^S(\mathbf{z}_{t-1})} = \beta_0 + \gamma^\top \mathbf{z}_{t-1} + \sigma_k S_{kt} \tag{14.11}$$

where the left-hand side represents the logits of the default frequencies. Now let us consider the default rates for a given sector. The default rates are given by

$$p_t = \frac{\sum_{A=1}^{K_t} \mathbf{1}_{At}}{K_t} \text{ for } t = 1, \ldots, T.$$

In analogy to (14.11) we can define the model

$$ln\frac{p_t}{1 - p_t} = \beta_0 + \gamma^\top \mathbf{z}_{t-1} + \sigma_k S_{kt} + \varepsilon_t^* \tag{14.12}$$

with the error term

$$\varepsilon_t^* = ln\frac{p_t}{1 - p_t} - ln\frac{p_{At}^S(\mathbf{z}_{t-1})}{1 - p_{At}^S(\mathbf{z}_{t-1})} \text{ for } t = 1, \ldots, T$$

so that model (14.12) follows from model (14.11) when only default rates and not individual default frequencies, are considered. Since the error terms are nonlinear transformations of the default rates, we only have "asymptotic" formulas for the moments. The asymptotic variance is given by

$$\text{var}(\varepsilon_t^*) = \frac{1}{K_t p_{At}^S(\mathbf{z}_{t-1})[1 - p_{At}^S(\mathbf{z}_{t-1})]}. \tag{14.13}$$

From (14.13) it can be seen that the errors ε_t^* are heteroscedastic. Combining the "error terms" $\sigma_k S_{kt}$ and ε_t^* to

$$\varepsilon_t = \sigma_k S_{kt} + \varepsilon_t^*, \tag{14.14}$$

(14.12) is a linear regression model

$$y_t = \beta_0 + \gamma^\mathsf{T} z_{t-1} + \varepsilon_t^* \text{ for } t = 1, \ldots, T, \tag{14.15}$$

where $y_t = ln\frac{p_t}{1-p_t}$. Now we extend the approach to N sectors. The sectors have to be defined in advance, i.e. industry or country by industry. For each sector we have a linear regression model for the transformed default rates like (14.15) where the risk factors z_{t-1} may differ between sectors. We obtain a system of N individual regression models that can be written in matrix notation

$$\mathbf{y}_k = \mathbf{Z}_k \gamma_k + \varepsilon_k \text{ for } k = 1, \ldots, N.$$

$\mathbf{y}_k$ is the $(T \times 1)$ vector of transformed default rates for sector k, $\mathbf{Z}_k$ is a suitably defined $(T \times L_k)$ matrix with the values of the risk factors and ones in the first column, and ε_k is the $(T \times 1)$ vector of random errors (14.14). T denotes the number of time series observations. γ_k is a $(L_k \times 1)$ vector of regression parameters.

The seemingly unrelated regression (SUR) is an extension of the ordinary least squares regression (OLS). Rather than estimating parameters for each of the N models individually, SUR includes information of contemporaneous correlation into the estimates. Combining N OLS equations into one big model yields

$$\begin{bmatrix} \mathbf{y}_1 \\ \mathbf{y}_2 \\ \cdots \\ \mathbf{y}_N \end{bmatrix} = \begin{bmatrix} \mathbf{Z}_1 & 0 & \cdots & 0 \\ 0 & \mathbf{Z}_2 & \cdots & 0 \\ \cdots & \cdots & \cdots & \cdots \\ 0 & 0 & \cdots & \mathbf{Z}_N \end{bmatrix} \begin{bmatrix} \gamma_1 \\ \gamma_2 \\ \cdots \\ \gamma_N \end{bmatrix} + \begin{bmatrix} \varepsilon_1 \\ \varepsilon_2 \\ \cdots \\ \varepsilon_N \end{bmatrix} \tag{14.16}$$

or alternatively

$$\mathbf{y} = \mathbf{Z}\gamma + \varepsilon$$

where the definitions of $\mathbf{y}, \mathbf{Z}, \gamma$ and ε are obvious from (14.16) and where their dimensions are respectively $(NT \times 1), (NT \times L), (L \times 1)$ and $(NT \times 1)$ with $L = \sum_{k=1}^N L_k$. Let ε_{kt} be the error for the k-th equation in the t-th time period. We make the assumption of absence of correlation between residuals over different points in time:

$$\mathrm{cov}(\varepsilon_{kt}, \varepsilon_{ju}) = \mathbb{E}(\varepsilon_{kt} \cdot \varepsilon_{ju}) = \begin{cases} \sigma_{kj} & \text{for } t = u \\ 0 & \text{otherwise} \end{cases}.$$

The covariance stucture of the segments' regression residuals is then given by the $(N \times N)$ covariance matrix Σ:

$$\Sigma = \begin{bmatrix} \sigma_{11} & \cdots & \sigma_{1N} \\ \cdots & & \cdots \\ \sigma_{N1} & \cdots & \sigma_{NN} \end{bmatrix}.$$

If Σ is a diagonal matrix there is no gain in efficiency using SUR instead of OLS (see [8, Ch. 11]. The complete covariance structure for the residuals of all observations is then given by

$$\text{cov}(\varepsilon) = \mathbb{E}(\varepsilon\varepsilon^\top) = \Sigma \otimes \mathbf{I}_T$$

where $\mathbf{I}_T$ is the $(T \times T)$ identity matrix and $\otimes$ denotes the Kronecker product. In practice the variances and covariances σ_{kj} are unknown and have to be estimated consistently in the first step from the OLS regressions. Let $\hat{\beta}_k^{OLS}$ denote the estimated OLS coefficients for the k-th model:

$$\hat{\beta}_k^{OLS} = (\mathbf{Z}_k^\top\mathbf{Z}_k)^{-1}\mathbf{Z}_k^\top\mathbf{y}_k$$

are computed for the N segments. Then, in the second step, we compute the OLS residuals

$$\hat{\varepsilon}_k = \mathbf{y}_k - \mathbf{Z}_k\hat{\beta}_k^{OLS}$$

and the (consistent) estimator $\hat{\Sigma}$ has elements given by

$$\hat{\sigma}_{kj} = \frac{\hat{\varepsilon}_k^\top\hat{\varepsilon}_j}{T} \qquad k,j = 1,\ldots,N.$$

Finally, the SUR estimator is

$$\hat{\beta} = \left[\mathbf{Z}^\top(\hat{\Sigma}^{-1} \otimes \mathbf{I}_T)\mathbf{Z}\right]^{-1}\mathbf{Z}^\top(\hat{\Sigma}^{-1} \otimes \mathbf{I}_T)\mathbf{y}$$

and an estimator of its covariance matrix is given by

$$\text{cov}(\hat{\beta}) = [\mathbf{Z}^\top(\hat{\Sigma}^{-1} \otimes \mathbf{I}_T)\mathbf{Z}]^{-1}.$$

It should be noted that there are further estimators of Σ and further estimation procedures for the regression parameters, for example maximum likelihood estimation. For more details see, for example [7, Ch. 17.2]. For an extension of the SUR model to the case with unequal number of observations for the regression equations see, for example [1]. For sector analysis it is of interest to test whether Σ is a diagonal matrix, i.e different sectors are independent. A test suggested by [4] is the Lagrange multiplier statistic,

$$\lambda_{LM} = T\sum_{k=2}^{N}\sum_{j=1}^{k-1}r_{kj}^2,$$

where r_{kj}^2 is the estimated correlation based on the residuals $\hat{\varepsilon}_k$. Asymptotically, this statistic is distributed as chi-squared with $N(N-1)/2$ degrees of freedom.

14.4 Example for SUR-based Sector Analysis

The following example shows how SUR-based sector derivation can improve portfolio risk analysis under three aspects:

- The background factors responsible for systematic risk are identified as macroeconomic variables.
- Static averages of default probabilities are replaced by dynamic forecasts.
- Sector standard deviations can be inferred from forecast errors and therefore provide the measure for default rate uncertainty.

We use the term "segments" to identify industries within a country. The aim of the analysis is to group segments with correlated defaults into CreditRisk$^+$ sectors such that defaults between segments of different sectors remain uncorrelated.

14.4.1 Data description

The regressions were performed on a dataset of insolvency rates including a range of countries and industries. Table 14.1 provides an overview of the time series data underlying the study. Regressors have been taken from a wide range of publicly available macroeconomic indices for the areas capital market, consumption, employment, foreign investments, governmental activity, income, prices and value added to capture systematic influences on default rates as completely as possible. A more detailed description of the data and data validation can be found in [3].

14.4.2 Default rate dynamics

In a typical parameter setting, default probabilities are calculated from historical averages of default rates. Scaling those default probabilities with a random variable of unit expectation as is done within CreditRisk$^+$ can be interpreted as forecasting that default probabilities over the next period will be equal to their historical average. Under this interpretation the standard deviation is a measure for the error associated with this forecast. In the model documentation it is suggested that the sectors' standard deviations be set to one, reflecting the empirical result that the historical volatility of default rates has been roughly as high as their historical average. This average-of-the-past approach to forecasting generally yields very poor results for any particular year, indicating a high degree of model risk. Alternatively, the forecasts from the SUR analysis can be used for scaling factors of default rates, given the current values of the corresponding macroeconomic risk drivers.

Segment	Time of observation	Number of quarterly defaults				Total number of companies			
		Min	Max	Mean	Std dev	Min	Max	Mean	Std dev
Canada Construction	1990–1999	327	595	439.25	71.54	106641	109848	108232.27	1047.73
Canada Manufacturing	1990–1999	156	335	244.98	37.05	62330	73843	67531.87	4293.19
⋯									
Germany Construction	1981–1998	283	1248	689.49	248.55	187242	245928	207225.61	18699.33
Germany Manufacturing	1981–1998	303	705	531.64	110.22	250949	297292	282836.49	17025.76
⋯									
GB Construction	1980–1999	56	1353	506.65	324.21	167575	264032	210165.88	28263.39
GB Manufacturing	1980–1999	26	415	163.10	105.72	144630	169656	157538.58	6487.98
⋯									
USA Agriculture	1984–1999	278	1015	587.45	166.83	60247	123070	92787.05	19333.36
USA Commerce	1984–1999	2542	6812	4556.33	863.22	1450697	1525844	1485535.18	24105.37
USA Construction	1984–1999	1509	3348	2206.09	510.55	544257	699472	605435.82	45999.53
USA Manufacturing	1984–1999	495	2203	1210.95	328.75	312855	357176	339423.73	16697.92
USA Services	1984–1999	3053	8285	5422.59	1307.82	1796207	2338809	2007393.18	171772.74

Table 14.1. Data description (selection)

14.4.3 Sector analysis

The CreditRisk$^+$ assumption of independence between sectors poses a constraint for the implementation of sector analysis with regard to the way sectors are defined. When different sectors are constructed for groups of obligors whose default rates show a high degree of correlation – for example because they all depend on the same risk factors, such as industries within a country – portfolio losses will be underestimated. For other approaches to tackling this problem also see Chapters 9, 10 and 15. The alternative to sector analysis is the one-sector approach where all obligors are attributed to the same sector, which leads to positive correlations between default events of all obligors. This conservative approach, however, obstructs potential risk concentrations and the benefits of diversifying across different systematic risk sources. Table 14.2 shows the correlation matrices of default rates compared with the correlation of SUR residuals for a limited set of data.

Residuals generally show a much lower degree of correlation than default rates, reflecting the fact that once the factors influencing default rate movements are identified and their values are known, a lot of "co-movement" formerly perceived as correlation, is already explained. From this result sectors can be constructed by grouping those segments where correlation of residuals is significant and separating groups of segments with insignificant correlations between each other. The shaded areas show the two sectors "North America" and "Western Europe" resulting from the test procedure for conditional independence after [4] of segments' default rates as described in the previous section. For all other arrangements of segments that were tested, like one sector for each country and/or industry the null hypothesis of conditional independence of the sectors, i.e. insignificant correlation of the residuals, is rejected. For those arrangements, correlations between the residuals of the segments persist, even after the systematic risk factors have been accounted for in the default rates. This suggests country-specific systematic risk factors not identified by the model. Sector analysis based on countries or even industries alone would therefore underestimate systematic risk in the portfolio. From the SUR analysis, estimates for the development of segment-specific default rates, i.e. the systematic component within individual default probabilities, are obtained. The associated forecast standard errors are a measure for default probability uncertainty given the current values of the corresponding systematic risk drivers and can be used to derive the sectors' standard deviations.

Correlation of residuals

	Canada Constr.	Canada Manuf.	USA Constr.	USA Manuf.	GB Constr.	GB Manuf.	Sweden Constr.	Sweden Manuf.	France Constr.	France Manuf.	Germany Constr.	Germany Manuf.
Canada Construction	1.0											
Canada Manufacturing	0.49	1.0										
USA Construction	0.49	0.45	1.0									
USA Manufacturing	0.26	0.64	0.42	1.0								
Great Britain Construction	0.17	0.04	-0.09	0.07	1.0							
Great Britain Manufacturing	0.18	0.19	-0.02	0.24	0.77	1.0						
Sweden Construction	0.24	0.01	0.11	0.19	0.45	0.52	1.0					
Sweden Manufacturing	0.13	0.26	0.12	0.27	0.18	0.45	0.15	1.0				
France Construction	0.28	0.01	0.05	0.14	0.14	-0.01	0.18	0.01	1.0			
France Manufacturing	0.42	0.30	-0.07	0.19	0.40	0.40	0.33	-0.02	0.25	1.0		
Germany Construction	0.16	0.05	0.16	-0.15	-0.14	0.03	0.08	0.06	-0.03	-0.10	1.0	
Germany Manufacturing	0.22	-0.04	-0.30	-0.04	-0.02	-0.07	0.02	-0.29	0.18	0.31	0.09	1.0

Correlation of default rates

	Canada Constr.	Canada Manuf.	USA Constr.	USA Manuf.	GB Constr.	GB Manuf.	Sweden Constr.	Sweden Manuf.	France Constr.	France Manuf.	Germany Constr.	Germany Manuf.
Canada Construction	1.0											
Canada Manufacturing	0.58	1.0										
USA Construction	0.56	0.61	1.0									
USA MManufacturing	0.36	0.86	0.79	1.0								
Great Britain Construction	0.39	0.63	0.68	0.78	1.0							
Great Britain Manufacturing	0.27	0.71	0.71	0.90	0.90	1.0						
Sweden Construction	0.43	0.59	0.67	0.76	0.90	0.82	1.0					
Sweden Manufacturing	0.43	0.69	0.72	0.87	0.88	0.92	0.90	1.0				
France Construction	0.56	0.21	0.44	0.19	0.47	0.24	0.50	0.29	1.0			
France Manufacturing	0.43	0.44	0.52	0.54	0.80	0.63	0.83	0.65	0.75	1.0		
Germany Construction	-0.09	-0.60	-0.39	-0.77	-0.66	-0.83	-0.61	-0.76	0.01	-0.45	1.0	
Germany Manufacturing	-0.02	-0.55	-0.46	-0.72	-0.45	-0.73	-0.36	-0.62	0.18	-0.11	0.86	1.0

Table 14.2. Correlation coefficients of default rates and SUR residuals

14.4.4 Example

The following examples illustrate how sector analysis based on SUR modelling improves the perception of risk within a sample portfolio of credits.[1] The example is constructed to reflect a typical situation where an institution's measure of counterparty risk is largely based on the long-term assessment of individual risks, condensed into rating categories ("through the cycle rating"). The expected default rate over the reference period is then typically assumed to be the historical average of default rates for the same rating, averaging over previous states of the economy. In the example the systematic component of risk is now included into the analysis by multiplying the obligors' historical default rates by a scaling factor. This factor reflects the segment-specific forecasted development of default rates relative to the historical average. The uncertainty associated with the resulting default probability is expressed in the forecast standard error. Table 14.3 shows scaling factors and forecast standard errors for the segments used in the example and the resulting sector standard deviations. The 2003 forecast for the segment Germany Construction, for example, predicts a 78% increase in the one-year default rate compared with the historical average. The associated forecast standard error is 16.43% of the forecasted rate, which is significantly lower than the 100% associated with the average-of-the-past forecast. CreditRisk[+] requires one standard deviation for each random variable representing a sector. This number can be calculated in different ways, for example by averaging over the standard deviations of the default rates of obligors attributed to the sector, as suggested in the model documentation. This approach implies perfect correlation between default rates within the sector. In our example we calculated the sector standard deviations from the forecast standard errors of the SUR residuals. The last column of Table 14.3 displays the sectors' forecast standard errors in percent of the average forecasted default rate, i.e. the average of the forecasted default rate of the segments composing the sector.

In Table 14.4 we compare the perceptions of the risk distribution within the portfolio under the one-sector approach with SUR sector analysis. A segment's contribution to risk is measured as the decrease in the 99.9% percentile of the portfolio loss distribution when all obligors from the considered segment are removed from the portfolio.[2] In the example the inclusion of SUR sector analysis reveals a higher than average risk contribution of the Germany Construction segment due to the differences in expected default rates.

---

[1] The sample portfolio consists of 800 unit size exposures, which are distributed evenly over the segments used for the example. The probability of default for all exposures is 1%. The loss given default is set to 100%.

[2] Here the overall reduction is normalized to 100%. Note that the higher quantiles in the model are not subadditive such that the sum of all reductions would be less then the chosen quantile. Here we are only interested in the relative reductions, the results and conclusions are valid for other measures of risk contribution. For a detailed discussion of risk measures and risk contributions see Chapter 3.

Segment	Sector	Scaling factor according to forecast	Forecast standard error in % of forecasted default rate	Sector standard deviation
Canada Construction		0.86	11.76%	14.06%
Canada Manufacturing	North	0.87	12.41%	
USA Construction	America	0.98	16.23%	
USA Manufacturing		1.29	15.82%	
Germany Construction		1.78	16.43%	20.51%
Germany Manufacturing	Western	0.86	13.35%	
GB Construction	Europe	1.01	30.57%	
GB Manufacturing		1.37	21.96%	

Table 14.3. Segment-specific scaling factors and forecast standard error deviations

The previous contribution of 12.5% under the one-sector approach increases to 23.99% when SUR sector analysis is included. Germany's combined share of risk contribution rises from 25% to 35.58%.

Segment	Static default rates, one sector	Dynamic default rates, SUR sectors Risk contribution	Country	Risk contribution	Sector	Risk contribution
Canada Constr.	12.50%	6.95%	Canada	13.99%		
Canada Manuf.	12.50%	7.03%			North	
USA Constr.	12.50%	7.92%	USA	18.36%	America	32.34%
USA Manuf.	12.50%	10.43%				
Germany Constr.	12.50%	23.99%	Germany	35.58%		
Germany Manuf.	12.50%	11.59%			Western	
GB Constr.	12.50%	13.61%	GB	32.08%	Europe	67.66%
GB Manuf.	12.50%	18.46%				

Table 14.4. Risk distributions, diversified portfolio

For the second set of calculations, exposure in the sample portfolio is shifted from the segment USA Manufacturing to Great Britain Construction in order to generate a concentration risk scenario.

As shown in Table 14.5 under the one-sector implementation, risk is simply distributed according to the exposures. SUR sector analysis reveals the added concentration risk resulting from increased dependency on fewer correlated risk factors in the sector Western Europe. The segment Great Britain Construction now includes 25% of the exposure but concentrates 27.66% of the risk contributions. Note that this concentration effect manifests although

Segment	Static default rates, one sector	Dynamic default rates, SUR sectors	Country		Sector	
Segment	Risk contribution	Risk contribution	Country	Risk contribution	Sector	Risk contribution
Canada Constr.	12.50%	5.53%	Canada	11.12%		
Canada Manuf.	12.50%	5.59%			North	
USA Constr.	12.50%	6.30%	USA	6.30%	America	17.42%
USA Manuf.	0%	0.00%				
Germany Constr.	12.50%	24.38%	Germany	36.15%		
Germany Manuf.	12.50%	11.78%			Western	
GB Constr.	25%	27.66%	GB	46.42%	Europe	82.58%
GB Manuf.	12.50%	18.76%				

Table 14.5. Risk distributions, concentrated portfolio

the exposures have been moved to a segment with a more favourable expected default rate. Shifting 12.5% of exposure to the Sector Western Europe substantially increases this sector's risk contribution from 67.66% to 82.58%.

14.5 Conclusions

Default rates show a high degree of variability, which can be partially explained by the dynamics of macroeconomic risk factors. The SUR estimation procedure includes information about the correlation structure into the estimated parameters and therefore provides a suitable tool for the estimation and prediction of default rates, which show a high degree of correlation across industries and countries. When the explanatory effect of the SUR model is taken into account, correlations are reduced massively and become insignificant for certain combinations of segments, suggesting independence of default rates conditional on the macroeconomic risk factors. These results can be used to include default rate dynamics and sector analysis into the assessment of portfolio risk within CreditRisk$^+$.

References

1. B.H. Baltagi. *Econometric Analysis of Panel Data*. John Wiley, 2nd edition, 2001.
2. D. M. Bates and J. C. Pinheiro. Approximations to the log-likelihood function in the nonlinear mixed-effects model. *Journal of Computational and Graphical Statistics*, 4:12–35, 1995.

3. L. Boegelein, A. Hamerle, R. Rauhmeier, and H. Scheule. Modelling default rate dynamics in the CreditRisk$^+$ framework. *Risk*, 15(10):S24-S28, 2002.

4. T. Breusch and A. Pagan. The Lagrange multiplier test and its applications to model specification in econometrics. *Review of Econometric Studies*, 47:239–253, 1980.

5. Credit Suisse Financial Products. CreditRisk$^+$: A credit risk management framework. London, 1997. Available at http://www.csfb.com/creditrisk.

6. M. B. Gordy. A comparative anatomy of credit risk models. *Journal of Banking and Finance*, 24:119-149, 2000.

7. W. Greene. *Econometric Analysis*. Macmillan, 2nd edition, 1993.

8. G. Judge, W. Griffiths, R. Hill, H. Lütkepohl and T. Lee. *Introduction to the Theory and Practice of Econometrics*. John Wiley, 2nd edition, 1988.

15

Estimation of Sector Weights from Real-World Data

Michael Lesko, Frank Schlottmann and Stephan Vorgrimler

Summary. We discuss four different approaches to the estimation of sector weights for the CreditRisk$^+$ model from German real-world data. Using a sample loan portfolio, we compare these approaches in terms of the resulting unexpected loss risk figures.

Introduction

The CreditRisk$^+$ model in its general sector analysis formulation assumes the presence of common independent systematic risk factors that influence the default rates of all obligors in a given credit portfolio in order to incorporate default dependencies. The levels of implicit default correlation depend on volatilities of the risk factors and on the factor loadings, also called risk factor weights, of the obligors. We discuss different estimation methods for these model parameters from historical macro-economic data and investigate the empirical differences concerning the resulting loss distribution and the risk measures (e.g. unexpected loss/credit-value-at-risk) using a sample portfolio.

To avoid unnecessary complexity, we assume a one-period calculation for a given portfolio of K corporate obligors using N systematic risk factors (sectors) and an idiosyncratic risk factor for a time period of one year as introduced in Chapter 2. For multiple-period calculations (hold-to-maturity) see [3].

In the following sections we assume given parameters α_k for each random scalar factor S_k, such that the respective second parameter β_k of the gamma distribution, see Chapter 2, for sector k is then determined by

$$\beta_k := \frac{1}{\alpha_k}.$$

When parameterizing from observable risk factors one could choose

$$\alpha_k := \frac{1}{\text{var}[S_k]}.$$

(recall that $\mathbb{E}[S_k] \equiv 1$). If the risk factors are not directly observable the choice of the α_k values leads to N additional degrees of freedom (since there are N systematic risk factors). This topic is addressed later in this chapter.

15.1 Methods for the Estimation of the Risk Factor Weights

15.1.1 Plain vanilla industry sectors as risk factors

An obvious approach that is motivated by the choice of countries as risk factors in [3, A.7] is to introduce a separate risk factor for each industry within a given country. This is a common approach in real-world applications that allows a straightforward choice of the risk factor weights: each obligor A is allocated to the risk factor k_A that matches his industry classification best, i.e.

$$w_{Ak_A} := 100\%, \qquad \forall k \neq k_A : w_{Ak} := 0\%.$$

Since the respective industry classification is quite easy to obtain either from the obligor's data stored in the usual databases of financial lending institutions or from publicly available sources, the simplicity of this approach with respect to an IT-based data providing process seems very appealing.

However, the CreditRisk$^+$ model assumes the risk factors to be independent, and within most countries, this will most probably not be a valid assumption for the different industries. For instance, we have analysed macroeconomic data provided by the German federal statistical office (Statistisches Bundesamt) and found correlation coefficients of about 80% between pairs of the annual historical default rates within different West German industries in the years 1962–1993.[1] In the empirical section the consequences of neglecting these industry default rate correlations will be analysed. Therefore, such a simplified approach might lead to significant underestimation of unexpected losses. A sample result illustrating this fact is presented later in this chapter.

We now discuss different alternatives for the estimation of risk factor weights that incorporate correlations between industries.

15.1.2 The approach of Bürgisser, Kurth, Wagner and Wolf

Bürgisser et al. [2] developed an approach that incorporates correlations between the risk factors, see Chapter 9. The authors calculate the standard deviation of total portfolio losses assuming correlated risk factors (denoted by $\sigma_{X,\text{corr}}$ below) using the following formula (cf. [2, Eq. (12)] or (9.11)):

---

[1] These coefficients are estimated in a simplified manner by measuring the default rate correlations. A more refined estimation procedure would consist of estimating correlations of forecast errors. In this case the estimated correlations are lower, nonetheless they still have values significantly above zero.

$$\sigma^2_{X,\text{corr}} = \sum_k \sigma^2_k EL^2_k + \sum_{k,l \wedge k \neq l} \text{corr}(S_k, S_l) \sigma_k \sigma_l EL_k EL_l + \sum_A p_A \nu^2_A \qquad (15.1)$$

where $\text{corr}(S_k, S_l)$ denotes the correlation between two distinct sector variables.

The total portfolio loss distribution is then calculated using the standard CreditRisk$^+$ algorithm for the given portfolio data under the assumption of one systematic risk factor $\widehat{S_1}$. The parameters of the gamma distribution for $\widehat{S_1}$ are calibrated such that the resulting standard deviation of portfolio loss σ_X obtained by application of the standard CreditRisk$^+$ model matches the standard deviation specified in (15.1):

$$\sigma_X = \sigma_{X,\text{corr}}.$$

After calibrating $\widehat{S_1}$, the desired unexpected loss figures can be computed from the output of the standard CreditRisk$^+$ algorithm, e.g. the unexpected loss at the 99th percentile level (denoted by $UL_{0.99}$).

For a sample portfolio, Bürgisser et al. reported a higher standard deviation of about 25% concerning the total portfolio loss, and the corresponding unexpected loss at the 99th percentile level was about 20% higher compared with the results obtained by using the approach from Section 15.1.1.

This approach is very interesting for real-world applications, since it directly incorporates the observed correlation parameters into a transparent and easy-to-implement calibration process, and no additional information or modelling assumption is required. However, from the perspective of the model-theoretic assumptions, it has to be kept in mind that the resulting distribution of total loss from the given portfolio is not obtained under the assumption of correlated risk factors, but merely a distribution that has the same standard deviation as the total losses that result from the portfolio if the risk factors are correlated. In other words, this avoids the problem of using correlated risk factors in the CreditRisk$^+$ calculations in an appealing way, but e.g. the percentile figures calculated by the calibrated one risk factor approach might not be very precise, and nothing is known about the true level of unexpected loss.

15.1.3 The approach of Lesko, Schlottmann and Vorgrimler

In Lesko et al. [6] we have proposed the estimation of risk factor weights by principal component analysis (PCA), see e.g. [7] for a comprehensive introduction of this method. The basis for this approach are abstract, independent risk factors (as assumed by standard CreditRisk$^+$) which explain the evolution of default rates within the different industries. All obligors that have the same industry classification are assumed to share the same risk factor weight within this industry.

The first goal for the risk factor weight estimation under these assumptions is to identify appropriate independent risk factors. Secondly, the influence of these risk factors on the respective industry (and therefore, on the obligors that belong to this industry) has to be estimated. The PCA method provides a framework that allows the simultaneous identification of independent risk factors and the calculation of the corresponding risk factor weights. Using this method, the abstract risk factors and their influences on the respective industry are obtained directly from the volatilities between default rates in the considered industries and the correlations between these default rates. In general, the number of systematic risk factors can be chosen arbitrarily between 1 and the number of given industries, which we denote by J below. When choosing this parameter, it should be kept in mind that for a higher number of abstract risk factors, the accuracy of the resulting risk factor weights is also higher in terms of the implicit correlation structure represented by these weights.[2]

Our algorithm works basically as follows:

Algorithm 1. *PCA approach for CreditRisk*[+]

Input: *positive semidefinite covariance matrix M_C of dimension $J \times J$*

1:   Calculate the eigenvalues and the corresponding eigenvectors
     (column vectors) of the covariance matrix M_C
2:   Build diagonal matrix M_D containing the eigenvalues
3:   Build matrix M_E containing the eigenvectors as columns
4:   Choose the number N of systematic risk factors
5:   Truncate the matrix of eigenvectors to the number of systematic
     risk factors by deletion of the eigenvector columns having the
     smallest associated eigenvalues, also delete the columns and rows of
     these eigenvalues from M_D (final result of this operation: $J \times N$
     matrix $\widetilde{M_E}$ and $N \times N$ matrix $\widetilde{M_D}$)
6:   Incorporate idiosyncratic risk factor by adding a column to $\widetilde{M_E}$ and
     a column of zeros as well as a row of zeros to $\widetilde{M_D}$ (result:
     $J \times (N+1)$ matrix $\widehat{M_E}$, $(N+1) \times (N+1)$ matrix $\widehat{M_D}$)

Output: *risk factor weight matrix $\widehat{M_E}$, risk factor variances as diagonal elements of $\widehat{M_D}$*

The input covariance matrix M_C is estimated from the given correlations and the given volatilities of the industry default rates. The diagonal matrix M_D contains the eigenvalues of M_C, and each column of the matrix M_E contains the eigenvector associated with the eigenvalue in the corresponding column of M_D. These matrices satisfy the property

$$M_C = M_E \times M_D \times (M_E)^T \tag{15.2}$$

---

[2] Note that the numerical stability of the calculations using the Panjer recursion in CreditRisk[+] has to be watched carefully if the number of abstract risk factors is raised, cf. [8, 611 ff.]. When using a large number of abstract risk factors, one of the numerically more stable calculation schemes presented in the second part of this volume is preferable.

where $(M_E)^T$ is the transposed matrix of eigenvectors and $\times$ denotes the usual product of matrices.

The number of systematic abstract risk factors N is determined by analysing the eigenvalues of M_C first and choosing N depending on these eigenvalues. For instance, a single large eigenvalue indicates that only one systematic risk factor explains most of the systematic risk in the economy specified by the given industries' data.

In the next step (cf. line 5) the risk factor allocation based on the chosen number of systematic risk factors is derived by deletion of the columns from the matrix of eigenvectors $\widehat{M_E}$ that correspond to the $J - N$ smallest eigenvalues in M_D. These eigenvalues are also removed from M_D by deletion of the respective row and column vectors.

After this construction of the matrices $\widetilde{M_D}$ and $\widetilde{M_E}$ the resulting covariance matrix $\widetilde{M_C}$ can be checked in analogy to (15.2) to validate the calculated risk factor allocation parameters:

$$\widetilde{M_C} = \widetilde{M_E} \times \widetilde{M_D} \times (\widetilde{M_E})^T. \tag{15.3}$$

In line 6, an idiosyncratic risk factor is included by adding a column to the truncated matrix of eigenvectors $\widetilde{M_E}$ such that the sum for each row of the resulting matrix $\widehat{M_E}$ equals 1. Moreover, an additional row of zeros and an additional column of zeros for the specific sector have to be added to the truncated matrix $\widetilde{M_D}$. Then, the final output of Algorithm 1 is the resulting risk factor weight matrix $\widehat{M_E}$ (each row i contains the risk factor weights of an obligor belonging to industry $i \in \{1, \ldots, J\}$) and the matrix $\widehat{M_D}$ containing the volatilities of the risk factors as diagonal elements.

We consider the following sample correlation matrix in our calculations below:

Industry	Agricult.	Manufact.	Constr.	Trade	Transport.	Services
Agriculture	100%	70%	95%	94%	50%	96%
Manufacturing	70%	100%	72%	84%	90%	78%
Construction	95%	72%	100%	95%	45%	98%
Trade	94%	84%	95%	100%	64%	96%
Transportation	50%	90%	45%	64%	100%	51%
Services	96%	78%	98%	96%	51%	100%

Table 15.1. Sample correlations between industry sectors in Germany

The values in Table 15.1 reflect typical correlations between historical default rates of selected industries in West Germany. Obviously, the historical default rates are highly correlated, which underlines the criticism concerning the approach described in Section 15.1.1 that neglects the correlation between the risk factors.

Assuming three abstract risk factors, an application of our approach based on principal component analysis on a small sample portfolio consisting of three obligors yields the results shown in Table 15.2.

A	Industry	p_A	σ_A	ν_A	w_{A0}	w_{A1}	w_{A2}	w_{A3}
1	Construction	1.2%	0.9%	1,200,000	5.0%	67.7%	22.4%	4.9%
2	Trade	0.4%	0.3%	3,200,000	23.4%	70.9%	6.1%	-0.4%
3	Construction	2.8%	2.0%	400,000	5.0%	67.7%	22.4%	4.9%

Table 15.2. Input data for a sample portfolio using PCA output

There are different possible variants of our approach, since the given volatility of the historical default rates within the industries can be respected either by the calibration of the risk factor weights or by the calibration of the abstract risk factors' parameters α_k, β_k. Both variants lead to the same standard deviation of total losses from the portfolio, but the resulting unexpected loss figures are different in general, cf. the corresponding remarks by Gordy [5]. We followed the first path and assumed fixed parameters of the S_k random scalar factors in our above example, which lead to the shown calibration of the risk factor weights.

15.1.4 Single risk factor for the whole economy

Compared with the second and the third approach, which are motivated by mathematical considerations, the following approach is suggested by empirical observation of historical macro-economic default rates. As already mentioned in the above sections, the statistical analysis of historical default rates within German industries yields correlations of about 80% and above. Using the CreditRisk$^+$ approach of abstract background risk factors, this suggests that there is only one systematic background risk factor (e.g. the general state of the German economy) that influences the default rates of all German obligors uniformly.[3] Under this assumption, a single risk factor approach seems appropriate, the risk factor weights of all obligors are assumed to be $w_{A1} \equiv 100\%$, and the free parameter of the single risk factor is obtained e.g. by calibration of β_1 according to an observable variance of the German all-industries corporate default rate.

This is a straightforward approach that is somewhat similar to the approach described in Section 15.1.1, but this time, the implicit error moves the unexpected loss results into the opposite direction: the above choice of the risk

---

[3] The empirical data considered in the above analysis contains only defaults of corporate obligors. Nonetheless, the single risk factor approach is also a straightforward approach for retail portfolios or mixed corporate/retail portfolios consisting solely of obligors that are assumed to depend on a single systematic background risk factor.

factor weights and the calculation of a standard deviation for the single risk factor is similar to an implicit correlation between the default rates of 100% for each pair of industries, so this leads to an overestimation of the portfolio risk if the true correlation is lower, and this is e.g. the case for the historical West German data. In other words, this approach ignores possible diversification by investing in different industries. Moreover, if the correlation structure between industries is inhomogeneous, the accuracy of the CreditRisk$^+$ calculations under this approach might not be appropriate.

In the next section, we compare the empirical results for a sample portfolio by application of the different approaches discussed so far.

15.2 Empirical Comparison

We now apply the four approaches of sector weight estimation from real-world data described in the previous sections to a typical German middle-market loan portfolio consisting of 1000 obligors within the six industry sectors shown in Table 15.1. The sum of net exposures of all obligors is approximately 1,000,000,000 EUR, whereas the individual net exposures (ν_A) are between 1000 EUR and 35,000,000 EUR. The mean default probability of each obligor is in the range of $0.1\% \leq p_A \leq 8\%$.

The results of the CreditRisk$^+$ model calculations depending on the chosen risk factor weight estimation approach are shown in Table 15.3. We use EL as an abbreviation for expected loss, SD for the standard deviation of total portfolio loss, and UL_ϵ is again the unexpected loss at the confidence level ϵ.

Approach	Description	EL	SD	$UL_{0.99}$	$UL_{0.999}$
1	Industry sector→risk factor	21,365	15,080	49,370	75,894
2	Bürgisser et al.	21,365	19,441	66,195	104,231
3	PCA	21,365	19,445	68,963	112,450
4	Single systematic sector	21,365	19,844	67,936	107,328

Table 15.3. Comparison of CreditRisk$^+$ results for real-world portfolio

Looking at the different results depending on the chosen method of risk factor weight estimation, the first observation is that the choice of industry sectors as risk factors yields significantly lower measures of risk. This is due to the fact that the correlations between the industry sectors are ignored in this first approach of risk factor weight estimation, while the correlations are in fact significantly positive as shown in our example from Table 15.1. Hence, this approach truly underestimates the risk in our example and also in any other case where the correlations between industry sectors are all non-negative.

In contrast to that, the results of the other three approaches are quite similar concerning the standard deviation as a first risk measure. The standard deviation of the second and third approach is approximately the same, while the corresponding risk measure of the fourth approach is slightly higher. This is a result that is consistent with our considerations in the preceding sections, since the Bürgisser et al. and the PCA approach respect the pairwise correlations between the industry sectors that are < 100% between different sectors in Table 15.1 while the single systematic sector approach ignores this diversification possibility. Hence, the latter approach overestimates the standard deviation.

The results concerning the UL_ϵ values are somewhat more different: the PCA yields the highest values. Comparing the second and the third approach, the result is not surprising, since there is a difference between the parameter estimation based on calibration of the standard deviation by Bürgisser et al. compared with the PCA transformation of the risk factor weights. The more interesting observation is that the UL_ϵ value of the single systematic sector approach (which overestimates the risk in general) is lower than the corresponding UL_ϵ value of the PCA approach.

A more detailed analysis reveals that the difference lies in the distinct risk factor weight estimation procedure. To achieve a higher default rate volatility (and therefore, higher default correlations) an important choice can be made within both the estimation of the risk factor weights and the estimation of the default rate volatilities: either higher default rate volatilities or higher risk factor weights can be chosen.[4] In our sample calculations, we have chosen a standard deviation of $\sigma_1 := 0.72$ for the single systematic risk factor (fourth approach) while we assumed a default rate standard deviation equal to 1 in the PCA approach. The choice of a higher default rate volatility leads to a higher UL_ϵ in many cases, and this effect even increases for $\epsilon \to 1$, see also the corresponding results obtained by Gordy [5]. Hence, this is the main reason for the difference between both approaches concerning the percentiles computed by CreditRisk$^+$.

Finally, to illustrate the differences between the two simplest methods of risk factor estimations presented in the previous sections, the two probability functions of portfolio loss are displayed in Figure 15.1, and their tails are shown in Figure 15.2. All loss values on the x-axis are in millions of euros.

Clearly, the probabilities of large losses are higher for the fourth approach (cf. the tail of the dotted line in Figure 15.2). Hence, the corresponding risk measures are higher than its counterparts calculated using the first approach of risk factor weight estimation (cf. the solid line in Figure 15.2).

15.3 Conclusion

In the previous sections, we have discussed different alternative methods of risk factor weight estimation for CreditRisk$^+$.[5] Summarizing our remarks and

---

[4] This has already been pointed out by Gordy [5].

[5] Cf. also the alternative approach by Boegelein et al. in Chapter 14.

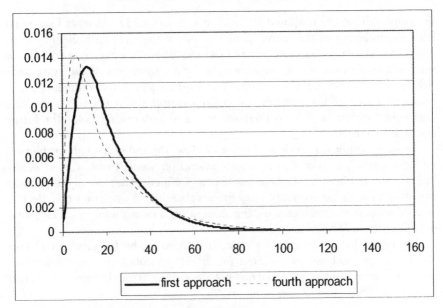

Fig. 15.1. Comparison of probability function of portfolio loss for the first approach (solid line) and the fourth approach (dotted line)

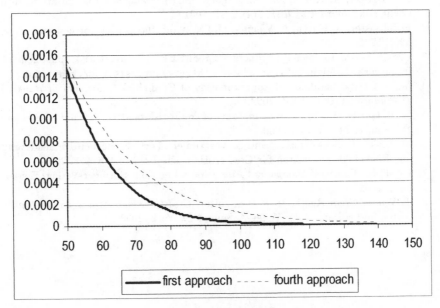

Fig. 15.2. Comparison of the tail probability function of portfolio loss for the first approach (solid line) and the fourth approach (dotted line)

the empirical results obtained for a West German middle-market loan portfolio, the influence of the correlations between industry sectors should not be neglected. Moreover, we have pointed out that there is a degree of freedom in the joint estimation of risk factor weights and parameters of the systematic risk factor random variables, which can lead to significant differences concerning the tails of the resulting portfolio loss probability distributions. This degree of freedom and its consequences are an interesting point for future research activities.

From an application-oriented point of view, the use of a single systematic sector approach of risk factor weight estimation seems to be a good choice in many cases where banks hold portfolios of obligors that mainly depend on the state of a single economy (e.g. firms operating in one country). Besides general economic considerations, this is supported by our analysis of historical default rates in West Germany.

Moreover, from our point of view, the approach by Bürgisser et al. and the PCA approach are well-suited for credit portfolios that are diversified at least over two different countries not depending strongly upon each other concerning their state of the economy.

References

1. P. Bürgisser, A. Kurth, and A. Wagner. Incorporating severity variations into credit risk. *Journal of Risk*, 3(4):5–31, 2001.
2. P. Bürgisser, A. Kurth, A. Wagner, and M. Wolf. Integrating correlations. *Risk*, 12(7):57–60, 1999.
3. Credit Suisse Financial Products. CreditRisk[+]: A credit risk management framework. London, 1997. Available at http://www.csfb.com/creditrisk.
4. M. B. Gordy. Saddlepoint approximation of CreditRisk[+]. *Journal of Banking & Finance*, 26:1335–1353, 2002.
5. M. B. Gordy. A comparative anatomy of credit risk models. *Journal of Banking & Finance*, 24:119–149, 2000.
6. M. Lesko, F. Schlottmann, and S. Vorgrimler. Fortschritte bei der Schätzung von Risikofaktorgewichten für CreditRisk[+]. *Die Bank*, 6:436–441, 2001.
7. I. Jolliffe. *Principal Component Analysis*, 2nd edition. Springer-Verlag, Heidelberg, 2002.
8. T. Wilde. CreditRisk[+]. In S. Das, editor, *Credit Derivatives and Credit Linked Notes*, 2nd edition. John Wiley & Sons, New York, 2000.

Risk-Return Analysis of Credit Portfolios

Frank Schlottmann, Detlef Seese, Michael Lesko and Stephan Vorgrimler

Summary. We consider a problem of real-world risk-return analysis of credit portfolios in a multi-objective function setting with respect to additional constraints. For the approximation of a set of feasible, risk-return-efficient portfolio structures in this setting we discuss a flexible approach that incorporates multi-objective evolutionary and local search methods as well as specific features of the CreditRisk$^+$ model. We apply the hybrid approach to a sample loan portfolio to illustrate its working principle.

Introduction

The intensive development of quantitative portfolio credit risk models like CreditRisk$^+$ and the increasing trade in financial instruments for transferring credit risk like credit default swaps, collateralized loan obligations etc. are major reasons for a growing importance of credit portfolio risk-return analysis and optimization. Beyond that, there will possibly be more demand for credit portfolio optimization, as soon the supervisory capital requirements for banks will be changed due to proposals of the Basel Committee, e.g. by setting new capital weights on some credit risk exposure types and providing supervisory capital relief for risk mitigation (cf. e.g. Basel Committee for Banking Supervision [1]).

In the following sections, we will focus on the approximation of risk-return efficient sets[1] for credit portfolios with respect to constraints, e.g. imposed by changes of supervisory capital regulations or internal reallocation of risk capital. This kind of portfolio management is of great importance, especially for, but not limited to, many German and European banks, since the typical largest exposures to credit risk for small and medium-size universal banks are loans given to companies or private households not having direct access to the

---

[1] Risk-return efficient sets consist of Pareto-optimal solutions with respect to the objective functions portfolio risk and portfolio return. A formal definition is given later.

capital market. Such exposures are well-suited for risk measurement within the CreditRisk$^+$ model framework.

In contrast to the methods for the computation of the efficient frontier for a given set of alternative stock market investments based on the portfolio's variance and related measures, usually a non-linear, non-convex downside risk measure like the credit-value-at-risk is preferred for portfolio credit risk-return analysis, therefore requiring a different method of computation. Moreover, e.g. Lehrbass [6] has pointed out that this computational problem often cannot be modelled using real-valued variables, since typically neither the decision alternatives allow an arbitrary amount of each credit risk exposure to be traded nor is it possible to obtain a short position providing a hedge for each arbitrarily chosen exposure from a given portfolio. In addition to that, e.g. the capital requirements for credit risk exposures imposed by the banking supervision authorities are an important constraint to be considered in the computation of efficient credit portfolio structures.

For our considerations, the concept of Pareto-optimality is essential, i.e. efficient structures are Pareto-optimal concerning the two distinct (and usually contrary) objective functions specifying the aggregated risk and the aggregated return of each potential credit portfolio structure for a given set of alternatives. Therefore, we are interested in multiple, feasible non-dominated solutions to a constrained portfolio credit risk-return optimization problem that are comparable to the efficient frontier in stock portfolio investment analysis. However, in our case we deal with a discrete search space having many local optima and particularly using multiple target functions not required to be linear, quadratic or convex. In this context, a feasible non-dominated solution is a portfolio structure that does not violate the constraints, and for which we cannot find any other feasible solution being better in all two target function values.

The rest of this chapter is organized as follows. In the first section, we specify our portfolio credit risk optimization problem. Afterwards, we compare this problem to a related problem considered by Lehrbass [6]. Then we give an overview of our hybrid evolutionary algorithm framework for the portfolio credit risk optimization problem. The next section shows how the CreditRisk$^+$ model is integrated into our framework. In a sample application we illustrate the working principle of the hybrid algorithm.

16.1 Notation and Problem Definition

In this section, we will present the basic terminology of the constrained discrete credit portfolio risk-return analysis problem to be solved.

Definition 1. *Given is a credit portfolio containing $K > 1$ obligors. Each investment alternative (obligor) $A \in \{1, \dots, K\}$ incorporates the risk of default and is characterized by the following data which is considered to be constant* *within the time period $[0, t_1]$ where t_1 is the chosen risk horizon (usually $t_1 = 1$ in practical applications):*

- *net exposure $\nu_A \geq 0$ (loss in monetary units if obligor A defaults)*
- *expected default probability $p_A \geq 0$*
- *standard deviation of default probability[2] $\sigma_A \geq 0$*
- *sector weights $w_{Ak}, k \in \{0, \ldots, N\} \geq 0$*
- *expected rate of return η_A in relation to ν_A (net of cost, particularly after cost of funding but before credit risk)*
- *supervisory capital requirement percentage $c_A \geq 0$ in relation to ν_A.*

The above variables are mainly the basic inputs for the CreditRisk$^+$ model as described in Chapter 2. Moreover, the expected return and the capital requirement for each obligor have been added. Both are expressed in relation to ν_A, which usually requires a simple rescaling step of real-world parameters like supervisory capital weights based on gross exposure but this avoids unnecessary additional variables in our problem specification. In the following text, we will abbreviate the respective set of scalar variables $\nu_A, p_A, \sigma_A, \eta_A, c_A$ of all obligors to vectors $\nu := (\nu_1, \ldots, \nu_K)^T$ and analogously η, p, σ and c. The sector weights are abbreviated to a matrix $w := (w_{Ak})_{A \in \{1,\ldots,K\}, k \in \{0,\ldots,N\}}$.

We assume that the investor has to decide about holding a subset of the obligors in her portfolio, consider e.g. a bank that wants to optimize its loan portfolio containing K different obligors. We fix the following notations according to the assumption that a supervisory capital budget for the investments is fixed and given by the bank's maximum supervisory capital.

- A supervisory capital budget of the investor is given by $B > 0$.
- A portfolio structure is given by a vector

$$x = (x_1, x_2, \ldots, x_K)^T, x_A \in \{0, \nu_A\}.$$

Since every x_A can only take the values 0 or ν_A, the investor has to decide whether to hold the whole net exposure ν_A in her portfolio. In many real-world portfolio optimization problems the decision is e.g. either keeping the obligor A in the credit portfolio or selling the entire net exposure of obligor A to a risk buyer. Or, alternatively stated, whether to add a new exposure to the credit portfolio. This is particularly true for formerly non-traded instruments like corporate loans in a bank's credit portfolio. Even if there are more than two decision alternatives for each potential investment in obligor A, the decision variables will often consist of a finite, discrete number of choices.

Facing these decision alternatives, an investor has to consider two conflicting objective functions: the aggregated return and the aggregated risk from her portfolio. Usually, there is a tradeoff between both objectives, since any rational investor will ask for a premium (additional return) to take a risk.

---

[2] Alternatively, the variation coefficient of the gamma-distributed random scalar factors in the CreditRisk$^+$ model can be given, see the remarks in Chapter 2. In this case, no obligor-specific standard deviation of default probability is required.

Definition 2. *The aggregated expected return from a portfolio structure x is calculated by*

$$ret(x,p,\eta) := \sum_A \eta_A x_A - \sum_A p_A x_A = \sum_A (\eta_A - p_A)\, x_A.$$

This is a common net risk-adjusted return calculation since the aggregated expected losses are subtracted from the portfolio's aggregated expected net return before cost of credit risk.

Definition 3. *The aggregated downside risk from the portfolio structure x for the investor is calculated by*

$$risk(x,p,\sigma,w) := UL_\epsilon(x,p,\sigma,w)$$

where $UL_\epsilon(x,p,\sigma,w)$ denotes the resulting credit-value-at-risk to the chosen confidence level ϵ for the given portfolio structure represented by the vector x.

The above choice of the risk measure is a common choice for measuring credit risk or economic capital in banks, cf. the remarks by Gundlach in Chapter 2. For our considerations in this section, we do not need to choose the calculation procedure for the cumulative distribution function of aggregated losses or the approximation procedure for the risk measure. A summary of Panjer's recursion algorithm can be found in the appendix to Chapter 2 and more recent approaches are discussed in Part II of this volume.

Definition 4. *The required supervisory capital of a given portfolio structure x is*

$$cap(x,c) := \sum_A x_A c_A.$$

Definition 5. *A portfolio structure x is feasible if and only if*

$$cap(x,c) \leq B.$$

The following definition is essential for the concept of Pareto-optimality.

Definition 6. *Given two distinct feasible portfolio structures x and y, x dominates y if and only if one of the following cases is true:*

$$ret(x,p,\eta) > ret(y,p,\eta) \wedge risk(x,p,\sigma,w) \leq risk(y,p,\sigma,w)$$

$$ret(x,p,\eta) \geq ret(y,p,\eta) \wedge risk(x,p,\sigma,w) < risk(y,p,\sigma,w)$$

If x dominates y, we will denote this relationship by $x >_d y$.

This means that a feasible portfolio structure x is better than a feasible portfolio structure y if and only if x is better in at least one of the two criteria and not worse than y in the other criterion. It is obvious that a rational investor will prefer x over y if $x >_d y$.

Definition 7. *Let the set S of all possible portfolio structures for the specified data from Definition 1 and the subset $S' \subseteq S$ of all feasible structures in S be given. A solution $x \in S'$ is a feasible Pareto-optimal portfolio structure if and only if it satisfies the following condition:*

$$\forall y \in S' : \neg \, (y >_d x) \, .$$

To choose between the best combinations of obligors using her preferences or utility function, a rational investor is interested in finding the set of feasible Pareto-optimal portfolio structures that has maximum cardinality. This set is comparable to the efficient frontier of Markowitz [7], but in a constrained, discrete decision space.

Problem 1. The problem of finding the set of feasible Pareto-optimal portfolio structures having maximum cardinality for the set of investment alternatives S can be formulated as: calculate the set

$$PE^* := \{x \in S' : \forall y \in S' : \neg(y >_d x)\}.$$

Problem 1 is a computationally hard problem in the sense of **NP**-hardness – see Seese and Schlottmann [12] for a formal analysis of such and further portfolio credit risk problems.[3] This means that unless the computational complexity classes **P** and **NP** satisfy the condition **P** = **NP**, which is still an open problem besides many attempts to prove this relationship, there is no deterministic algorithm that calculates PE^* within polynomial computing time (measured by the number of the obligors K). For practical implementations, these results on computational complexity imply that the constrained search space for feasible risk-return efficient portfolio structures grows exponentially with the numbers of obligors, i.e. the number of potential solutions is 2^K. Given limited computational resources (e.g. a fixed number of executable operations per second) and a bounded runtime, Problem 1 cannot be solved exactly by simple enumeration of all possible solutions even when dealing with rather small problem dimensions.

16.2 The Approach by Lehrbass Using Real-Valued Decision Variables

In his article on constrained credit portfolio risk-return optimization, Lehrbass [6] maximized a single target function using real-valued variables, i.e. he considered a relaxation of our Problem 1 in a single objective function setting.

---

[3] The computational complexity class **P** contains all problems that can be solved by a deterministic algorithm in polynomial time (measured by the size of the input). **NP** contains all problems for which a deterministic algorithm can verify in polynomial time whether a given solution candidate actually is a solution. A problem is **NP**-hard if all problems in the class **NP** can be reduced to it.

Using our notation, the following optimization problem is an example for a two-obligor problem discussed by Lehrbass:

Problem 2. Given an arbitrary, but fixed bound $B > 0$ (capital budget) and an arbitrary, fixed bound $V > 0$ on the downside risk (in [6]: risk-adjusted capital), solve the following optimization problem:

$$\max_{x_{A_1}, x_{A_2}} \{\eta_{A_1} x_{A_1} + \eta_{A_2} x_{A_2}\}, \tag{16.1}$$

$$c_{A_1} x_{A_1} + c_{A_2} x_{A_2} \leq B, \tag{16.2}$$

$$risk(x, p, \sigma, w) \leq V, \tag{16.3}$$

$$x_{A_1} \leq \nu_{A_1}, x_{A_2} \leq \nu_{A_2}, \tag{16.4}$$

$$x_{A_1}, x_{A_2} \in \mathbb{R}_+^0. \tag{16.5}$$

Note that in contrast to the formulation in [6] the above problem has upper bounds on the decision variables to keep consistent with the decision alternatives considered so far, but this does not change the main idea. In his study, Lehrbass applied standard methods for solving convex optimization problems to Problem 2, see [6] for more details.

Assuming that there is a solution satisfying (16.2), (16.3), (16.4) and (16.5), one obtains a single optimal solution $x_{A_1}^*, x_{A_2}^*$ by solving an instance of Problem 2 once for the given parameters. This optimal solution has the maximum return for the given upper bound V on the risk and for the given upper bound B on the required capital.

By solving Problem 2 several times using different upper bounds V in the *risk* constraint (16.3) one obtains a set of Pareto-optimal solutions that represent the tradeoff between portfolio risk and return. Since Problem 2 has real-valued decision variables, there are usually infinitely many solutions, hence this set of Pareto-optimal solutions is an approximation for the tradeoff between portfolio risk and return.

In contrast to the above formulation of Problem 2, when returning to our Problem 1 we have to deal with a discrete optimization problem consisting of a fixed number of distinct choices. This is due to real-world restrictions like integral constraints and market frictions, e.g. transaction cost or round lots. Moreover, portfolio optimization problems based on downside risk measures incorporate non-linear objective functions. If the credit-value-at-risk is used in the *risk* objective function as in Problem 1, this objective function becomes even non-convex.[4] Thus, the necessary conditions for the application of standard algorithms for convex optimization problems are not satisfied for Problem 1. Therefore, we apply an algorithm to this problem that searches for discrete solutions and does not assume linearity or convexity of the objective functions or constraints.

---

[4] This is due to the non-convexity of the value-at-risk downside risk measure. See Chapter 3 for properties of downside risk measures.

16.3 A Multi-Objective Evolutionary Approach Combined with Local Search

16.3.1 Description of the algorithmic framework

Since the first reported implementation and test of a multi-objective evolutionary approach, the vector-evaluated genetic algorithm (VEGA) by Schaffer [9] in 1984, this special branch of evolutionary computation has attracted many researchers dealing with non-linear and non-convex multi-objective optimization problems. After the introduction of VEGA, many different evolutionary algorithms (EAs) have been proposed for multi-objective optimization problems, see e.g. Deb [3] for an overview and a detailed introduction.

In general, a multi-objective evolutionary algorithm (MOEA) is a randomized heuristic search algorithm reflecting the Darwinian "survival of the fittest" principle that can be observed in many natural evolution processes, cf. e.g. Holland [5]. At each discrete time step t, an MOEA works on a set of solutions $P(t)$ called population or generation. A single solution $x \in P(t)$ is an individual. To apply an MOEA to a certain problem the decision variables have to be transformed into genes, i.e. the representation of possible solutions by contents of the decision variables has to be transformed into a string of characters from an alphabet Σ. The original representation of a solution is called phenotype, the genetic counterpart is called genotype. For our portfolio credit risk optimization problem (Problem 1), we assume that the decision variables x_A will be arranged in a vector to obtain gene strings representing potential solutions. The resulting genotypes consist of real-valued genes that are connected to strings and take either the value 0 or ν_A depending on the absence or presence of obligor A in the current solution. So we obtain strings of length K that represent some of the 2^K combinations of possible (but neither necessarily feasible nor necessarily optimal) portfolio structures. Examples showing such gene strings are given later in Figure 16.1 and Figure 16.2.

The following hybrid multi-objective evolutionary algorithm (HMOEA) combining MOEA concepts and local search methods computes an approximation of PE^* mainly by modifying individuals using so-called variation operators, which change the contents of genes, by evaluating individuals based on the given objective functions and the constraints, and by preferring individuals that have a better evaluation than other individuals in $P(t)$. More details on these subjects and the other elements of the HMOEA are given below.

Algorithm 1. *HMOEA basic algorithm scheme*

1: $t := 0$
2: Generate initial population $P(t)$
3: Initialize elite population $Q(t) := \emptyset$
4: Evaluate $P(t)$

5:   **Repeat**
6:       *Select individuals from $P(t)$*
7:       *Recombine selected individuals (variation operator 1)*
8:       *Mutate recombined individuals (variation operator 2)*
9:       *Apply local search to mutated individuals (variation operator 3)*
10:      *Create offspring population $P'(t)$*
11:      *Evaluate joint population $J(t) := P(t) \cup P'(t)$*
12:      *Update elite population $Q(t)$ from $J(t)$*
13:      *Generate $P(t+1)$ from $J(t)$*
14:       $t := t + 1$
15: **Until** $Q(t) = Q(\max\{0, t - t_{diff}\}) \vee t > t_{max}$
16: **Output**: $Q(t)$

At the start of the algorithm, the initial population $P(0)$ will be generated by random initialization of every individual to obtain a diverse population in the search space of potential solutions.

We use an elite population $Q(t)$ in our algorithm that keeps the best feasible solutions found so far at each time step t. Rudolph and Agapie [8] have shown under weak conditions that in an MOEA having such an elite population, the members of $Q(t)$ converge to elements of PE^* with probability 1 for $t \to \infty$, which is a desirable convergence property. Furthermore, the algorithm can be terminated at any time by the user without losing the best feasible solutions found so far. This is particularly important for real-world applications. At the start of the algorithm, $Q(t)$ is empty. When the algorithm terminates, $Q(t)$ contains the approximation for PE^*.

The evaluation of $P(t)$ in line 4 and $J(t)$ in line 11 is based on the non-domination concept proposed by Goldberg [4] and explicitly formulated for constrained problems, e.g. in Deb [3, p. 288]. In our context, it leads to the following type of domination check, which extends Definition 6.

Definition 8. *Given two distinct portfolio structures x and y, x constraint-dominates y if and only if one of the following cases is true:*

$$cap(x, c) \leq B \wedge cap(y, c) \leq B \wedge$$
$$ret(x, p, \eta) > ret(y, p, \eta) \wedge risk(x, p, \sigma, w) \leq risk(y, p, \sigma, w) \qquad (16.6)$$
$$cap(x, c) \leq B \wedge cap(y, c) \leq B \wedge$$
$$ret(x, p, \eta) \geq ret(y, p, \eta) \wedge risk(x, p, \sigma, w) < risk(y, p, \sigma, w) \qquad (16.7)$$

$$cap(x, c) \leq B \wedge cap(y, c) > B \qquad (16.8)$$

$$cap(x, c) > B \wedge cap(y, c) > B \wedge cap(x, c) < cap(y, c). \qquad (16.9)$$

If x constraint-dominates y, we will denote this relationship by $x >_{cd} y$.

The first two cases in Definition 8 refer to the cases from Definition 6 where only feasible solutions were considered. Case (16.8) expresses a preference for feasible over infeasible solutions and case (16.9) prefers the solution that has lower constraint violation.

The non-dominated sorting procedure in our HMOEA uses the dominance criterion from Definition 8 to classify the solutions in a given population, e.g. $P(t)$, into different levels of constraint-domination. The best solutions, which are not constraint-dominated by any other solution in the population, obtain domination level 0 (best rank). After that, only the remaining solutions are checked for constraint-domination, and the non-constraint-dominated solutions among these obtain domination level 1 (second best rank). This process is repeated until each solution has obtained an associated domination level.

In line 6 of Algorithm 1, the selection operation is performed using a binary tournament based on the domination level. Two individuals x and y are randomly drawn from the current population $P(t)$, using uniform probability of $p_{sel} := \frac{1}{|P(t)|}$ for each individual. The domination levels of these individuals are compared, which implicitly yields a comparison between the individuals' dominance relation according to Definition 8. If, without loss of generality, $x >_{cd} y$ then x wins the tournament and is considered for reproduction. If they cannot be compared using the constraint-domination criterion (i.e. they have the same domination level) the winning individual is finally determined using a draw from a uniform distribution over both possibilities.

The first variation operator is the standard one-point crossover for discrete decision variables, i.e. the gene strings of two selected individuals are cut at a randomly chosen position and the resulting tail parts are exchanged with each other to produce two new offspring. This operation is performed with crossover probability p_{cross} on individuals selected for reproduction. The main goal of this variation operator is to move the population through the space of possible solutions. In the example displayed in Figure 16.1, two individuals are shown on the left, each representing a possible portfolio structure for the given investment alternatives. Each individual has five genes, which code the decision variables. The randomly chosen cut position is between the second and the third gene, such that the contents of the third, fourth and fifth gene are exchanged between the two individuals to obtain two offspring.

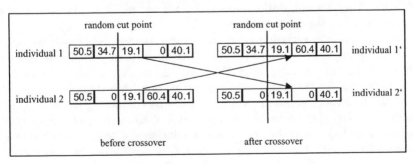

Fig. 16.1. One-point crossover variation operator

In analogy to natural mutation, the second variation operator changes the genes of the obtained offspring individuals randomly with probability p_{mut} (mutation rate) per gene to allow the invention of new, previously undiscovered solutions in the population. Its second task is the prevention of the HMOEA stalling in local optima as there is always a positive probability leaving a local optimum if the mutation rate is greater than zero. Figure 16.2 gives an example for this variation operator where the fifth gene of the offspring individual 1' from Figure 16.1 is mutated.

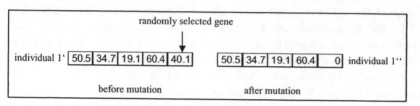

Fig. 16.2. Mutation variation operator

Our third variation operator in line 9 of Algorithm 1 represents a problem-specific local search procedure that is applied with probability p_{local} to each selected solution x after crossover and mutation. This local search procedure can exploit the structure of a given solution x to perform an additional local optimization of x towards elements of PE^*, e.g. by using an algorithm that changes x according to local information about our objective functions in the region around x. We consider this to be a significant improvement compared with a standard, non-hybrid MOEA, since the randomized search process of the MOEA can be guided a bit more towards the elements of PE^* and there-fore, such a local search operator can improve the convergence speed of the overall algorithm towards the desired solutions. This is particularly important for real-world applications, where speed matters when large portfolios are to be considered. In addition to these arguments, portfolio credit risk models like CreditRisk$^+$ provide additional local structure information for a current solution x beyond the objective function values that can be exploited very efficiently from a computational complexity's perspective. This is described in more detail in the next subsection below.

By applying the variation operators to the selected individuals we obtain an offspring population $P'(t)$. The members of the joint population $J(t)$ containing all parent solutions from $P(t)$ and all offspring solutions from $P'(t)$ are evaluated using the non-dominated sorting procedure described above. After that, the elite population $Q(t)$ is updated by comparing the best feasible solutions in $J(t)$ (i.e. having domination level 0) to the solutions in $Q(t)$: if a solution $x \in J(t)$ constraint-dominates a solution $y \in Q(t)$, the solution y is removed from $Q(t)$. Furthermore, if $x \in J(t)$ is not constraint-dominated by any solution from $Q(t)$, x is added to $Q(t)$. After this update procedure, $Q(t)$ contains the best feasible solutions found so far.

Before finishing the population step t and setting $t \to t + 1$ the members of the new parent population $P(t + 1)$ have to be selected from $J(t)$, since $|J(t)| > |P(t + 1)|$ by definition of $J(t) := P(t) \cup P'(t)$. Since elitist EAs, which preserve the best solutions from both parents and offspring, usually have better convergence properties, we also use this mechanism in our algorithm. Hence, the parents of the new parent population $P(t + 1)$ are selected from $J(t)$ according to their domination level. This means the solutions from $J(t)$ having domination level 0 are copied to $P(t + 1)$, the remaining capacity of $P(t+1)$ is filled with solutions from $J(t)$ having domination level 1 and so on, until not all solutions from $J(t)$ having a certain domination level k can be copied to $P(t + 1)$ due to the bounded capacity of $P(t + 1)$. Of course, if the number of solutions from $J(t)$ that have domination level 0 already exceeds the capacity of $P(t + 1)$ then $k = 0$ in the above example.

Let $J'(t) \subseteq J(t)$ denote the subset of solutions having domination level k that cannot be copied entirely to $P(t+1)$. To solve the selection problem and to obtain a good distribution of the solutions in the two-dimensional objective function space, an additional criterion is used to choose the solutions from $J'(t)$ to be copied to $P(t + 1)$: we incorporate the concept of crowding-sort described e.g. in Deb [3], which selects the solutions from $J'(t)$ by considering the distance between the objective function values of these solutions. Here, the perimeter of the largest possible rectangle that can be drawn around a single solution $x \in J'(t)$ in the objective function space without adding a second solution $y \in J'(t), y \neq x$, to the rectangle's interior serves as a distance measure. The solution $x \in J'(t)$ that has the largest value of this distance measure is copied to $P(t + 1)$, and afterwards x is removed from $J'(t)$. This process is repeated until $P(t+1)$ is filled up. As a consequence, at this step the algorithm prefers solutions that belong to less crowded regions of the objective function space.

Finally, the algorithm terminates if $Q(t)$ has not been improved for a certain number t_{diff} of population steps or if a maximum number of t_{max} population steps has been performed.

16.3.2 Incorporating CreditRisk$^+$ into the algorithmic framework

In the remaining sections of this chapter, we assume the bank has implemented the standard CreditRisk$^+$ model. For the corresponding calculations no explicit default correlations are required, since the volatilities of the obligors' default probabilities in conjunction with the common risk factors of all obligors replace a direct modelling of the default correlation $\rho(A_1, A_2)$ for two obligors A_1, A_2. However, in [2, 56 ff.], see also (10.1), the following implicit default correlation formula is provided (below we use the assumption that all random scalar factors S_k are normalized to an expectation $\mu_k = 1$):

$$\rho(A_1, A_2) \approx \sqrt{p_{A_1} p_{A_2}} \sum_{k=1}^{N} w_{A_1 k} w_{A_2 k} \sigma_k$$

where σ_k is the variance of the random scalar factor for sector k.

The dependence structure between obligors can be exploited to provide an adequate genetic modelling of the decision variables for the given portfolio data by sorting the obligors in ascending order according to the level of default dependency. This can be performed using a simple sorting procedure that exploits the structure of the sector weights and the variation coefficients of the sector random variables, see Schlottmann and Seese [10, 10–11] for a detailed example and further explanations.

To create a local search operator required by an implementation of the HMOEA scheme, we use the following local search target function, which uses the quotient between aggregated net return and aggregated risk to evaluate a given portfolio structure x:

$$f\left(x, p, \sigma, w, \eta\right) := \frac{ret(x, p, \eta)}{risk(x, p, \sigma, w)}$$

This is a common definition of a risk-adjusted performance measure like RAROC and similar concepts. Considering the specification of the return function in Definition 2 as well as the chosen risk measure UL_ϵ we obtain

$$f(x, p, \sigma, w, \eta) := \frac{\sum\limits_{A=1}^{K} x_A(\eta_A - p_A)}{UL_\epsilon\left(x, p, \sigma, w\right)}. \tag{16.10}$$

If we maximize this function f, we will implicitly maximize the net return and minimize UL_ϵ, and this will drive the portfolio structure x towards the set of global Pareto-optimal portfolio structures (cf. the domination criteria specified in Definition 6). The local search variation operator shown in Algorithm 2 exploits this property and besides that, it also respects the given supervisory capital constraint.

Algorithm 2. *Local search variation operator*

1:   **For each** $x \in P(t)$ *execute the following lines with probability* p_{local}
2:     **If** $cap(x, c) > B$ **Then**
3:         $D := -1$
4:     **End If**
5:     **If** $cap(x, c) \le B$ **Then**
6:         *Choose D between* $D := 1$ *or* $D := -1$ *with uniform probability 0.5*
7:     **End If**
8:     $\forall A : \widehat{x_A} := x_A, Step := 0$
9:     **Do**
10:       $\forall A : x_A := \widehat{x_A}, Step := Step + 1$
11:       $ret_{old} := \sum\limits_{A=1}^{K} x_A(\eta_A - p_A)$ *and* $risk_{old} := UL_\epsilon(x, p, \sigma, w)$
12:       **For each** x_J *calculate partial derivative* $d_J := \frac{\partial}{\partial x_J} f(x, p, \sigma, w, r)$
13:       **If** $D = -1$ **Then**

14:          *Choose* $\widehat{A} := \arg\min_J \{d_J \,|\, x_J > 0\}$

15:          *Remove this exposure from portfolio:* $x_{\widehat{A}} := 0$

16:     **Else**

17:          *Choose* $\widehat{A} := \arg\max_J \{d_J \,|\, x_J = 0\}$

18:          *Add this exposure to portfolio:* $x_{\widehat{A}} := \nu_{\widehat{A}}$

19:     **End If**

20:     $ret_{new} := \sum_{A=1}^{K} \hat{x}_A(\eta_A - p_A)$ *and* $risk_{new} := UL_\epsilon(\hat{x}, p, \sigma, w)$

21:     **While** $Step \leq Step_{max} \wedge (\exists A : \hat{x}_A > 0) \wedge (\exists J : \hat{x}_J = 0) \wedge$
          $\hat{x} \notin P(t) \wedge \hat{x} \notin Q(t) \wedge$
          $((D = -1 \wedge cap(\hat{x}, c) > B) \vee (D = 1 \wedge cap(\hat{x}, c) \leq B \wedge$
          $(ret_{new} > ret_{old} \vee risk_{new} < risk_{old})))$

22:     *Replace x in P(t) by its optimized version*

23: **End For**

24: **Output:** $P(t)$

If the current solution x from $P(t)$ to be optimized with probability p_{local} (which is a parameter of the algorithm) is infeasible because the capital restriction is violated (cf. line 2 in Algorithm 2), the algorithm will remove the obligor having the minimum gradient component value from the portfolio (lines 14 and 15). This condition drives the hybrid search algorithm towards feasible solutions. In the case of a feasible solution that is to be optimized, the direction of search for a better solution is determined by a draw of a uniformly distributed (0,1)-random variable (cf. lines 5 and 6). This stochastic behaviour helps in preventing the local search variation operator from stalling into the same local optima.

The partial derivative d_J for obligor J required in line 12 of Algorithm 2 can be obtained by evaluation of the following formula (a proof is provided in the appendix):

$$d_J = \frac{x_J\,(\eta_J - p_J)\,UL_\epsilon\,(x, ...) - \sum_{A=1}^{K} x_A\,(\eta_A - p_A) r_J(UL_\epsilon\,(x, ...))}{x_J\,(UL_\epsilon\,(x, ...))^2} \qquad (16.11)$$

where we abbreviated $UL_\epsilon\,(x, ...) := UL_\epsilon\,(x, p, \sigma, w)$ and $r_J(UL_\epsilon\,(x, ...))$ is the marginal risk contribution for obligor J to the unexpected loss, cf. the corresponding remarks in Chapter 2.

Remembering the fact that the risk contributions, and therefore, the partial derivatives d_J can be calculated efficiently (e.g. in KN computational steps for K obligors and N sectors if the approximation suggested by [2] is used[5]) for an individual that has already a valid fitness evaluation, this yields a very fast variation operator.

---

[5] Using the common O-notation the computational effort is $O(KN)$, hence the required number of operations grows linearly in the number of obligors.

Besides the proposed use of the risk contributions suggested by CSFP, the above algorithm can easily be adapted to other types of risk contributions or calculation schemes, see Chapter 3 by Tasche for more details.

The local search algorithm terminates if at most $Step_{max}$ iterations have been performed (parameter $0 < Step_{max} \leq K$), if the current solution cannot be modified further, if it is already included in the populations $P(t)$ or $Q(t)$ or if no improvement considering the violation of constraints or the target function can be made.

16.3.3 Application to a sample loan portfolio

We now focus on a small sample portfolio containing middle-market loans to illustrate the proposed method of risk-return analysis in the CreditRisk$^+$ model framework. Table 16.1 gives a short overview of the characteristics of the data and the chosen parameters.[6]

Parameter	Value
K (# obligors)	20
N (# systematic risk factors)	1
p_A (probability of default)	2% to 7%
Variation coefficient of default probability	0.75
$\frac{(\sum_{A=1}^{K} v_A)^2}{\sum_{A=1}^{K} v_A^2}$ (granularity of portfolio)	15.70
L_0 (loss unit)	$\frac{\max_{A \in \{1,\ldots,K\}} \{v_A\}}{100}$
ϵ (confidence level)	0.99
$\frac{B}{\sum_{A=1}^{K} c_A v_A}$ (constraint level)	$\frac{2}{3}$
$|P(t)|$ (population size)	30
p_{cross} (crossover probability)	0.95
p_{mut} (mutation probability)	$\frac{1}{20}$
p_{local} (local search probability)	0.10
$Step_{max}$ (max. local search iterations)	4
t_{diff} (termination condition)	100

Table 16.1. Overview of data and parameters

In our example, we used $|P(t)| = 30$ individuals in each population step t, which is a compromise between developing a diverse set of solutions in $P(t)$ and the corresponding computational effort spent per step t. Due to the elite population and the local search variation operator in the HMOEA, the choice of this parameter is not very crucial for this algorithm. In other tests using different portfolio sizes, we chose $20 \leq |P(t)| \leq 100$, and this parameter range is a common choice in other evolutionary algorithm applications, too. The

---

[6] The scheme for choosing L_0 was taken from the CreditRisk$^+$ reference implementation on http://www.csfb.com/institutional/research/credit_risk.shtml. The influence of this choice on the algorithm is discussed later in this section.

common parameter setting of $p_{cross} := 0.95$ and $p_{mut} := \frac{1}{K}$ is reported to work well in many other studies using elitist evolutionary algorithms, and this was also supported by test results during development of the HMOEA.

The choice of p_{local} and $Step_{max}$ can be made by the respective user of the HMOEA depending on her preferences: if one is interested in finding better solutions in earlier populations, then both the probability and the number of local search iterations given an application of the respective variation operator will be set higher, and in this case more computational effort is spent by the algorithm on the local improvement of the solutions. However, the local search optimization pressure should not be too high, since one is usually also interested in finding a diverse set of solutions. A choice of $0 < p_{local} \leq 0.1$ and $Step_{max} << K$ (which means the parameter is significantly lower than K) is also preferable concerning the additional computational effort to be spent by the local search variation operator. In our tests, we use a parameter set of $p_{local} := 0.1$ and $Step_{max} := 4$ for the HMOEA, which also yielded promising results in other test cases consisting of significantly different loan portfolio data sets.

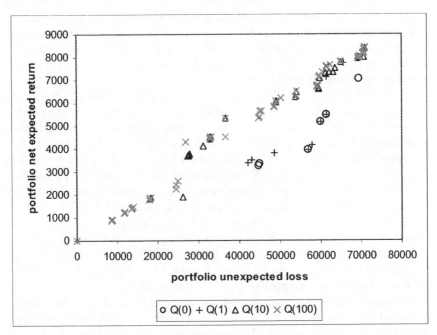

Fig. 16.3. Evolution of individuals within $Q(t)$

Figure 16.3 illustrates the evolution of the individuals within $Q(t)$ during a run of the HMOEA until $t = 100$. The objective function values of the six individuals within the first elite population $Q(0)$ are indicated by circles in Figure 16.3. The elite population $Q(1)$ obtained after the following iteration of lines 5 to 15 of Algorithm 1 contains the individuals that have the objective function values marked by a plus sign. For instance, the two individuals from $Q(0)$ whose objective function values are approximately (45000,3500) are dominated by the two individuals from $Q(1)$ that are close to them but have higher portfolio return and/or less portfolio unexpected loss values. Hence, the former two members of $Q(0)$ are no longer members of $Q(1)$. On the other hand, three elements from $Q(0)$ are not dominated within the next iteration of the evolution process, hence they remain in $Q(1)$. This is indicated by the three circles that contain a plus sign.

Until $t = 10$ the individuals in the elite population have moved significantly in the relevant direction towards the upper left corner of the displayed risk-return space, and the cardinality of the elite population has been raised to 33. Several members of $Q(10)$ even cannot be improved until $t = 100$, which can be seen in the upper right corner of Figure 16.3. However, the members of $Q(100)$ have a larger spread over the two-dimensional objective function space due to individuals that have been found in the lower left area of this figure. The cardinality of $Q(100)$ is 68.

Summarizing the elite populations shown, Figure 16.3 displays typical convergence properties during a run of the hybrid algorithm: in early iterations, the algorithm mainly improves the domination-related objective function values. Over time, more solutions are discovered and the diversity of the individuals in the objective function space is raised due to the crowding-sort procedure described in Section 16.3.1.

Our test data set is small enough to allow a complete enumeration of the search space, i.e. the feasible Pareto-optimal set PE^* can be computed within reasonable computing time to verify the solution quality of the approximation set computed by the HMOEA after a total of $t = 581$ population steps.

Figure 16.4 displays a comparison between the objective function values of the 100 elements within PE^* and those of the 97 members of the final elite population $Q(581)$. The hybrid approach required 43 seconds for the computation of the approximation set $Q(581)$ on a standard PC (2 GHz single CPU), whereas the enumeration took 1318 seconds (approx. 22 minutes) of computation time for PE^*. A visual inspection reveals that $Q(581)$ is a good approximation set for PE^* since the points of PE^* (indicated by "x") are approximated by mostly identical or at least very close points of $Q(581)$, which are marked by a respective circle in Figure 16.4.

In many MOEA applications, actually a single run of the algorithm is performed to obtain an approximation of the feasible Pareto-optimal solutions as shown in the above example. We performed a total of 50 independent runs of the HMOEA on the test problem using different pseudorandom number sequences to obtain 50 approximations of PE^*. The average runtime over 50 independent runs of the HMOEA was only 33 seconds, and no single run required more than 54 seconds. In each independent run, the HMOEA found the two boundary solutions, which is a desirable result for the stochastic search algorithm. The cardinality of the approximation set was 96.3 on average and

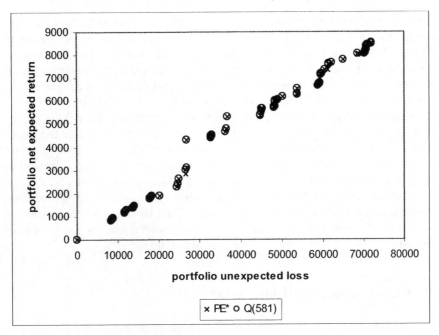

Fig. 16.4. Comparison of PE^* to $Q(581)$

≥ 91 in all runs, and the spread of the approximation solutions in the objective function space was similar to Figure 16.4 in each single run.

Remembering that adding one obligor to the portfolio doubles the size of the search space, which means that the enumeration effort increases exponentially with the number of obligors, we expect that the gap between the runtime of the enumeration and the runtime for obtaining an approximation using the HMOEA increases significantly. For a suitably large portfolio, a complete enumeration cannot be performed within reasonable time given a bounded runtime on a specific computer. Hence, the hybrid approach is particularly useful to compute approximations for PE^* in these cases. For instance, we have computed an approximation set for a non-homogeneous portfolio containing 386 corporate obligors of a German bank on the specified PC within approximately 1 hour.

Concerning larger portfolio dimensions, a few issues have to be kept in mind. First, the computational effort to calculate or approximate UL_ϵ strongly influences the runtime of the HMOEA. Thus, the choice of the calculation procedure for the loss distribution and the corresponding percentile has to be made with respect to the tradeoff between accuracy of the results and the required computation time. In our case, the extent of loss discretization in the standard CreditRisk$^+$ model (i.e. choice of L_0) strongly influences the computational effort to be spent on the calculation of the risk objective function. A finer discretization yields more precise results but at the cost of a higher runtime, whereas less precise UL_ϵ results can be achieved in less runtime. We have to point out that this issue is not specific to our algorithm but to all algorithms that require a large number of risk objective function evaluations during the search for risk-return efficient solutions. Alternative methods of approximating UL_ϵ offer potentially faster approximations that might be interesting for larger portfolios, see Chapters 5 to 8 for more details.

Secondly, the usual convergence results for evolutionary algorithms assume $t \rightarrow \infty$, whereas in practical applications our hybrid algorithm has to be stopped after a finite number of population steps t. Therefore, it is important to remember that $Q(t)$ is an approximation and not necessarily identical to PE^*, particularly for large portfolios having a huge search space.

Thirdly, since evolutionary algorithms are well-suited for distributed computation or parallel implementation (see e.g. [11]) there is a good prospect of improving the speed of our approach by using more than one CPU at least for some tasks to process large-dimensional portfolios.

16.4 Conclusion and Outlook

We formally defined a constrained multi-objective problem of risk-return analysis for a credit portfolio based on binary decision variables. The aggregated expected net return from a potential portfolio structure and the corresponding aggregated downside risk measured by unexpected loss were considered as objective functions of a bank also having an additional capital budget restriction imposed e.g. by supervisory capital.

For the approximation of the set of feasible Pareto-optimal solutions to our problem of portfolio risk-return analysis, we discussed a hybrid approach that combines concepts from different multi-objective evolutionary algorithm schemes and a problem-specific local search operator based on a risk-adjusted performance measure. This hybrid approach is not restricted to linear or convex objective functions and is also flexible concerning the constraints. We implemented the CreditRisk$^+$ portfolio credit risk model into the algorithm and derived a local search operator that exploits model-specific features. Although our implementation was executed on a single standard desktop PC, the hybrid algorithm found approximations of almost all feasible, Pareto-optimal solutions within seconds for our small test portfolio. Moreover, we have pointed out important issues for applications of our proposed hybrid method to larger problem instances.

Besides the integration of another variant of CreditRisk$^+$ (cf. the chapters in the first three parts of this volume), further research from the viewpoint of risk measurement could focus on alternative objective functions like expected shortfall, tail conditional expectation and related measures (see e.g. Chapter 3), and in this case an appropriate local search operator can easily be implemented. Moreover, the multi-objective evolutionary approach supports the incorporation of more than two objective functions without significant

changes, which might e.g. be very interesting for hold-to-maturity calculations as proposed in the CreditRisk$^+$ manual [2].

Due to the flexibility of our algorithm, many further constraints of practical interest can be considered, for instance the simultaneous use of different
capital budgets or unexpected loss limits on subsets of obligors (e.g. depending
on obligor-specific criteria like country or industry) in the risk-return analysis. Even more sophisticated restrictions can be handled, e.g. restrictions on
the structure of the parts of a portfolio to be sold in an asset-backed security transaction that is itself calculated using a non-linear pricing model as
described in the Chapters 18 by Kluge and Lehrbass and 19 by Hellmich and
Steinkamp.

Appendix: Proof of Formula (16.11)

Given a portfolio structure specified by x, p, σ, w, η, the partial derivative of
the function f given by (16.10) is calculated using the quotient rule and the
abbreviation $UL_\epsilon(x, \ldots) := UL_\epsilon(x, p, \sigma, w)$:

$$
\begin{aligned}
d_J &:= \frac{\partial}{\partial x_J} f(x, p, \sigma, w, \eta) \\
&= \frac{(\eta_J - p_J) UL_\epsilon(x, \ldots) - \left(\sum_{A=1}^{K} x_A (\eta_A - p_A) \right) \frac{\partial}{\partial x_J}(UL_\epsilon(x, \ldots))}{(UL_\epsilon(x, \ldots))^2}.
\end{aligned}
$$

Assuming $x_J \neq 0$ yields

$$
d_J = \frac{x_J (\eta_J - p_J) UL_\epsilon(x, \ldots) - \left(\sum_{A=1}^{K} x_A (\eta_A - p_A) \right) x_J \frac{\partial UL_\epsilon(x, \ldots)}{\partial x_J}}{x_J (UL_\epsilon(x, \ldots))^2}.
$$

Using the relationship

$$
r_J(UL_\epsilon(x, p, \sigma, w)) \equiv x_J \frac{\partial}{\partial x_J}(UL_\epsilon(x, p, \sigma, w))
$$

(cf. Chapters 2 and 3) we finally obtain

$$
d_J = \frac{x_J (\eta_J - p_J) UL_\epsilon(x, \ldots) - \left(\sum_{A=1}^{K} x_A (\eta_A - p_A) \right) r_J(UL_\epsilon(x, \ldots))}{x_J (UL_\epsilon(x, \ldots))^2}.
$$

278   Frank Schlottmann, Detlef Seese, Michael Lesko and Stephan Vorgrimler

References

1. Basel Committee for Banking Supervision. The new Basel Capital Accord. Bank for International Settlements, Basel, http://www.bis.org/bcbs/cp3full.pdf, 2003.
2. Credit Suisse Financial Products. CreditRisk$^+$: A credit risk management framework. London, 1997. Available at http://www.csfb.com/creditrisk.
3. K. Deb. *Multi-Objective Optimisation Using Evolutionary Algorithms.* John Wiley & Sons, Chichester, 2001.
4. D. Goldberg. *Genetic Algorithms for Search, Optimisation and Machine Learning.* Addison-Wesley, Reading, 1989.
5. J. Holland. *Adaptation in Natural and Artificial Systems.* Michigan University Press, Ann Arbor, 1975.
6. F. Lehrbass. Rethinking risk-adjusted returns. *Credit Risk Special Report, Risk,* 4:35–40, 1999.
7. H. Markowitz. Portfolio selection. *Journal of Finance,* 7:77ff, 1952.
8. G. Rudolph and A. Agapie. Convergence properties of some multi-objective evolutionary algorithms. In A. Zalzala, editor, *Proceedings of the 2000 Congress on Evolutionary Computation,* pages 1010–1016, IEEE Press, Piscataway, 2000.
9. J. Schaffer. *Some experiments in machine learning using vector evaluated genetic algorithms.* PhD thesis, Vanderbilt University, Nashville, 1984.
10. F. Schlottmann and D. Seese. A hybrid genetic-quantitative method for risk-return optimisation of credit portfolios. In C. Chiarella and E. Platen, editors, *Quantitative Methods in Finance 2001 Conference abstracts,* page 55, University of Technology, Sydney, 2001. Full paper available under: www.business.uts.edu.au/finance/resources/qmf2001/ Schlottmann_F.pdf
11. H. Schmeck, J. Branke and U. Kohlmorgen. Parallel implementations of evolutionary algorithms. In A. Zomaya, F. Ercal and S. Olariu, editors, *Solutions to Parallel and Distributed Computing Problems,* pages 47–68, John Wiley & Sons, Chichester, 2001.
12. D. Seese and F. Schlottmann. The building blocks of complexity: a unified criterion and selected problems in economics and finance. Presented at the Sydney Financial Mathematics Workshop 2002, www.qgroup.org.au/SFMW, 2002.

Numerical Techniques for Determining Portfolio Credit Risk

Sandro Merino and Mark Nyfeler

Summary. Two numerical algorithms for the risk analysis of credit portfolios are presented. The first one determines the distribution of credit losses and is based on the fast Fourier transform. The algorithm has a strong analogy to the CreditRisk$^+$ approach, since both the Poisson approximation as well as the technique of forming homogeneous exposure bands are being used. An application to the analysis of collateralized debt obligations is also given. The second algorithm makes use of an importance sampling technique for allocating credit risk contributions according to the risk measure expected shortfall. The coherent risk spectrum that is obtained by varying the loss exceedance level is introduced and its properties are discussed.

Introduction

The risk analysis of credit portfolios rises two main questions:

- What is the probability distribution of total credit losses in the portfolio over a given time horizon?
- How can the total portfolio credit risk be allocated down to the single counterparty exposures?

We will present two numerical algorithms that can be applied to these problems. These generalize the CreditRisk$^+$ approach by allowing for a higher flexibility regarding the dependence and distribution assumptions for the underlying systematic risk factors. In fact, one disadvantage of CreditRisk$^+$ (and probably of analytical techniques in general) is that the Panjer algorithm, and its generalizations, impose restrictions on the assumptions that can be made on the distribution of the systematic risk factors and their joint copula function. With a numerical approach such restrictions fall and the choice of the systematic factors and of their copula can be made, focusing purely on the available empirical data without the necessity to impose artificial ex ante restrictions that are merely due to tractability considerations.

Computing the portfolio's loss distribution

In Sections 17.1 to 17.6 we explain a numerical algorithm for determining the portfolio loss distribution, first presented in [9]. It makes use of the Poisson approximation and of the idea of introducing exposure bands, which are also two of the CreditRisk$^+$ cornerstones. In contrast to CreditRisk$^+$ we apply the fast Fourier transform (FFT) to determine conditional loss distributions. Analytical and numerical methods are thus combined to obtain a flexible, accurate, and computationally effective analysis of credit portfolios. Let us point out that Oliver Reiß in his chapters discusses more general applications of the Fourier transform in the CreditRisk$^+$ context both for discrete and continuous formulations of the model, which e.g. cause integrability problems absent in our approach.

The saddlepoint approximation is yet another combination of analytical and numerical techniques that has been shown to provide an extremely efficient and accurate approach to determine the tail of loss distributions (see [8] or Chapter 7) within the conditional independence framework. However, if information on both the body and the tail matters then the saddlepoint approximation cannot be applied, since – at least in general – it fails to produce accurate results for the distribution's body (see Table 17.2). This issue is particularly important for the application to the risk transfer analysis for CDOs that is discussed in Section 17.6.

The algorithm we describe to determine the portfolio's loss distribution uses techniques from numerical mathematics and actuarial science. We have implemented the algorithm in a mathematical software package and are able to compute the loss distribution of credit portfolios containing half a million individual counterparties within 4 hours with adequate accuracy (on a standard personal computer). We will find that to achieve this performance it is necessary to simplify neither the credit risk model, i.e. the structure of the underlying risk factors, nor the portfolio structure.

Risk allocation

Once the total portfolio credit risk is understood we may need to allocate the total risk down to the individual counterparties or segments in the credit portfolio. For example, for performance measurement, pricing or business strategy considerations such an allocation plays an important role. Sections 17.7 to 17.11 are devoted to these issues. An approach increasingly being adopted in practice is the allocation of risk by means of coherent risk measures. Most prominently expected shortfall contributions provide a conceptually convincing mechanism for breaking down the total risk of a credit portfolio to the level of its individual counterparties or any other portfolio segmentation of interest. In contrast to widely used VaR contributions, risk allocation based on expected shortfall possesses a number of desirable properties and coincides with the intuition and experience of possibly not very quantitatively inclined credit portfolio managers. In fact, in contrast to VaR, expected shortfall is a risk measure permitting a better detection and consistent quantification of the risk arising from concentrating credit exposure on single counterparties. Thereby expected shortfall can be calibrated to the institution's risk preferences by choosing an appropriate loss exceedance level.

Unfortunately all good things come at a price, which in this case is paid in the form of the considerable computational effort required for computing expected shortfall contributions by standard Monte Carlo simulation. In fact, naive attempts to determine such risk contributions with acceptable accuracy fail, since the number of simulations increases to exorbitant and thus computationally prohibitive levels.

We describe a solution to this problem, which is based on an importance sampling technique and dramatically increases the rate of convergence of the simulation. We will show that the effort for computing expected shortfall contributions at any loss exceedance level is not significantly higher than the rapidly converging simulation of expected loss by naive Monte Carlo. Furthermore, apart from the assumption that a conditional independence framework is being used, we do not impose any restrictions either on the dependence structure or on the composition of the portfolio. Therefore the method we describe is applicable to a very wide range of credit risk models including all current conditional independence models.

17.1 The Conditional Independence Framework

In Chapter 4 it was shown that CreditRisk$^+$ and CreditMetrics are very similar (see also [5, 6]). In particular they share the conditional independence framework: conditional on a small set of latent variables $S = (S_1, \ldots, S_N)$, default events among obligors are independent. These latent variables represent the systematic default risk factors and mimic the future state of the world at a pre-defined time horizon T (usually one year). Given a specific future state of the world $S = s$, counterparty A's default probability is denoted by p_A^s.

The relationship between systematic risk factors and default probabilities at time T is given by

$$p_A^S = f_A(S), \qquad S \sim F$$

for some functions f_A and some distribution function F. In most applications it is assumed that the expected default frequency $p_A = \mathbb{E}[p_A^S]$ is known for all counterparties, e.g. inferred from a rating process. Recall from (2.9) that the default rate volatility model of CreditRisk$^+$ without specific risk factor is given by

$$p_A^S = p_A \sum_{k=1}^{N} w_{Ak} S_k$$

where $(S_k)_k$ are independent and gamma distributed with mean 1, standard deviation σ_k and $\sum_{k=1}^{N} w_{Ak} = 1$, $w_{Ak} \geq 0$ for all A, k.

In contrast, we recall from Chapter 4 that credit risk models based on the Merton approach (such as CreditMetrics) model the asset returns $(Z_A)_A$ of companies by a (standard) multivariate normal distribution. Default occurs if the corresponding return falls below a certain threshold. Parameter estimation is done by calibrating a factor model

$$Z_A = \sum_{k=1}^{N} w_{Ak} S_k + \sigma_A \epsilon_A.$$

Here $S = (S_1, \ldots, S_N)$ is assumed to be multivariate (standard) normal and $(\epsilon_A)_A$ represents independent, firm-specific standard normally distributed risk. Since p_A is assumed to be known, the asset return factor model can be written as a default rate volatility model

$$p_A^S = \Phi \left(\frac{\Phi^{-1}(p_A) - \sum_{k=1}^{N} w_{Ak} S_k}{\sigma_A} \right),$$

where Φ denotes the standard normal cumulative distribution function.

Now we recall that the total portfolio loss up to a specific future date T is modelled by the random variable

$$X = \sum_A \mathbf{1}_A \nu_A,$$

where $\mathbf{1}_A$ stands for the default indicator and ν_A for the constant net exposure (exposure net expected recovery) to the A-th counterparty. The variable $\mathbf{1}_A$ takes the value 1 with probability p_A and the value 0 with probability $1 - p_A$.

The conditional independence framework assumes that given the state s of the latent variables $S = (S_1, \ldots, S_N)$ the indicator variables $\mathbf{1}_A$ are independent. This means that the remaining risk, given the systematic part, is considered as purely idiosyncratic (obligor-specific). Consequently, we can define a credit model based on a conditional independence framework as

$$p_A^S = f_A(S), \qquad S \sim F \tag{17.1}$$

$$(\mathbf{1}_A \mid S = s)_A \sim \text{Bernoulli}\,(p_A^s), \text{ independent} \tag{17.2}$$

$$X = \sum_A \mathbf{1}_A \nu_A. \tag{17.3}$$

A note on our notation: we denote by $(X \mid S = s)$ a random variable with distribution defined by the conditional probability $\mathbb{P}[(X \mid S = s) = x] := \mathbb{P}[X = x \mid S = s]$. If $\mathbb{P}[S = s] > 0$, then this is well-defined by elementary probability theory. If $\mathbb{P}[S = s] = 0$, then one needs to resort to measure theory to define $\mathbb{P}[X = x \mid S = s]$, cf. e.g. to [2, Theorem 5.3.1.] on product measures. It is interesting to note that the first few pages of Chapter 5 in [2] give a rigorous mathematical justification for the conditional independence framework. In particular, it is shown that there are two equivalent points of view for the conditional independence framework: one may regard the conditional density of X given $S = s$ as being derived from the joint density of X and S or equivalently, one may derive X by considering a two-stage observation based on the conditional values of X given S. In either case, probabilities involving X and Y coincide. Our notation thus stresses the interpretation of X as being derived by observations of $(X|S = s)$.

Next, we describe a numerical algorithm for generating the loss distribution of any credit risk model that can be written in the form of (17.1) to (17.3), i.e. that belongs to the conditional independence framework.

Additionally, we assume all net exposures to be integer, which amounts to expressing net exposures in terms of integer multiples of a pre-defined base unit of loss L_0. We will now turn to the task of deriving the probability mass function p^X of the loss variable X, given in terms of $p_m^X = \mathbb{P}[X = m]$ for $m = 0, 1, \ldots$ up to the level $m = M$.

17.2 A General Representation of the Portfolio Loss Distribution

The unconditional probability $\mathbb{P}[X = m]$ can be obtained (see [2]) from the conditional probabilities by integration with respect to the distribution F of $S = (S_1, \ldots, S_N)$:

$$\mathbb{P}[X = m] = \int_{\mathbb{R}^N} \mathbb{P}[X = m \mid S = s]\, dF(s). \qquad (17.4)$$

Furthermore, the conditional independence property of the default indicators allows for a closed-form solution for the integrand. By definition of the model we have $(X \mid S = s) = \sum_A (\mathbf{1}_A \nu_A \mid S = s)$, a sum of independent random variables. By elementary probability theory it follows that the distribution of $(X \mid S = s)$ is equal to the convolution of the distributions of $(\mathbf{1}_A \nu_A \mid S = s)_A$; or in mathematical terms

$$p^{X|S=s} = \left(\bigotimes_A p^{\mathbf{1}_A \nu_A | S=s} \right), \qquad (17.5)$$

where $x \otimes y$ denotes the convolution of two discrete densities x, y. Rewriting (17.4) in vector notation yields

$$p^X = \int_{\mathbb{R}^N} \left(\bigotimes_A p^{\mathbf{1}_A \nu_A | S=s} \right) dF(s). \qquad (17.6)$$

Consequently, the task of deriving the loss mass function reduces to:

- analytical derivation of the integrand (given in (17.5));
- numerical integration of the integrand in N dimensions (as in (17.6)).

In order to reduce the computational burden for the convolutions and the N-dimensional integration the application of the following techniques from numerical and actuarial mathematics is essential in our approach:

- Poisson approximation;,
- fast Fourier transform;,
- numerical integration using quasi-Monte Carlo methods.

17.3 Numerical Computation of the Portfolio Loss Distribution

We will now explain how the above-mentioned techniques can be applied to obtain an efficient numerical algorithm for evaluating the formula (17.6).

17.3.1 Evaluating the integrand $p^{X|S=s}$

A solution for fast and efficient evaluation of the integrand (17.5) is obtained via the ideas of standard CreditRisk$^+$. Recall from Chapter 2 or [4] that we can assume the indicators $(\mathbf{1}_A \mid S = s)$ to be Poisson distributed with parameter $f_A(s)$ rather than Bernoulli distributed. We group counterparties with equal exposures into exposure bands $B_j := \{A \mid \nu_A = \bar{\nu}_j\}$ for some $\bar{\nu}_j$, $j = 1, \ldots V$. We denote by X_j the loss variable of the subportfolio generated by band B_j and rewrite the conditional portfolio loss variable as a sum of the exposure bands

$$(X \mid S = s) = \sum_{j=1}^{V} \left(\bar{\nu}_j \sum_{A \in B_j} (\mathbf{1}_A \mid S = s) \right) = \sum_{j=1}^{V} (X_j \mid S = s).$$

Note that the conditional loss variables $(X_j \mid S = s)_j$ are independent. Since the sum of independent Poisson random variables is still Poisson distributed it follows immediately that

$$(X_j \mid S = s) \sim \text{Poisson}(\lambda_j(s)), \qquad \lambda_j(s) := \sum_{A \in B_j} f_A(s). \qquad (17.7)$$

Consequently, the conditional probability mass function of a subportfolio B_j is easily calculated by

$$\mathbb{P}[X_j = m \mid S = s] = \begin{cases} \frac{\lambda_j(s)^n}{n!} \exp\left(-\lambda_j(s)\right) & \text{if } m = n \cdot \bar{\nu}_j\,,\, n \in \mathbb{N} \\ 0 & \text{else} \end{cases} \qquad (17.8)$$

for $m = 0, 1, \ldots, M$. The number of distributions involved in the convolution problem has now reduced from K (obligors) to V (exposure bands), since the conditional independence framework allows us to rewrite (17.5) as

$$p^{X|S=s} = \left(\bigotimes_{j=1}^{V} p^{X_j|S=s} \right). \tag{17.9}$$

The number of exposure bands required to keep the approximation error small mainly depends on the homogeneity of the distribution of exposures within the portfolio. Even for portfolios with an inhomogeneous exposure distribution, 200 exposure bands should suffice to achieve very high accuracy.

For efficient evaluation of (17.9) we use the discrete Fourier transform,[1] denoted by $\mathcal{F}$. The Fourier transform can be applied for computing convolutions, since for two vectors x, y we have

$$x \otimes y = \mathcal{F}^{-1}(\mathcal{F}(x) \odot \mathcal{F}(y)), \tag{17.10}$$

where $\odot$ denotes the componentwise multiplication $(x \odot y)_k := x_k \cdot y_k$ of two vectors with equal number of components. More precisely, (17.10) only holds if the vectors x, y on the right-hand side of the above equation are extended by "zero-pads". By slight abuse of notation we use the same symbol for vectors and their zero-padded extension. We will discuss this issue together with the choice of vector length M in Section 17.4.

We numerically process (17.10) by the fast Fourier transform (FFT), see also Chapter 8. This reduces the number of multiplications required to determine $\mathcal{F}(x)$ to $\frac{1}{2}N \log_2(N)$ if $\log_2(N) \in \mathbb{N}$, whereas the standard convolution algorithm for computing $x \otimes y$ requires $\frac{1}{2}N(N-1)$ multiplications.

Remarks on the Panjer recursion

Alternatively to our proposed method, the portfolio conditional loss distribution $p^{X|S=s}$ can be generated directly with a recursion proposed by Panjer [10]. For the classical CreditRisk$^+$ model, the Panjer recursion scheme can even be used to directly obtain the unconditional loss distribution p^X, which is due to a very specific choice of the risk factor distribution F.

For calculating the conditional loss distribution $p^{X|S=s}$ for a portfolio with V exposure bands the Panjer recursion only requires multiplications of order $O(VN)$ to determine a conditional loss distribution $p^{X|S=s}$ compared with $O(VN \log_2(N))$ for FFT. Unfortunately, the Panjer recursion suffers from a well-known underflow problem. See Section 12.5.4 of Chapter 12 by Binnenhei for the description of this problem and a possible solution.

The algorithm suggested there is as fast as Panjer's, and admits an "almost" arbitrary value for starting the recursion. For large portfolios the FFT method still looks favourable compared with Binnenhei's recursion. Binnenhei argues that if the starting value is of order 100^{-400}, then the algorithm can be invoked with numbers in the order of 100^{-300} and all resulting loss

---

[1] The n-dimensional discrete Fourier transform $\mathcal{F}_n$ is a linear isomorphism on $\mathbb{C}^n$. When no confusion seems likely we will suppress the dimension index n.

probabilities can be rescaled by a factor 100^{-100} at the end of the recurrence scheme. Now if the conditional probability of zero loss takes a value smaller than say 100^{-700}, then a likely result is an overflow problem for some probabilities, which a priori cannot be foreseen and not be corrected, unless rescaling is done at the beginning and repeatedly within the recurrence scheme. At present the final clarification of the advantages of these recent alternatives for optimizing the technical implementation of the algorithms is still incomplete and its detailed discussion is therefore postponed to future work.

17.3.2 Numerical integration in N dimensions

For numerically integrating (17.4) we apply the strong law of large numbers which yields

$$\int_{\mathbb{R}^N} \mathbb{P}[X = m \mid S = s] \, dF(s) = \lim_{I \to \infty} \frac{1}{I} \sum_{i=1}^{I} \mathbb{P}[X = m \mid S = s^i],$$

where $(s^i)_i$ is a sequence of independent draws from the distribution F of S. Hence for sufficiently large I the loss probabilities are approximated by

$$\mathbb{P}[X = m] \approx \frac{1}{I} \sum_{i=1}^{I} \mathbb{P}[X = m \mid S = s^i], \qquad m = 0, 1, \dots . \qquad (17.11)$$

To increase the rate of convergence of the Monte Carlo estimates when simulating in several dimensions $(N > 1)$ quasi-random or low-discrepancy numbers usually outperform standard pseudo-random numbers, as is shown in [7]. Quasi-random numbers are more evenly scattered in the $[0, 1]^N$-cube. Since in applications the integrand is a smooth function of S the rate of convergence can increase substantially, especially for $N < 100$.

By the linearity of the Fourier transform we can rewrite (17.11) as follows:

$$p^X \overset{(17.11)}{\approx} \frac{1}{I} \sum_{i=1}^{I} p^{X|S=s^i} \overset{(17.9),(17.10)}{=} \frac{1}{I} \sum_{i=1}^{I} \left(\mathcal{F}^{-1} \left(\bigodot_{j=1}^{V} \mathcal{F} \left(p^{X_j|S=s^i} \right) \right) \right)$$

$$= \mathcal{F}^{-1} \left(\frac{1}{I} \sum_{i=1}^{I} \left(\bigodot_{j=1}^{V} \mathcal{F} \left(p^{X_j|S=s^i} \right) \right) \right). \qquad (17.12)$$

Hence the inverse Fourier transform needs to be carried out only once rather than at every simulation step, providing a simple but effective reduction of computational time. Therefore, as shown by (17.12), the numerical integration is carried out in the Fourier space.

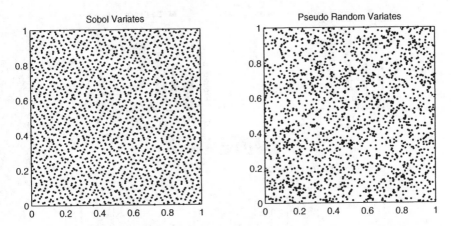

Fig. 17.1. 2000 Sobol (quasi-) random numbers vs standard pseudo-random variates for $N = 2$ dimensions

17.3.3 The algorithm in a nutshell

To provide a concise illustration of the algorithm we consider a simple case of a portfolio consisting of just four obligors with an exposure distribution of $\{1, 2, 4, 8\}$. Consequently, we have four exposure bands B_j, $j = 1, \ldots 4$, each containing one obligor. We assume a default rate volatility model with a single binary latent variable S ($N = 1$). Here S describes the state of the economy with two possible outcomes, *good* or *bad*.

By the Poisson approximation, the number of defaults in each subportfolio, conditional on the state of the economy, is Poisson distributed (see (17.7)). The remaining task, namely processing (17.12) from right to left is represented in Figure 17.2. For either of the two given states of the economy, the conditional loss distributions of each exposure band are (fast) Fourier transformed and componentwise multiplied. As a result we retrieve the two conditional Fourier-transformed portfolio loss densities for each state of the economy. By performing a weighted average of the two sequences, where each weight reflects the relative frequency of the corresponding state of the economy, we arrive at the unconditional Fourier-transformed loss distribution. Finally, we apply the inverse (fast) Fourier transform to obtain the portfolio loss distribution.

17.4 Accuracy of the Method

Since the exposure bands can be chosen in a fine-grained way and the number of draws from the distribution of systematic risk factors can be increased, the quality of the method heavily relies on the accuracy of the Poisson approximation, applied in (17.7). Since this approximation is used to obtain the

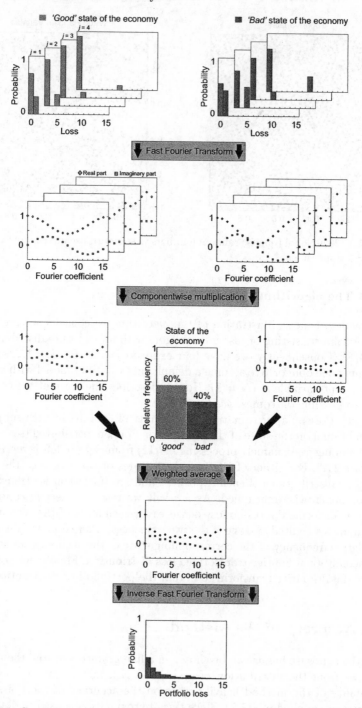

Fig. 17.2. Graphical representation of the algorithm for deriving the portfolio loss distribution

conditional portfolio loss distribution, we investigate the error for each simulation step, conditional on the systematic risk factors. Averaging the error encountered in each simulation step gives the error for the loss distribution.

17.4.1 The Poisson approximation

Rigorous error estimates for the approximation can be found in [3]. The error is assessed by the measure of total variation

$$d(G, H) := \sup_{A \subset \mathbb{N}_0} [G(A) - H(A)].$$

Conditional on a state of the latent variable $S = s$, let G_{js} and H_{js} denote the probability measures induced by the number of defaults in exposure band B_j and its Poisson approximation respectively.

$$S = s : \sum_{A \in B_j} 1_A \sim G_{js}, \quad (1_A)_{A \in B_j} \sim \text{indep. Bernoulli}(p_A^s)$$

$$S = s : \sum_{A \in B_j} 1_A \sim H_{js}, \quad (1_A)_{A \in B_j} \sim \text{indep. Poisson}(p_A^s)$$

In this setting the optimal upper error bound (EB) is given by

$$d(G_{js}, H_{js}) \leq \left(1 - e^{-\sum_{A \in B_j} p_A^s}\right) \frac{\sum_{A \in B_j} (p_A^s)^2}{\sum_{A \in B_j} p_A^s}.$$

Table 17.1 provides error bounds for exposure bands of different size. Each obligor default probability p_A^S is uniformly distributed in an interval of type $[0, \epsilon]$. That is $p_A^s := \hat{U}_A$, $U_A \sim \text{Uniform}[0, \epsilon]$, independent. Two observations

Size	p_A^s interval	EB	Size	p_A^s interval	EB
2000	$[0, 1\%]$	0.0066	20	$[0, 3\%]$	0.0056
2000	$[0, 3\%]$	0.0198	200	$[0, 3\%]$	0.0197
2000	$[0, 10\%]$	0.066	2000	$[0, 3\%]$	0.0198

Table 17.1. Error bounds for the Poisson approximation

are important: the error is uniformly bounded with respect to the number of counterparties K and the approximation error decreases with smaller counterparty default probabilities $(p_A^s)_A$.

An efficient yet very accurate approximation method for determining the tail of credit loss distributions is the saddlepoint method. Therefore we will benchmark the results obtained with our method to both the saddlepoint method and the true loss distribution. We will again present the results at the level of the conditional loss distribution in order to allow for a direct comparison with the results for the saddlepoint method obtained by Martin,

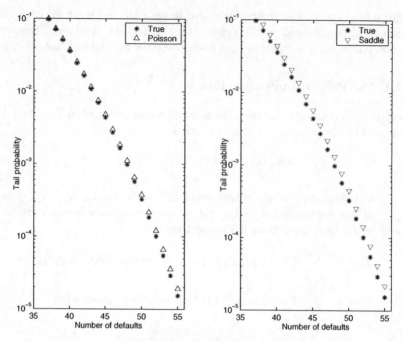

Fig. 17.3. Poisson approx. vs saddlepoint method in the tail: $(p_A^s)_A$ independent and uniformly sampled from $[0, 3\%]$, 2000 counterparties. The logarithmic scale highlights the very high accuracy of both methods

Thompson and Browne [8]. Since the method we propose makes use of the Poisson approximation in each exposure band, we assume all counterparties in the next test portfolio to have unit exposure.

An appropriate graphical representation of tail probabilities is given by the logarithmic plot in Figure 17.3, the so-called loss exceedance curve (VaR against tail probability).

In the following table it is shown that in contrast to the Poisson approximation the saddlepoint method fails to produce accurate values for the body of the loss distribution.

		$P(X \leq x)$		Δ	
x	True	Saddle	Poisson	Saddle	Poisson
20	4.12%	3.19%	4.27%	22.42%	3.59%
24	17.73%	14.83%	17.99%	16.37%	1.45%
28	43.80%	39.03%	43.91%	10.91%	0.25%
32	71.88%	67.84%	71.72%	5.62%	0.22%
36	90.07%	88.12%	89.85%	2.16%	0.24%
40	97.54%	96.93%	97.42%	0.63%	0.12%

Table 17.2. Poisson approx. vs saddlepoint method for the body of the distribution: $(p_A^s)_A$ chosen uniformly on $[0, 3\%]$, 2000 counterparties

17.4.2 The Fourier transform and the maximal loss level M

As already mentioned, (17.10) only holds if the vectors are extended by zero-pads. More precisely, the convolution of two vectors $x, y \in \mathbb{C}^{n+1}$ is given by

$$x \otimes y = \mathcal{F}^{-1} \left(\mathcal{F} \left((x_0, \ldots, x_n, \underbrace{0, \ldots, 0}_{n}) \right) \odot \mathcal{F} \left((y_0, \ldots, y_n, \underbrace{0, \ldots, 0}_{n}) \right) \right).$$

Note that the minimal length of the zero-pads is n, but they can also be chosen longer. From (17.8) we know that the probabilities in the tail of the conditional loss vectors $\left(p^{X_j | S = s} \right)_j$ decay exponentially towards zero. Hence if we choose M sufficiently large, we can ensure that these conditional loss vectors are already padded with numbers that are extremely close to zero[2]). In other words, the vectors are already zero-padded and we can work with a constant vector length throughout the algorithm.

Since the maximum loss level M, expressed as a multiple of the loss unit L_0, defines the length of the vectors in the algorithm it needs to be fixed when the numerical method is initialized. Hence we need criteria for choosing M: recall, that for a given portfolio both the choice of L_0 and M jointly determine the computational cost of the numerical method. To achieve highest possible efficiency for the FFT, M should always be chosen such that $\log_2(M + 1)$ is an integer. The choice of M also depends on the structure of the underlying portfolio:

- For small, poorly diversified portfolios choose M as large as the total portfolio exposure, i.e. $M \geq \sum_A \nu_A$.
- For well-diversified portfolios M can be chosen considerably lower (diversification effect). Our experience shows that the Chebyshev inequality with a confidence level of 1% provides a good rule of thumb. That is, choose M such that

$$\mathbb{P}[X > M] \leq \frac{1}{M^2} \left(\mathbb{E}[X]^2 + \mathrm{var}[X] \right) \overset{!}{\approx} 1\%.$$

Expectation and variance of the portfolio loss can be derived analytically or simply be estimated conservatively.

Since we were not able to derive rigorous error estimates in terms of M, it is commendable to backtest the accuracy in the tail obtained for a certain choice of M against the saddlepoint method (as illustrated in Figure 17.3). However, with the above choice of M, we have never observed any relevant deviations from results obtained with the saddlepoint method.

17.5 Application of the Method

We provide loss distribution results obtained by our proposed method for a credit portfolio containing 15,000 obligors with the following underlying

---

[2] $\mathcal{F}$ and $\mathcal{F}^{-1}$ are continuous linear maps. Therefore small input errors can be controlled.

default rate volatility model:

$$p_A^S = p_A \cdot S, \qquad S \sim \text{Gamma} \ (\mu_S = 1, \sigma_S = 0.8).^3$$

Counterparty exposures $(\nu_A)_A$ for this particular portfolio are generated by stretching variates by a factor of 100 drawn from a beta distribution. Additionally we allocate all obligors in the portfolio evenly among three rating classes. According to a counterparty's rating, it is assumed to have one of the following expected default frequencies: 0.5%, 1% and 1.5%. The portfolio loss distribution is provided in Figure 17.4.

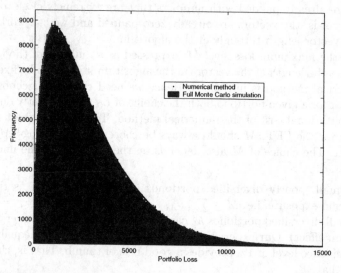

Fig. 17.4. Numerical method vs full Monte Carlo simulation. Number of simulations: 100,000 and 1,000,000, respectively

Figure 17.4 shows that the match of the distribution obtained with our algorithm with the distribution obtained by full Monte Carlo simulation is almost perfect! To underline the accuracy we additionally provide some statistics of the two distributions.

17.6 Risk Analysis of CDO Structures

The algorithm described in the previous sections turns out to be particularly useful for analysing and designing CDO structures. This is due to the fact that the method can be applied to accurately determine the risk mitigation resulting from hedging loss tranches on a given subportfolio. In fact, the risk

---

[3] Corresponds to the single factor CreditRisk$^+$ default rate volatility model.

	NM	fMC	Δ		NM	fMC	Δ
Mean	2932	2923	0.31%	$VaR_{80\%}$	4540	4518	0.49%
Std. dev.	2351	2359	-0.34%	$VaR_{90\%}$	6050	6067	-0.28%
Skewness	1.589	1.607	-1.12%	$VaR_{95\%}$	7530	7544	-0.19%

Table 17.3. Numerical method (NM) vs full Monte Carlo simulation (fMC): statistics of the loss distribution

mitigation can be determined from the underlying portfolio without making any additional ad-hoc assumptions.

In this section we will describe how the conditional independence framework can be applied to quantify the risk transfer involved in collateralized debt obligations (CDO) structures. In a simple CDO structure a bank buys credit protection on a specified subportfolio of its total credit portfolio. Usually a self-retention is involved (equity tranche) and losses exceeding the self-retention up to a specified maximum are covered by the protection provider on a quota share basis (quota share on the transferred tranche). One key question for the buyer of the credit protection is: How does the hedge on the subportfolio change the overall credit loss distribution? The advantage of the following treatment of this question is that the approach is fully consistent with the bank's internal credit risk model (as long as a conditional independence framework is chosen) and no separate and potentially inconsistent model for the risk analysis of CDO structures needs to be introduced ad hoc.

17.6.1 The risk transfer function to a subportfolio

The effect of the credit protection on the subportfolio can be determined by applying a risk transfer function that contains the specific features of the credit protection agreement (attachment points, quota share). A typical risk transfer function $\mathcal{R}$ is shown in Figure 17.5. The impact of the risk transfer function on the subportfolio can easily be determined for any given state of the latent variable $S = s$. In fact, if $p^{X_B|S=s}$ is the conditional loss distribution of a subportfolio B, then the distribution of the transformed random variable $\mathcal{R}(X_B^{S=s})$ is the conditional loss distribution of the subportfolio incorporating the hedging effect of the credit protection. Since $\mathcal{R}$ is usually a piecewise linear and increasing function it is not necessary to resort to a Monte Carlo simulation to determine the mass function of $\mathcal{R}(X_B^{S=s})$ since it can easily be derived from the conditional loss distribution of the subportfolio $p^{X_B|S=s}$.

17.6.2 The impact of the CDO on the bank's total credit risk

Note that for each given state of the latent variable the transformed loss distribution of the subportfolio and the loss distribution of the rest of the

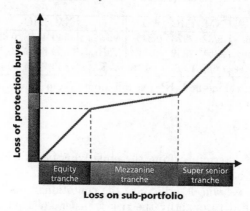

Fig. 17.5. Risk transfer function $\mathcal{R}$

bank's credit portfolio can be simply aggregated by convolution, since they are (conditionally) independent.[4]

By repeating the above aggregation step for each state of the latent variable we can determine the unconditional loss distribution of the whole portfolio fully taking into account the credit protection on the subportfolio. The process of applying the risk transfer function $\mathcal{R}$ is illustrated in Figure 17.6. This is just a simple modification of the general algorithm illustrated in Figure 17.2.

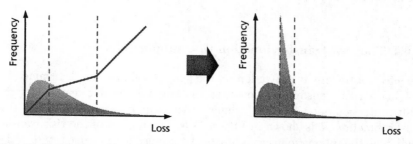

Fig. 17.6. Applying the risk transfer function to the conditional loss distribution of the subportfolio

17.7 Coherent Risk Allocation

The previous sections were devoted to numerical techniques for the determination of the probability distribution of total credit losses of a portfolio.

---

[4] Here we need to assume that each counterparty is either a member of the subportfolio or not, i.e. that the exposure of a counterparty is not partially hedged.

Different methods are required for the problem of allocating risk to the individual obligors in the portfolio. In the following sections we address this second main topic and refer to Chapter 3 for the introduction of coherent risk measures. Let us mention that the main result in [1] characterizes coherent risk measures as being scenario-based risk measures, i.e. any coherent risk measure expresses expectations on losses conditional on one or more different scenarios. Thereby the scenarios used for determining expectations uniquely characterize the coherent risk measure. For practical applications the expected shortfall, defined as $\mathbb{E}[X \mid X \geq c]$, is of particular importance since it answers the question: How much do we expect to lose if we assume that portfolio losses exceed a certain level c? Thus the loss level c needs to be chosen. This can be interpreted as follows: by specifying c the portfolio manager is forced to specify the loss scenario for which he is defining his risk management policies and actions. Thus a more conscious discussion of the firm's risk appetite and risk limits is required before resorting to mathematical techniques for implementing these choices at a more granular level. We consider the following function $\varphi_A(\cdot)$, which we shall call the coherent risk spectrum of obligor A. To define it on the unit interval $[0, 1]$ we normalize by expressing c as a percentage of the total portfolio volume $\nu = \sum_A \nu_A$, i.e., we define

$$\varphi_A : [0, 1] \to [0, 1] \,, \; \varphi_A(c) := \frac{\mathbb{E}[X_A \mid X \geq c\nu]}{\mathbb{E}[X \mid X \geq c\nu]} \,.$$

Hence $\varphi_A(c)$ is the relative contribution of obligor A to the total expected shortfall of the portfolio determined at the loss level $c\nu$, where $c \in [0, 1]$. We note that for $c = 0$ and $c = 1$ we can immediately determine $\varphi_A(c)$:

- For $c = 0$ the expected shortfall contribution reduces to (unconditional) expected loss since
$$\mathbb{E}[X_A \mid X \geq 0] = \mathbb{E}[X_A].$$

- At the other extreme $c = 1$ we are conditioning with respect to the scenario $[X \geq \nu] = [X = \nu]$, i.e., for $c = 1$ the expected shortfall contribution of any obligor reduces to its exposure contribution
$$\mathbb{E}[X_A \mid X = \nu] = \nu_A.$$

Therefore at $c = 0$ the coherent risk spectrum $\varphi_A(c)$ always starts with the value $\varphi_A(0) = \mathbb{E}[X_A]/\mathbb{E}[X]$ and connects this value with the value at $c = 1$ where $\varphi_A(1) = \nu_A/\nu$. We thus find that risk allocation according to expected loss contributions and with respect to exposure contributions are at the two opposite extremes of the coherent risk spectrum (in Section 17.10 we explicitly determine the risk spectrum of a specific portfolio). Portfolio managers will need to discuss the role and interpretation of the level c between those two extremes and it will be a key parameter in their risk management framework.

We will see later that the risk spectrum is an instrument that is able to analyse the interplay of rating quality, correlation level, and assumed loss level scenarios in a concise and explicit way, thereby unveiling the complex interaction of these three factors for the given portfolio.

We only consider credit risk models that are embedded in the conditional independence framework, as defined in Section 17.1. In contrast to previous sections, we neither express net exposures in terms of a pre-defined basic loss unit nor do we make use of the Poisson approximation.

17.8 Importance Sampling

We first discuss the difficulties encountered when trying to use standard Monte Carlo simulation to compute the coherent risk spectrum of a credit portfolio.

17.8.1 Standard Monte Carlo, a dead end

The inadequacy of the standard Monte Carlo approach for estimating expected shortfall contributions becomes evident by considering the following situation: assume that for independent default events we wish to estimate $\mathbb{E}[X_A \mid X \geq c]$ by using I independent draws $(x^i)_{1 \leq i \leq I}$ of the portfolio loss variable X. Let n_c denote the number of portfolio loss draws that are greater than or equal to the exceedance level c, i.e. $n_c = \sum_{i=1}^{I} 1_{\{x^i \geq c\}}$, where 1 denotes the indicator function taking values in $\{0, 1\}$. The standard Monte Carlo shortfall estimator for obligor A is then given by

$$\mathbb{E}[X_A \mid X \geq c] \approx \frac{1}{n_c} \sum_{i=1}^{I} x_A^i 1_{\{x^i \geq c\}} \,.$$

The inefficiency of this estimator is striking: for fixed I and increasing threshold level c, fewer and fewer portfolio loss variates x^i remain for estimating expected shortfall. For example, if c is set to the 90% quantile of the portfolio loss distribution, then on average only 10% of the drawn variates x^i can be used for estimating $\mathbb{E}[X_A \mid X \geq c]$. For determining the coherent risk spectrum with increasing c this naive approach becomes more and more computationally expensive, since the number of simulations I required to observe acceptable convergence will increase very rapidly. In practice, even small portfolios and values of c not even close to 1 constitute an unsurmountable task for this simple approach. Fortunately a better solution does exist.

17.8.2 Importance sampling, a viable alternative

Let Y be a random variable having a probability density function $f(y)$. Suppose we are interested in estimating

$$\Theta := \mathbb{E}_f[h(Y)] = \int h(y)f(y)dy. \tag{17.13}$$

A direct simulation of the random variable Y, so as to compute values of $h(Y)$, might be inefficient possibly because it is difficult to simulate from the distribution f, or the variance of $h(Y)$ is large.

Imagine we had another distribution function $g(y)$, which is equivalent to $f(y)$ (meaning that $g(y) = 0 \iff f(y) = 0$), then we could alternatively estimate Θ by

$$\Theta = \mathbb{E}_g \left[\frac{h(Y)f(Y)}{g(Y)} \right] = \int \frac{h(y)f(y)}{g(y)} g(y) dy . \tag{17.14}$$

It follows from (17.14) that Θ can be estimated by successively generating values of a random variable Y having a density function $g(y)$. If such a density function $g(y)$ can be chosen so that the random variable $h(Y)f(Y)/g(Y)$ has a small variance, then this approach – referred to as importance sampling – can result in an efficient estimator of Θ.

17.8.3 Tilted probability densities

For an estimation problem of the form (17.13) there is no straightforward way to determine a density $g(y)$ that efficiently reduces the variance of the estimator given in (17.14). A very useful approach for important special cases is, however, described in [5], a special one-parameter family of probability densities, so-called tilted densities, is discussed. These are particularly adapted to the efficient estimation of exceedance probabilities of the form $\mathbb{P}\left[\sum_i Y_i \geq c\right]$ for independent Bernoulli random variables Y_i.

Given the probability density $f(y)$ of a random variable Y its associated one-parameter family of *tilted densities* $(f_t)_{t \in \mathbb{R}}$ is defined by

$$f_t(y) := \frac{e^{ty} f(y)}{M(t)} ,$$

where $M(t) := \mathbb{E}_f[e^{tY}]$ is the moment-generating function of Y, which is assumed to exist. This definition is easily extended to the multivariate case: if $Y = (Y_1, \ldots, Y_m)$ has independent components with respective densities $f_1, \ldots, f_m$, then the tilted densities $(f_t)_{t \in \mathbb{R}}$ associated with $f(y_1, \ldots, y_m) = \prod_{i=1}^m f_i(y_i)$ are defined by

$$f_t(y_1, \ldots, y_m) := \prod_{i=1}^m f_{i,t}(y_i),$$

where $f_{i,t}$ denotes the tilted density of the component Y_i.

17.9 Shortfall Contributions for Independent Default Events

We only consider independent default events and show how expected short-fall contributions can be expressed in terms of exceedance probabilities. The generalization to dependent defaults is straightforward within the conditional independence framework and will be explained in a later section.

For independent default events our credit risk model given by (17.1) to (17.3) reduces to (17.3) and

$$(\mathbf{1}_A)_A \sim \text{Bernoulli } (p_A), \text{ independent.}$$

In accordance with Chapter 3 we get

$$\mathbb{E}[X_A \mid X \geq c] = \nu_A \frac{\mathbb{P}[\mathbf{1}_A = 1, X \geq c]}{\mathbb{P}[X \geq c]}$$

$$= \frac{\nu_A}{\mathbb{P}[X \geq c]} p_A \mathbb{P}\left[\sum_{A' \neq A} X_{A'} \geq c - \nu_A\right].$$

We thus conclude that the portfolio's expected shortfall contributions can be expressed by the exceedance probability $\mathbb{P}[X \geq c]$ and the restricted exceedance probabilities $\mathbb{P}\left[\sum_{A' \neq A} X_{A'} \geq c_A\right]$ determined at exceedance levels $c_A := c - \nu_A$. Before explaining how these probabilities can be estimated by means of tilted probability densities we apply the general definition to the case of Bernoulli variables.

17.9.1 Tilted Bernoulli mass functions

The probability mass function f of a Bernoulli variable Y with parameter p is defined by

$$f(y) = p^y (1 - p)^{1-y}, \qquad y = 0, 1.$$

An elementary calculation yields that a tilted Bernoulli mass function $f_t(y)$ is again a Bernoulli mass function with parameter

$$p_t = \frac{pe^t}{pe^t + 1 - p}. \tag{17.15}$$

In other words, tilted Bernoulli variables are again Bernoulli variables. Two important properties of the tilted distributions are:

1. $t \mapsto p_t$ is strictly monotone;[5]
2. $p_t = p$ for $t = 0$.

Accordingly, we recover the original mass function for $t = 0$, i.e. $f_0(y) = f(y)$. Obviously for any $t \in \mathbb{R}$ the mass function $f_t(y)$ is equivalent to $f(y)$.

---

[5] We assume that $p \in]0, 1[$.

17.9.2 A family of importance sampling estimators

To estimate any exceedance probability $\Theta_c := \mathbb{P}[X \geq c]$ we use the representation $\mathbb{P}[X \geq c] = \mathbb{E}[1_{\{X \geq c\}}]$. This allows us to write Θ_c in terms of tilted Bernoulli mass functions by

$$\Theta_c = \mathbb{E}_t \left[1_{\{X \geq c\}} \prod_A \frac{f_A(\mathbf{1}_A)}{f_{A,t\nu_A}(\mathbf{1}_A)} \right], \tag{17.16}$$

which holds for any given $t \in \mathbb{R}$. Here $\mathbb{E}_t$ denotes the expectation with respect to the importance sampling measure. This is just the application of the general idea (17.14) with the choice $g(x_1, \ldots, x_m) = \prod_{A=1}^{m} f_{A,t\nu_A}(x_A)$. From (17.16) we directly obtain the following family of unbiased importance sampling estimators for Θ_c:

$$\hat{\Theta}_{c,t} := 1_{\{X \geq c\}} \prod_A \frac{f_A(\mathbf{1}_A)}{f_{i,t\nu_A}(\mathbf{1}_A)} = 1_{\{X \geq c\}} M(t) e^{-tX} \tag{17.17}$$

with $M(t) := \prod_A [p_A e^{t\nu_A} + 1 - p_A]$. In contrast, the standard Monte Carlo estimator is given by

$$\hat{\Theta}_c = 1_{\{X \geq c\}} = \hat{\Theta}_{c,0}.$$

From the strong law of large numbers it follows that for any t-value $\hat{\Theta}_{c,t}$ is a consistent estimator of Θ_c, i.e. for independent $\hat{\Theta}_{c,t}^{(i)}$, $i \in \{1, 2, \ldots\}$ we have

$$\Theta_c = \lim_{I \to \infty} \frac{1}{I} \sum_{i=1}^{I} \hat{\Theta}_{c,t}^{(i)}.$$

Since any Monte Carlo simulation will only draw finitely many variates, the question arises, which estimator (out of the infinitely many) should we choose?

17.9.3 Choosing an appropriate family member

We now need to decide which member of the family we should choose to simulate from, i.e. which t minimizes the variance of $\hat{\Theta}_{c,t}$. Finding the answer to this question turns out to be very tricky. By the definition of variance

$$\text{var}[\hat{\Theta}_{c,t}] = \mathbb{E}_t \left[1_{\{X \geq c\}} M^2(t) e^{-2tX} \right] - \Theta_c^2.$$

Both the random variable of which the expectation is taken and the expectation itself depend on t. Finding this minimum is computationally very burdensome. Alternatively, we shall look for a different but also very suitable family member, which is much easier to compute.

An obvious but important observation is that for any $t \geq 0$

$$1_{\{X \geq c\}} e^{-tX} \leq e^{-tc}. \tag{17.18}$$

Combining (17.17) with (17.18) we conclude that $0 \leq \hat{\Theta}_{c,t} \leq M(t)e^{-tc}$, if $t \in [0, \infty)$. Hence a good candidate for the importance sampling distribution may be the one that yields a minimal range for the values of $\hat{\Theta}_{c,t}$, which is the case for

$$t^* := \max \left(\arg\min_t M(t)e^{-tc}, 0 \right).$$

Backing out t^* turns out to be very easy. The function $G(t) := M(t)e^{-tc}$ is strictly positive for all values of t. Hence

$$t^* \text{ minimizes } G(t) \implies \frac{d\log(G(s))}{ds}(t^*) = 0. \tag{17.19}$$

Evaluating the above derivative yields

$$\frac{d\log(G(s))}{ds}(t^*) = 0 \iff \sum_A \nu_A \frac{p_A e^{t\nu_A}}{p_A e^{t\nu_A} + 1 - p_A} - c = 0$$

$$\iff \mathbb{E}_{t^*}[X] = c. \tag{17.20}$$

Due to strict monotony of $t \mapsto \frac{d\log(G(s))}{ds}(t)$ it becomes apparent that there is always a solution to the minimization problem (17.19) and it is unique.

The interpretation of (17.20) is straightforward. Typically c is larger than the portfolio expected loss $\mathbb{E}[X]$ and the t^*-importance sampling distribution scales each counterparty's default probability p_A up to $p_{A,t^*\nu_A}$ according to (17.15) such that the portfolio expected loss under the importance sampling measure is exactly equal to the considered loss exceedance level c.

17.9.4 Example for independent default events

Let us consider a toy portfolio consisting of $K = 35$ counterparties. The portfolio contains three rating categories with 20, 10 and 5 obligors, respectively. The default probabilities and net exposures are chosen as follows.

No. obligors	Default prob. p_A	Net exposure ν_A
20	5%	{1,2,...,20}
10	1%	{13,16,...,40}
5	0.5%	{30,40,...,70}

Table 17.4. Set-up toy portfolio. Expected loss in percent of total portfolio volume is 1.99%

We benchmark the proposed importance sampling estimator $\hat{\Theta}_{c,t^*}$ with the standard Monte Carlo estimator $\hat{\Theta}_c$. For comparing the two estimators we use as statistics their coefficient of variation (CV).[6] Luckily, the CV for

[6] The CV of a non-negative random variable Y is defined as $\sigma(Y)/\mu(Y)$, i.e. standard deviation divided by mean. $CV(Y)$ measures the relative dispersion of Y around its mean.

the standard Monte Carlo estimator can be calculated analytically.[7] Not so for the importance sampling estimator, for which we use 10,000 simulations to determine its CV. This is done repeatedly for a number of loss levels c from 2% to 50% of total portfolio volume. As we see in Figure 17.7, the variance of

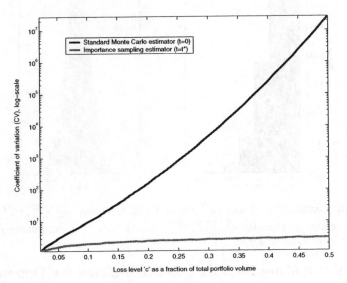

Fig. 17.7. Coefficient of variation (CV): importance sampling estimator $\hat{\Theta}_{c,t^*}$ vs standard Monte Carlo estimator $\hat{\Theta}_c$

the standard Monte Carlo estimator grows exponentially as a function of c, whereas the amount of variance in our proposed estimator is extremely small even for very high loss levels. In fact the statement in Figure 17.7 is very universal in the sense that the graphs remain alike for any kind of portfolio!

To give some further insight into the estimator's properties, we compare the distributions of the two estimators for a given loss level $c = 10\%$ of the total portfolio volume. For both methods we estimate the loss exceedance probability using 100 simulations. Repeating this step 10,000 times we arrive at the histogram charts in Figure 17.8. The discrepancy in performance is striking. In this example, the standard Monte Carlo estimator can, by definition, only take values on the grid $\{0, 0.01, 0.02, \ldots, 1\}$. Under the t^*-measure, conversely, loss exceedance events such as $[X \geq c]$ are much more frequent and the correction term $M(t)e^{-tX}$ prevents the estimator from being bound to any coarse grid.

---

[7] $\mathrm{CV}(\hat{\Theta}_c) = \sqrt{\frac{1 - \mathbb{P}[X \geq c]}{\mathbb{P}[X \geq c]}}$. For small portfolios the exceedance probability $\mathbb{P}[X \geq c]$ can be calculated analytically by convolution.

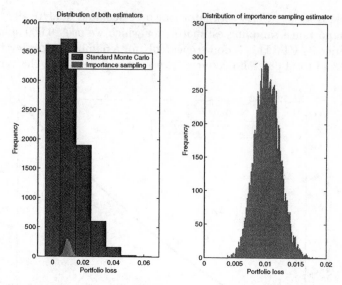

Fig. 17.8. Distribution of standard Monte Carlo estimator $\frac{1}{100}\sum_{i=1}^{100}\hat{\Theta}_c^{(i)}$ and importance sampling estimator $\frac{1}{100}\sum_{i=1}^{100}\hat{\Theta}_{c,t^*}^{(i)}$ for $c = 10\%$. The exceedance probability $\mathbb{P}[X \geq c]$ is equal to 1.03%

17.10 Estimating Shortfall Contributions for Dependent Default Events

We now turn back to the original question on how to estimate expected shortfall contributions for any credit portfolio model that is embedded within the conditional independence framework. The generalization from independent to dependent default events is straightforward:

$$
\mathbb{E}[X_A \mid X \geq c] = \nu_A \frac{\mathbb{P}[\mathbf{1}_A = 1, X \geq c]}{\mathbb{P}[X \geq c]}
$$

$$
= \lim_{I \to \infty} \frac{\nu_A}{\mathbb{P}[X \geq c]} \frac{1}{I} \sum_{i=1}^{I} \mathbb{P}\left[\mathbf{1}_A = 1, X \geq c \mid S = s^i\right] \quad (17.21)
$$

$$
= \lim_{I \to \infty} \frac{\nu_A}{I\mathbb{P}[X \geq c]} \sum_{i=1}^{I} f_A(s^i) \mathbb{P}\left[\sum_{A' \neq A} X_{A'} \geq c_A \mid S = s^i\right].
$$

$\left(s^{(i)}\right)_i$ are independent draws from the distribution F. The portfolio exceedance probability $\mathbb{P}[X \geq c]$ also depends on i, since it is estimated by

$$
\mathbb{P}[X \geq c] = \lim_{I \to \infty} \frac{1}{I} \sum_{i=1}^{I} \mathbb{P}\left[X \geq c \mid S = s^i\right]. \quad (17.22)
$$

The recipe for doing a Monte Carlo simulation for (17.21), (17.22) combined with the proposed importance sampling technique is as follows:

1. In each simulation step, generate a variate s of S from the distribution F of the systematic risk factors.
2. Determine each counterparty's conditional default probability $p_A^s = f_A(s)$.
3. Given $(p_A^s)_A$ find t^* for estimating $\mathbb{P}[X \geq c \,|\, S = s]$.
4. Given $(p_A^s)_A$ find for each counterparty A the value t_A^* for estimating

$$\mathbb{P}\left[\sum_{A' \neq A} X_{A'} \geq c_A \,|\, S = s\right].$$

In Steps 3 and 4 one usually uses between 10 and 1000 importance sampling simulations of the Bernoulli indicators for estimating the conditional exceedance probabilities.[8] In Section 17.11 we will provide a trick on how to cut the huge computational burden of Step 4 down to a minimum.

Our importance sampling technique does not impose any restrictions on either the portfolio structure (size, exposure distribution) or the distribution of systematic risk factors. So the method will work for any type of dependence structure that allows for efficient generation of random variates.

Example for dependent default events

We shall now estimate expected shortfall contributions on counterparty level for a larger portfolio counting 1200 obligors and for dependent default events. The analysis is presented in three parts. First we analyse the impact of the expected default rate level and the level of default rate volatility on the counterparty shortfalls $\varphi_A(c)$ for a fixed loss exceedance level c. Then we shall incorporate exposure inhomogeneity to get a glance at how exposure impacts $\varphi_A(c)$. Finally the coherent risk spectrum is discussed.

Dependence structure

For making default events dependent, we choose as an underlying model the classical CreditRisk$^+$ two-factor default rate volatility model, written as

$$p_A^S = p_A(w_A S_1 + (1 - w_A)S_2).$$

The factor loadings w_A are required to be in $[0,1]$ for all A. Both risk factors S_1, S_2 are assumed to be independent and gamma[9] distributed with $\mathbb{E}[S_1] = \mathbb{E}[S_2] = 1$. The risk factor volatilities are set to $\sigma_{S_1} = 0.4$ and $\sigma_{S_2} = 1.2$. A default rate volatility of 0.4 is generally considered as low, whereas 1.2 is rather high.

---

[8] A total of 100,000 simulations yields very accurate shortfall contribution estimates (see the upcoming example). So if we decide to draw $K = 10,000$ systematic risk factor variates, it suffices to use another 10 simulations in each Step 3 and 4.

[9] In this set-up the range of p_A is $[0, \infty)$ rather than $[0,1]$. If any p_A should take on a value larger than 1 (which is very unlikely to happen), we reset it to 1 during the simulation.

Portfolio composition

We divide the portfolio into six segments: three default rate volatility bands (indexed by numbers 1, 2, 3) and two rating bands (referred to by capital letters A, B). Each of the six segments shall contain 200 obligors. At first we assume that all counterparties have unit net exposure, i.e. $\nu_A = 1$ for all A.

Segment	A	w_A	p_A	ν_A
1B	1–200	0.8	1%	1
1A	201–400	0.8	0.3%	1
2B	401–600	0.5	1%	1
2A	601–800	0.5	0.3%	1
3B	801–1000	0.2	1%	1
3A	1001–1200	0.2	0.3%	1

Table 17.5. Portfolio composition: Risk segments, counterparty index A, risk factor loadings w_A, expected default frequencies p_A, net exposure ν_A

According to our set-up, default rates in the two segments 1A/B have a low, in 2A/B a medium and in 3A/B a rather high volatility. Each volatility segment contains two rating categories A and B, both with equal number of counterparties. The expected default probability of any obligor in band A is 0.3% and 1% in category B.

Simulation set-up

We simulate 10,000 variates from the distribution of the systematic risk factors $S = (S_1, S_2)$. Conditional on any given variate $S = s$, we use another 10 simulations to estimate the conditional portfolio exceedance probability $\mathbb{P}[X \geq c \,|\, S = s]$ and the restricted exceedance probabilities $\mathbb{P}[\sum_{A' \neq A} X_{A'} \geq c_A \,|\, S = s]$. Again the results are compared with the standard Monte Carlo estimation approach on a fair basis.

Results

The portfolio expected loss as a fraction of total portfolio volume is equal to 0.65%. The threshold level c is set at 2.25%, which is equivalent to the 98.8% quantile. Figure 17.9 again underpins the results obtained in the section on independent default events. The importance sampling estimator is extremely efficient and the Monte Carlo noise is practically zero.

From the point of view of a risk manager the above results are perfectly consistent with intuition: the coherent risk measure expected shortfall reacts as expected to different levels in default rate volatility (volatility segments) and expected default rate (rating segments).

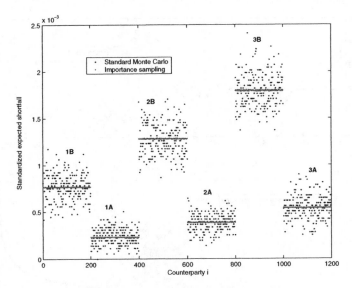

Fig. 17.9. Risk allocation by counterparty in terms of standardized expected short-fall contributions $\varphi_A(c)$ for a fixed level $c = 2.25\%$: Importance sampling versus standard Monte Carlo

Incorporating exposure granularity

Next we drop the assumption of unit exposures. For mimicking the exposure distribution of a typical credit portfolio, we randomly draw independent net exposures $(\nu_A)_{1 \le A \le 400}$ from a non-negative, skewed distribution. $\nu_A := (\hat{X})^4$, $X \sim \mathcal{N}(12,5)$.[10] These exposures are sorted in ascending order and applied to each volatility segment 1, 2, 3.[11] Hence for each volatility segment, the smallest 50% of the net exposures are assigned to the lower-rated ($p_A = 1\%$) and the largest 50% to the better-rated obligors ($p_A = 0.3\%$).

We provide a relative exposure impact analysis on expected shortfall for the highly volatile segment 3, meaning expected shortfall per dollar net exposure per counterparty. In the upper plot of Figure 17.10 the structural break in the curve from counterparty 1000 to 1001 is due to the change in rating category (from $p_A = 1\%$ to 0.3%). Apart from the discontinuity, the curve is strictly monotone increasing. This is due to the fact that we assigned sorted exposures to each volatility segment and that $\varphi_A(c)$ per dollar is sensitive to the overall portfolio exposure level and the exposure level of each counterparty. If the exposure sensitivity of $\varphi_A(c)$ were to be zero, then the two curves would both be a straight line (gradient of zero), preserving the discontinuity

---

[10] As throughout this chapter, exposures are assumed to be constant, not random. But for generating a skewed exposure distribution we used a procedure based on normal random numbers. $\mathcal{N}(\mu,\sigma)$ stands for the normal distribution with mean μ and standard deviation σ.

[11] This implies that $\nu_1 = \nu_{401} = \nu_{801}$, $\nu_2 = \nu_{402} = \nu_{802}$ etc.

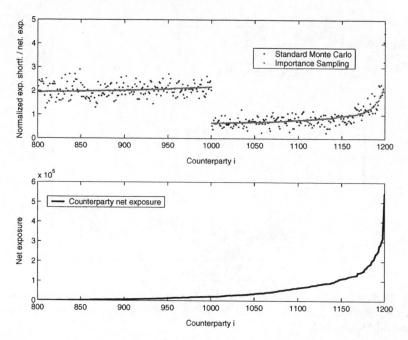

Fig. 17.10. Risk and exposure allocation by counterparty. Upper plot: Normalized expected shortfall contributions $\varphi_A(c)$ per dollar net exposure, importance sampling vs. standard Monte Carlo. Lower plot: Net exposure distribution for volatility segment 3

at 1001. Again this result is very much in line with our intuition on how risk "should" be allocated.

Due to its coherency expected shortfall is a very suitable measure for defining exposure limits on counterparty or portfolio segment basis from a risk management perspective.

The coherent risk spectrum

For large portfolios the probability that $X \geq c$ converges very quickly to 0 as c is increased. Thus to find an example where it is meaningful to plot the entire spectrum for $c \in [0,1]$ we reduce the current portfolio to 60 counterparties.[12] Again we distribute the obligors homogeneously among the six risk segments, for which we shall analyse the coherent risk spectrum. Last but not least, all net exposures are reset to 1. As mentioned in Section 17.7, the single curves in the spectrum start at the level of the respective segment's (normalized) expected loss and attain the right end at the level of the segment's relative

---

[12] Note that the portfolio exceedance probability at the 95% loss level is estimated to be $6.25 \cdot 10^{-86}$.

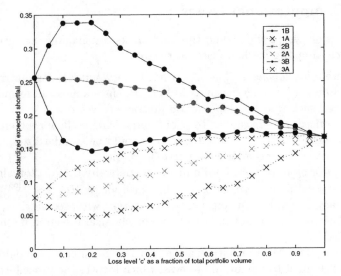

Fig. 17.11. Coherent risk spectrum of the six risk segments 1A to 3B. The curves were generated using $K = 10,000$ simulations for the systematic risk factors and 500 conditional simulation steps

exposure contribution. Since for this particular example all segments have equal exposure all the curves attain the same endpoint.

The risk spectrum displays the obvious risk characteristics of the portfolio, since segments with higher default probability and higher systematic risk also carry higher relative contribution to total expected shortfall, producing a risk ranking that is identical to the one shown in Figure 17.9. A more interesting observation is that the credit risk of segments 3A and 1B is at a similar level for high loss levels c, i.e. for stress loss scenarios, whereas for lower loss levels the risk assigned to segment 3A decreases as a consequence of its higher rating quality. Thus the higher systematic risk (correlation) present in segment 3A manifests itself in the event of stress scenarios. The interdependence of rating quality, correlation level, assumed loss level scenario and their impact on the risk allocation is therefore revealed by the coherent risk spectrum. In this framework, rather than being a static and quite arbitrary choice, the loss level c becomes a dynamic variable in itself and can be used to understand the relative risk contributions in the portfolio for different scenarios. The choice of the scenarios deserving consideration and the assessment of whether the corresponding risk contributions are acceptable or not is, however, subjective and is due to the portfolio management and cannot simply be delegated to a model. The model is, however, able to detect unexpected effects and to base decisions on more objective arguments. In contrast to a static choice of any particular risk measure the spectrum approach intrinsically delivers a rather comprehensive scenario analysis and fosters common sense where it is due.

17.11 Computational Issues

In this section we provide some tips and tricks for making the introduced importance sampling method fast and reliable.

1. *Programming language*: The algorithm is very suitable for matrix-based programming languages such as MATLAB, since almost all formulas can be coded such that they operate on matrices or vectors.
2. *Standardized net exposures*: To avoid numerical overflow problems when evaluating $e^{t\nu_A}$ all net exposures $(\nu_A)_A$ should be standardized, meaning expressed as a fraction of total portfolio volume.
3. *Numerical optimization*: For finding t^* numerically the Newton algorithm is a very suitable method for finding the root of $H(t) := \frac{d\log(\bar{G}(s))}{ds}(t)$. It is easy to implement and usually converges with geometric speed. In case Newton fails, it is handy to have the classical bisection method as a backup.

   To use as few Newton optimization steps as possible one should give a second thought to the initial t_0-guess, which invokes the procedure. Using Taylor's expansion theorem we have that

   $$0 = H(t^*) = H(0) + H'(0)(t^*) + \frac{1}{2}H''(0)(t^*)^2 + O\left((t^*)^3\right).$$

   Omitting all higher-order terms and solving directly for t^* we retrieve a very reliable and easy to compute starting point t_0. Starting at t_0, four to seven Newton steps are usually sufficient to find any t^* or t_A^* with very high accuracy.

4. t^*-*bands*: For estimating in simulation step i the conditional restricted exceedance probability $\mathbb{P}\left[\sum_{A'\neq A} X_{A'} \geq c_A \mid S = s^i\right]$ for each obligor A, there is no need to solve for each single $t_A^*(i)$. Instead, we can group obligors for which the corresponding $t_A^*(i)$ will take on very similar values. This bucketing is possible because of two facts: (a) t_A^* primarily depends on counterparty A's net exposure and (b) the importance sampling exceedance probability estimator is unbiased for any value $t \geq 0$.

   Hence by bucketing counterparties by their net exposure size, we may only slightly decrease the estimator's speed of convergence. The estimation of the restricted exceedance probabilities itself remains on counterparty level. Since the net exposure remains equal at every simulation step i, we only need to do the bucketing once at the beginning of the simulation. We have found that for most applications as little as two bands for almost any portfolio is sufficient!

5. *Runtime*: For the 1200 counterparty example, the calculation of each obligor's shortfall contribution using 2 t^*-bands (as described above) took 15 minutes on a standard PC (1.6 GHz) using MATLAB. In contrast, the results obtained using standard Monte Carlo were nowhere near the ones obtained by importance sampling, even after augmenting the number of simulations I from 10,000 to 5,000,000. Runtime increased to 10 hours.

6. *VaR*: The universality of the importance sampling method allows us to estimate VaR at any confidence level, for any portfolio type and for any conditional independence model. It is therefore a valuable alternative to the saddlepoint method, which can perform poorly for portfolios with a skewed exposure distribution.

References

1. P. Artzner, F. Delbaen, J.-M. Eber, and D. Heath. Coherent measures of risk. *Mathematical Finance*, 9(3):203–228, 1999.
2. R. B. Ash and C. A. Doléans-Dade. *Probability & Measure Theory*. Academic Press, San Diego, 2000.
3. A. D. Barbour. In D. N. Shanbhag and C. R. Rao, editors, *Topics in Poisson Approximation*, Elsevier Science, 2001.
4. Credit Suisse Financial Products. CreditRisk+: A credit risk management framework. London, 1997. Available at http://www.csfb.com/creditrisk.
5. M. B. Gordy. A comparative anatomy of credit risk models. *Journal of Banking & Finance*, 24:119–149, 2000.
6. H. Koyluoglu and A. Hickman. Reconcilable differences. *Risk*, 11(10):56–62, 1998.
7. P. Jäckel. *Monte Carlo Methods in Finance*. Wiley Finance, 2002.
8. Martin, R. Martin, K. Thompson, and C. Browne. Taking to the saddle. *Risk*, 14(6):91–94, 2001.
9. S. Merino and M. Nyfeler. Calculating portfolio loss. *Risk*, 15(8):82–86, 2002.
10. H: Panjer. Recursive evaluation of a family of compound distributions. *ASTIN Bulletin*, 12:22–26, 1981.

Some Remarks on the Analysis of Asset-Backed Securities

Daniel Kluge and Frank B. Lehrbass

Summary. In this chapter we discuss the analysis of asset-backed securities (ABS) in the environment of competitive risk-based pricing in the banking industry. We will cover the relevant aspects that need to be considered before investing in an ABS structure. When it comes to model-based pricing approaches a portfolio model is needed. The practitioner may either choose a simulation-based approach or an analytical model, where both have their advantages and shortcomings. We will focus on the usage of CreditRisk$^+$ in the context of ABS pricing, outline the prerequisites for running the model in practice and finally discuss the pricing of a simple ABS structure with CreditRisk$^+$.

Introduction

In recent years the presence of asset-backed securities (ABS) on the capital markets has increased dramatically. Depending on the structure they are employed mainly for generating liquidity, achieving capital arbitrage or realizing economic risk transfer. By no means is this list complete. Due to the virtually unlimited variety of ABS structures they offer a wide spectrum of investment alternatives to the capital market participants. Common to all ABS structures is, on the one hand, a reference pool of financial claims and on the other, issued notes linked to the performance of the reference pool. The reference pool may consist of all kinds of assets, from liquid ones like corporate bonds or credit default swaps to the illiquid end of credit products such as consumer loans or mortgage loans. The pool itself may be static or dynamic. The ABS structure will provide appropriate mechanisms to transfer the targeted risks from the reference pool to the noteholders. The transferred risks will always comprise risks associated with the credit of the underlying assets. When dealing with ABS structures the investor faces basically two types of securitizations, i.e. true sale structures or synthetic structures. In a true sale transaction a special purpose vehicle (SPV) will be set up to issue notes and directly acquire the assets in the reference pool from the proceeds. The issued notes are then served solely from the cash flow generated by the assets in the reference pool.

Contrary to that, in a synthetic deal the SPV guarantees the credit risk of the reference pool while at the same time issuing notes and buying appropriate collateral to be able to fulfil firstly its obligations under the guarantee for the reference pool and secondly its obligations towards the noteholders. The notes are served from the cash flow generated by the collateral pool and the premium income for the portfolio guarantee. The decision for either a true sale or a synthetic structure depends on various issues such as jurisdiction, purpose of securitization and types of assets in the reference pool. For example, a true sale generates liquidity for the securitising entity and reduces its balance sheet, whereas a synthetic deal is a pure risk transfer.

Out of an ABS structure there will be different tranches of notes issued with different ranking with respect to principal and interest as well as loss allocation. In the plain vanilla case the notes receive principal and interest in their order of seniority and loss allocations in the reverse order. As such, principal repayment of the most senior tranche has a higher priority than principal repayment for the other tranches. Moreover the most senior tranche is entitled to receive interest before the subordinated tranches receive interest, while having to bear losses only if all subordinated tranches were fully eaten up by former losses. Vice versa, the most junior tranche will bear the initial losses while receiving principal and interest only if all higher ranking notes have been served appropriately. The loss allocation mechanism can be viewed as standardized. However, we may often find other amortization patterns in an ABS structure than purely sequential. Each tranche can at most bear losses up to the outstanding tranche principal. The outstanding principal of subordinated tranches builds the credit enhancement for a given tranche. Moreover, in a synthetic deal the most senior tranche will often be a credit default swap, in which case the structure is only partially funded. Since we will focus on the pricing of credit risk we will from now on no longer distinguish whether a tranche is a credit default swap or an issued bond. Apart from the credit enhancement of an ABS tranche provided by the subordinated tranches there can be additional credit enhancement in the structure, e.g. overcollateralization, a reserve fund and/or excess spread as a first layer of protection against losses in the reference pool.

So far it should already have become clear that ABS structures can get fairly complex, especially regarding the fact that we did not elaborate all the risks inherent in a given structure. Risks we have not mentioned so far are, e.g. servicer risk in transactions backed by loan positions or the performance of the collateral pool in a synthetic transaction. Given the complexity and to some degree the uniqueness of each ABS structure a model-based pricing approach alone cannot be viewed as sufficient, when evaluating an ABS tranche. Moreover one should also carry out a thorough analysis of all risks and safeguards within a structure, ultimately to be able to get a feeling for the gap between reality and the employed model. We will discuss this more qualitative approach to a potential investment in Section 18.1.

Concerning a model-based pricing approach, we recall that an ABS tranche from a quantitative point of view is nothing more than a derivative whose payoff profile is linked to the performance of a reference portfolio of financial assets. Any ABS tranche pays a regular coupon to the noteholders, who in turn have to bear the losses allocated to the tranche. As such we may view the investment in an ABS tranche as a combination of a risk-free investment and a short position (i.e. selling protection) in a credit default swap (CDS). The risk premium incorporated in the coupon is the premium income from the CDS, while the allocated losses at each coupon date are the protection payments out of the CDS. Let $s < t$ be two adjacent coupon dates and $\xi(t)$ the loss allocated to the tranche at t. Let $N(s)$ be the outstanding tranche notional at s and $K_L(s)$ the respective credit enhancement at s. Then we have

$$\xi(t) = min(N(s); (\Delta X(t) - K_L(s))^+)^1$$

where $X(t)$ is the cumulative loss allocated to the structure at time t and $\Delta X(t) = X(t) - X(s)$ is the loss allocation to the structure at time t. Note that $K_L(t)$ decreases over time when losses are allocated to subordinated tranches. Assuming fixed coupon dates $0 < T_1 < \ldots < T_m = T$ with $\Delta_{j-1,j} = T_j - T_{j-1}$ and the existence of a pricing measure $\mathbb{Q}$ – we will discuss the latter assumption below – the fair risk premium for the ABS tranche must satisfy:

$$s^* = \frac{\sum_{j=1}^m B(0,T_j)\mathbb{E}_\mathbb{Q}[\xi(T_j)]}{\sum_{j=1}^m B(0,T_j)\Delta_{j-1,j}\mathbb{E}_\mathbb{Q}[N(T_{j-1})]}$$

with $B(0,t)$ being the price of a risk-free zero coupon bond with maturity t. In this expression we have implicitly assumed the independence between losses in the reference pool and interest rates. This assumption is not unusual when modelling defaultable claims. The simplicity of this expression is unfortunately a matter of notation. To determine $\mathbb{E}_\mathbb{Q}[N(T_{j-1})]$ we need to know the joint distribution of $(X(T_{j-1}), P(T_{j-1}))$ where $P(t)$ denotes the cumulative principal repayment in the ABS structure until time t.

In order to obtain $\mathbb{E}_\mathbb{Q}[\xi(T_j)]$ we need to know the joint distribution of $(X(T_{j-1}), P(T_{j-1}), K_L(T_{j-1}), \Delta X(T_j))$. In any case it is obvious that we need to employ some kind of a portfolio credit risk model to determine the expected values in the above pricing formula. With a sufficiently complex ABS structure there is little hope of solving for s^* analytically. In many cases Monte Carlo is therefore the method of choice. This will come at the price of the usual shortcomings of simulation approaches. Most fundamental is the inaccurate assessment of the probability of extreme events. This, however, is especially harmful in the evaluation of senior tranches where, e.g. only extreme losses in the reference pool will lead to losses in the tranche. As such, the pricing of a senior tranche will be sensitive to the obtained tail probabilities for the loss distribution of the reference pool.

---

[1] We denote by f^+ the non-negative part $\max\{f, 0\}$ of a function f.

CreditRisk$^+$ as an analytical model does not have these shortcomings. Thus we seek to identify ABS structures where we can employ CreditRisk$^+$ for the pricing of tranches. We will discuss a simple structure in Section 18.3 where this approach works. In our current practical work we strive to transform given structures into our simple structure. The task of finding an analytical extension of CreditRisk$^+$ to cope with more complex structures remains. Related work can be found in [3], where the pricing of a collateralized loan obligation (CLO) with CreditRisk$^+$ is discussed, and in Chapter 19 where the pricing of a synthetic structure with a reference pool consisting of single name CDS is considered. Crucial at this point is the determination of individual default probabilities of the names in the reference pool under the pricing measure $\mathbb{Q}$. In [3] these probabilities are implicitly assumed to be given; in Chapter 19 they are implied by CDS quotes available from the market. The latter should always be the method of choice when CDS quotes for the names in the reference pool are available.

This, however, will not apply to our simple structure, where we will consider a reference pool of mortgage loans. Having extracted default probabilities under the pricing measure $\mathbb{Q}$, it remains to specify the remaining risk parameters in the CreditRisk$^+$ model under $\mathbb{Q}$. These parameters comprise the variances of the systematic default risk factors that ultimately determine the correlation structure in the reference pool, as well as the variances of severity risk factors when using model extension as presented in [1] or Chapter 7. This will be discussed in our concluding remarks in Section 18.4. What about the pricing measure $\mathbb{Q}$? To be precise, we need the assumption that the assets in the reference pool have been priced under some pricing measure $\mathbb{Q}$. With a reference pool of traded CDS contracts as in Chapter 19 there is room for justifying the existence of a pricing measure, especially in the context of available market quotes. But as we consider a reference pool of loan positions the situation is different. Moreover, when an entity prices loans by using some pricing measure $\mathbb{Q}$, which is a reasonable assumption whenever a risk-based pricing system is applied, why should this be the measure used by other market participants to price credit risk?

From a practitioner's point of view there are the following arguments. Due to increasing market transparency and a clear trend towards economically reasonable determination of loan margins, we observe a substantial convergence in loan margins within the banking industry. Having said that, loan margins are increasingly dictated by the market. This applies especially to banks that are active in securitizing their assets. Alternatively, even if the pricing measure used by, e.g. a bank for the determination of loan margins is not the measure used by other investors to price credit risk, we may look at the bank considering an ABS investment as a surrogate for direct loan origination, and therefore it makes sense to price ABS and individual loan positions consistently under the same pricing measure. In the end this means that a price for an ABS tranche is not necessarily the fair market price, but identifies the ABS tranche as cheap, expensive or appropriately priced with respect to the bank's pricing of originated loans. Henceforth we assume the existence of a pricing measure $\mathbb{Q}$.

The rest of this contribution is organized as follows. In Section 18.1 we will discuss the qualitative approach to ABS structures as mentioned before, which we will call the pre-model analysis. Section 18.2 is dedicated to the essential steps that have to be taken before running the CreditRisk$^+$ model including some remarks on the estimation of systematic default risk factor variances. In Section 18.3 we will discuss the pricing of our simple structure with CreditRisk$^+$. Some concluding remarks can be found in Section 18.4. Finally the appendix shows how to compute the expected loss of an ABS tranche given our simple structure.

18.1 Pre-Model Analysis

For the pre-model analysis we will focus on safeguards and risks due to the structure embracing the reference pool in an ABS transaction. Due to structure related risks, two ABS transaction with identical reference pools may display very different risk profiles. In order to keep our explanations within a reasonable scope, we will focus on the analysis of mortgage-backed securities (MBS), which are ABS with a reference pool consisting of mortgage loans. However, the analysis presented can be applied with some modifications to many other ABS structures as well. We need the pre-model analysis to evaluate whether a transformation of a given structure into our simple structure is possible from a practical point of view. At the same time the discussed issues should be a building block in any analysis of an ABS structure. We propose a four-step analysis, where pricing considerations are deferred to Section 18.3.

The first step is the identification of structure-related risks. Potential risks are:

- Currency mismatch
- Interest rate mismatch
- Cash flow mismatch
- Servicer risk
- Collateral risk in a synthetic deal
- Replenishment risk
- Loss definition in a synthetic deal

The second step is the identification of the cash flow waterfall. The following questions need to be answered:

- Are principal and interest paid from a single or two distinct waterfalls?
- What is the pre-enforcement waterfall?
- What is the post-enforcement waterfall?
- Is pro rata amortization of tranches of different seniorities possible?
- What does the waterfall mean for the evolution of credit enhancements?

The third step is the identification of structural safeguards. The structure should be checked for the following:

- Liquidity facility and liquidity facility provider
- Interest rate hedge and hedge provider
- Currency hedge and hedge provider
- Mitigation of servicer risk (backup servicer)
- Mitigation of replenishment risk
- Mitigation of collateral risk in synthetic transactions
- Reserve fund
- Excess spread and excess spread trapping
- Transaction calls

In the context of structural safeguards it is important to figure out to what extent the risks identified in the first step are hedged and can be neglected in the further analysis. Besides that the impact of excess spread should be assessed. In fact, in some structures the excess spread can also be neglected. This is due to the fact that the excess spread is extracted from the structure, if not needed, and will be compressed considerably if needed most.

Step four is the identification of the legal risks and legal safeguards. These risks are often hard to quantify but should be kept in mind when deciding on an investment. Among others the following items need to be considered:

- Set-off risk and mitigation
- Effectiveness of the transfer of security over assets
- Representations and warranties for transferred assets
- Note events of default

This list is not exhaustive and step four often requires a thorough study of the transaction documents.

18.2 Getting Started with the Model

To run CreditRisk$^+$ in practice, especially in the context of pricing purposes, there are several data requirements that need to be fulfilled. Again we restrict our attention to MBS. The initial step needs to be the transformation of the portfolio data. We set up our portfolio model based on a normalized database with the elementary tables CLIENT, LOAN, PROPERTY, CONTRACT, TENANT (the latter two tables being needed only in the context of investment properties). Having filled the tables and established the relations the data is checked by applying mortgage bankers' expertise and knowledge. From this database the input for the CreditRisk$^+$ model is generated. Independent of the inclusion of model extensions we will always need the default probability and net exposure for each obligor in the reference pool. Our pricing approach as introduced in Section 18.3 requires furthermore the risk premium for each position in the reference pool. Depending on the available pool data risk premia need to be implied out of loan margins. As we will see in Section 18.3 we might also run into the situation where we have to estimate the variances of the systematic default risk factors from the variances of the individual default rates. For this case we want to propose a different estimation rule as given in [2].

To elaborate our ideas we need some notation in line with Chapter 2. Let $\mathbb{P}$ be the subjective probability measure. Denote obligors by $A = 1, \ldots, K$. Individual default probabilities are modelled as continuous random variables p_A^S satisfying

$$p_A^S = \sum_{k=0}^{N} p_A w_{Ak} \frac{S_k}{\mathbb{E}_{\mathbb{P}}[S_k]}. \tag{18.1}$$

S_0 is the degenerate zero-variance risk factor corresponding to the idiosyncratic default risk. The S_k, $k = 1, \ldots, N$, are the remaining systematic default risk factors, assumed to be mutually independent and gamma distributed with $\mathrm{var}_{\mathbb{P}}[S_k] = \sigma_k^2$.

Note that the risk factors S_k can be normalized to have any desired expected value. In [2] the risk factors are arranged to have $\mathbb{E}_{\mathbb{P}}[S_k] = \mu_k$, where $\mu_k = \sum_{A=1}^{K} p_A w_{Ak}$ is the mean default rate in the k-th sector. For the remainder of this section we will work within this special setting with respect to risk factor expectations. The factor loading $w_{Ak} \in [0, 1]$ indicates to what extent obligor A's default frequency is influenced by the k-th factor. Individual default frequency standard deviations are denoted by σ_A, i.e. $\mathrm{var}_{\mathbb{P}}[p_A^S] = \sigma_A^2$. The mutual independence of the risk factors together with (18.1) implies

$$\sigma_A^2 = \sum_{k=1}^{N} p_A^2 w_{Ak}^2 \frac{\sigma_k^2}{\mu_k^2}. \tag{18.2}$$

If the σ_k, $k = 1, \ldots, N$, are estimated directly, (18.2) gives a straightforward rule to derive each σ_A endogenously. In the case where each σ_A is given, the estimates of σ_k, $k = 1, \ldots, N$, should (at least approximately) be consistent with (18.2). Ideally, (18.2) needs to be satisfied for each obligor $A = 1, \ldots, K$, which is equivalent to the linear system

$$\begin{pmatrix} c_{11} & \cdots & c_{1N} \\ \vdots & & \vdots \\ c_{K1} & \cdots & c_{KN} \end{pmatrix} \begin{pmatrix} z_1 \\ \vdots \\ z_N \end{pmatrix} = \begin{pmatrix} b_1 \\ \vdots \\ b_K \end{pmatrix} \tag{18.3}$$

with $c_{Ak} = \frac{p_A^2 w_{Ak}^2}{\mu_k^2} \geq 0$, $z_k = \sigma_k^2 \geq 0$ and $b_A = \sigma_A^2 > 0$.

Since K (the number of obligors) will typically be much larger than N (the number of risk factors), the system (18.3) will in general have no solution. The best we can achieve from a mathematical point of view is to choose $z = (z_1, \ldots, z_N)^T$ s.t. its image under $C = (c_{Ak})_{A,k}$ minimizes the Euclidean distance to b, i.e. Cz must be the linear projection of b onto the image of C.

Although this linear projection is unique, as long as C does not have maximal rank, this will not imply a unique z. Hence the σ_k's obtained from (18.3) may not be unique.

A more pragmatic approach is to prescribe a simple rule to derive each $\sigma_k, k = 1, \ldots, N$, contingent on the given σ_A's. In [2] and Chapter 2 we find

$$\sigma_k = \sum_{A=1}^{K} w_{Ak} \sigma_A. \tag{18.4}$$

Consider the special case where the model is consistent with the data, i.e. where (18.2) is valid for each obligor. In this case we can compute the error in $\sum_{k=1}^{N} \sigma_k$, which we will call the total factor risk. Let σ_k denote the true factor standard deviation and $\hat{\sigma}_k$ the estimate of σ_k obtained from (18.4). Then we have

$$\sum_{k=1}^{N} \sigma_k - \sum_{k=1}^{N} \hat{\sigma}_k = \sum_{A=1}^{K} \left(\sum_{k=1}^{N} w_{Ak} \frac{\sigma_k}{\mu_k} - \sqrt{\sum_{k=1}^{N} \left(w_{Ak} \frac{\sigma_k}{\mu_k} \right)^2} \right) p_A. \tag{18.5}$$

Since the summands $\frac{\sigma_k}{\mu_k} w_{Ak}$ are all non-negative, (18.5) implies a strictly positive estimation error in the total factor risk, given the existence of at least one obligor with two or more nonzero factor loadings. Hence the model will underestimate the total factor risk. We propose the following rule to decrease this estimation error:

$$\sigma_k = \sum_{A=1}^{K} \sqrt{w_{Ak}} \sigma_A. \tag{18.6}$$

Going through the above calculations again, we obtain

$$\sum_{k=1}^{N} \sigma_k - \sum_{k=1}^{N} \hat{\sigma}_k = \sum_{A=1}^{K} \left(\sum_{k=1}^{N} w_{Ak} \frac{\sigma_k}{\mu_k} - \sqrt{\sum_{k=1}^{N} \left(w_{Ak} \frac{\sigma_k}{\mu_k} \right)^2} \left(\sum_{i=1}^{N} \sqrt{w_{Ai}} \right) \right) p_A \tag{18.7}$$

Since $\sum_{k=1}^{N} \sqrt{w_{Ak}} \geq 1$ with a strict inequality for at least one obligor, the estimation error in (18.7) is smaller than the one in (18.5). In extreme cases, it might even become negative, meaning that the model would overestimate the total factor risk. In any case, compared with estimation rule (18.4), estimation rule (18.6) results in a lower underestimation of the total factor risk and a higher estimate for each σ_k. The above proposed estimation rules are not directly transferable to the case with arbitrary $\mathbb{E}_\mathbb{P}[S_k], k = 1, \ldots, N$; e.g. for the case $\mathbb{E}_\mathbb{P}[S_k] = 1$ for all k, the respective estimators for the factor standard deviations are $\sigma_k = \frac{1}{\mu_k} \sum_{A=1}^{K} w_{Ak} \sigma_A$ and $\sigma_k = \frac{1}{\mu_k} \sum_{A=1}^{K} \sqrt{w_{Ak}} \sigma_A$.

18.3 Pricing of a Simple Structure with CreditRisk$^+$

Before turning to the pricing with CreditRisk$^+$ we briefly want to discuss the classical approach that the majority of investors follow, when evaluating ABS.

The evaluation is oriented towards the current spreads seen in the market and as such might be viewed as a relative value approach. In essence the investor will compare the considered ABS tranche to a similar note present in the market. He will seek to achieve the best possible match w.r.t. note principal, rating, asset type, jurisdiction and term (weighted average life is most common). The investor will then view the spread of the comparable note as the appropriate spread for the considered note.

This approach, however, has its shortcomings. In particular, finding a comparable note with a quoted spread in the market is most of the time unsuccessful unless the investor considers notes from generic liquid transactions (e.g. residential mortgage master trusts). Moreover, it may be worthwhile scrutinizing the ratings assigned by the rating agencies. One might question if the same rating really means the same credit risk. This should justify the use of a model-based pricing approach.

By the nature of CreditRisk$^+$ it is particularly well suited to reference pools of illiquid assets, and again we will focus on MBS as a representative of ABS. Constructing our simple structure for exemplification we need some notation. As discussed in the introduction we will assume the existence of a pricing measure $\mathbb{Q}$. Although we can use any model extension we will stick to the original version of CreditRisk$^+$ as described in Chapter 2. As such, we assume that each obligor A has a constant gross exposure $\mathcal{E}_A$, a constant loss given default LGD_A and hence a constant net exposure ν_A. We assume that the transaction has bullet maturity $T > 0$ and that all loans are also bullet loans with maturity T. The uniform maturity of the individual loan positions and the transaction itself is often artificially generated by a time call embedded into the structure. However, one then has to adjust the risk premia of the loans. Losses are allocated to the structure at time T. This would happen, e.g. if losses are recorded on principal deficiency ledgers during the life of the transaction and hence the loss allocation is done at the end of the life of the transaction. Let the risk premium for each loan position be given, denoted by s_A, $A = 1, \ldots, K$. For simplicity we assume that the individual loans have a single payment date at time T. This assumption is only for expository purposes and can be relaxed easily.

In order for our approach to work we must assume that each s_A is a risk-adjusted spread properly reflecting the credit risk of the respective mortgage loan. Since we consider each mortgage loan as a combination of a risk-free investment and a short position in a credit default swap providing protection against the respective credit losses, s_A should reflect the spread above the appropriate risk-free interest rate. It should therefore not cover overhead costs, as they apply to any investment and are not risk sensitive. Alternatively one could consider the credit protection within a mortgage loan as an insurance contract and the MBS tranche itself as a reinsurance contract, in which case we are in the setting of [4]. In essence that does not differ from the approach we take.

The loss to obligor A is given by

$$X_A(T) = \mathbf{1}_A \nu_A = \mathbf{1}_A \mathcal{E}_A LGD_A$$

where $\mathbf{1}_A$ is the default indicator of obligor A satisfying $\mathbb{E}_\mathbb{P}[\mathbf{1}_A] = p_A$. Define $p_A^* = \mathbb{E}_\mathbb{Q}[\mathbf{1}_A]$, $A = 1, \ldots, K$. Then under the assumption that interest rates are independent of defaults the following equality must hold true for all obligors:

$$\mathbb{E}_\mathbb{Q}[X_A(T)] = \mathbb{E}_\mathbb{Q}[(1 - \mathbf{1}_A)s_A \mathcal{E}_A]$$

which is equivalent to

$$p_A^* \mathcal{E}_A LGD_A = (1 - p_A^*)s_A \mathcal{E}_A.$$

After a few manipulations this yields

$$p_A^* = \frac{s_A}{LGD_A + s_A}.$$

This expression establishes the method of extracting individual default probabilities under the pricing measure $\mathbb{Q}$ from the risk premia of the respective mortgage loans in our simple structure. Note that p_A^* is strictly increasing in s_A. Assuming that the loan margins were calculated by a risk-averse economic agent, the expected payoff under $\mathbb{P}$ of the implicitly contained CDS contract in each loan position must have been strictly greater than zero. This means

$$(1 - p_A)s_A \mathcal{E}_A > p_A \mathcal{E}_A LGD_A$$

equivalent to

$$s_A > \frac{p_A}{1 - p_A} LGD_A. \tag{18.8}$$

Using that p_A^* is strictly increasing in s_A and (18.8) we immediately get

$$p_A^* > p_A$$

which is consistent with the results in [5].

Without loss of generality we will assume from now on that the risk factors S_k, $k = 1, \ldots, N$ all have mean 1. Then the remaining risk parameters of the CreditRisk$^+$ model under $\mathbb{Q}$ are given by the standard deviations of the default intensity risk factors S_k. In analogy to the notation under $\mathbb{P}$ we define $\sigma_k^{*2} = \text{var}_\mathbb{Q}[S_k]$. We will assume that the risk arising from the systematic factors is diversifiable, i.e. we assume

$$\sigma_k^* = \sigma_k, \quad k = 1, \ldots, N.$$

In addition we assume that the structure of independent risk factors is preserved under $\mathbb{Q}$. The latter two assumptions are ad hoc and should not be seen as more than a pragmatic approach. However, these kinds of assumptions are frequently used when applying model-based pricing. By (18.2) and the latter two assumptions the implied obligor-specific default intensity standard deviations under $\mathbb{Q}$ are

$$\sigma_A^* = \sum_{k=1}^{N} \left(p_A^* w_{Ak} \sigma_k\right)^2 .$$

Now we are in a position to compute the full distribution of the loss $X(T) = \sum_{A=1}^{K} X_A(T)$ in the reference portfolio under the pricing measure $\mathbb{Q}$ using CreditRisk$^+$. Let $\xi(T)$ be the loss in a specific tranche of the transaction at time T and assume for simplicity that all tranches have only one payment date at time T. Again under the assumption of independence between interest rates and defaults the implied risk premium s^* of the tranche is given by

$$s^* = \frac{\mathbb{E}_{\mathbb{Q}}[\xi(T)]}{N(0)}$$

where $N(0)$ is the notional of the tranche. The computation of $\mathbb{E}_{\mathbb{Q}}[\xi(T)]$ is straightforward once the full distribution of $X(T)$ under $\mathbb{Q}$ is known. The appropriate formulae for our simple structure can be found in the appendix.

18.4 Concluding Remarks

In the previous sections we have discussed the work with CreditRisk$^+$ in the context of the pricing of ABS. We derived the appropriate pricing formulae for a simple structure. In practice we get accurate loss distributions with CreditRisk$^+$ for the whole reference pool as well as the particular tranches, however at the cost of having to transform a given ABS structure into our simple structure, which obviously will distort reality to a certain extent. On the other hand, we may employ simulation-based approaches at the cost of an inaccurate assessment of tail probabilities. In the end it will depend on the taste of the user which approach she or he chooses. Common to both approaches is the problem of extracting the relevant risk parameters under $\mathbb{Q}$.

The assumption of diversifiable systematic risk that was proposed in Section 18.3 is somewhat driven by our original intention of pricing ABS positions consistently with the pricing of individual loan positions. If we take the view that $\mathbb{Q}$ is the pricing measure used by the market, it would be appropriate to calibrate the systematic factor variances, i.e. the risk-neutral correlation structure in the reference pool, to existing market prices. This approach is taken in Chapter 19.

Appendix: Computing the Expected Loss of a Tranche

Suppose the distribution of losses $X(T)$ in the reference portfolio of our simple structure has been computed under $\mathbb{Q}$. The loss incurred by the tranche compensating the losses in the reference portfolio between the lower threshold K_L and the upper threshold K_U $(0 \leq K_L < K_U)$ is given by

$$\xi(T) = (\min(X(T), K_U) - K_L)^+$$

If $X(T)$ is a discrete random variable with possible realizations $0 = V_0 < V_1 < V_2 < \ldots$ and $\pi^*(j) = \mathbb{Q}[X(T) = V_j]$, $j = 1, 2, \ldots$, then

$$\mathbb{E}_\mathbb{Q}[\xi(T)] = \sum_{\{j|K_L < V_j \leq K_U\}} \pi^*(j)(V_j - K_L) + (K_U - K_L) \sum_{\{j|V_j > K_U\}} \pi^*(j)$$

$$= \sum_{\{j|K_L < V_j \leq K_U\}} \pi^*(j)(V_j - K_L) + \left(1 - \sum_{\{j|V_j \leq K_U\}} \pi^*(j)\right)(K_U - K_L). \quad (18.9)$$

If $X(T)$ is an arbitrary random variable with distribution function F_X^* under $\mathbb{Q}$, then

$$\mathbb{E}_\mathbb{Q}[\xi(T)] = \int_{K_L}^{K_U} (x - K_L) dF_X^*(x) + (K_U - K_L)(1 - F_X^*(K_U)). \quad (18.10)$$

Results (18.9) and (18.10) are trivial, since

$$\xi(T) = \begin{cases} 0 & X(T) \leq K_L \\ X(T) - K_L & K_L < X(T) \leq K_U \\ K_U - K_L & X(T) > K_U \end{cases}.$$

For the senior tranche of our simple structure we know that it bears any losses above its lower threshold level K_L. Substituting K_L by $\mathcal{K}$ we may write the loss in the senior tranche as

$$\xi(T) = (X(T) - \mathcal{K})^+.$$

Distinguishing the discrete case and the arbitrary case as above we get, in the discrete case

$$\mathbb{E}_\mathbb{Q}[\xi(T)] = \sum_{\{J|V_J > \mathcal{K}\}} \pi^*(J)(X(T) - \mathcal{K}) \quad (18.11)$$

and in the arbitrary case

$$\mathbb{E}_\mathbb{Q}[\xi(T)] = \int_\mathcal{K}^\infty (x - \mathcal{K}) dF_X^*(x). \quad (18.12)$$

Let the number of tranches in our simple structure be $M > 0$. Then there are M threshold loss levels $0 = K_0 < K_1 < \ldots < K_{M-1} < \infty$, where the loss incurred by the m-th tranche is given by

$$\xi_m(T) = \begin{cases} (\min(X(T), K_m) - K_{m-1})^+ & m = 1, \ldots, M-1 \\ (X(T) - K_{m-1})^+ & m = M \end{cases}. \quad (18.13)$$

In this notation the loss of the senior tranche is $\xi_M(T)$. From (18.13) it is easy to deduce that $\xi_1(T) + \ldots + \xi_M(T) = X(T)$. Therefore we can compute the premium for the senior tranche according to

$$\mathbb{E}_Q[\xi_M(T)] = \mathbb{E}_Q[X(T)] - \sum_{m=1}^{M-1} \mathbb{E}_Q[\xi_m(T)].$$

Thus we can avoid computing the unwieldy expressions in (18.11) and (18.12), respectively.

References

1. P. Bürgisser, A. Kurth, and A. Wagner. Incorporating severity variations into credit risk. *Journal of Risk*, 3(4):5–31, 2001.
2. Credit Suisse Financial Products. CreditRisk$^+$: A credit risk management framework. London, 1997. Available at http://www.csfb.com/creditrisk.
3. F. B. Lehrbass. Bewertung von Basket-Kreditderivaten und Collateralized Loan Obligations. In H. P. Burghof, S. Henke, B. Rudolph, P. Schönbucher, D. Sommer, editors, *Handbuch Kreditderivate*, Schäffer Poeschel, Stuttgart, 2000.
4. D. Sondermann. Reinsurance in arbitrage-free markets. *Insurance: Mathematics and Economics*, 10:191–202, 1991.
5. O. A. Vasicek. Loan portfolio value. Risk 15(12):160–162, 2002.

Pricing and Hedging of Structured Credit Derivatives

Martin Hellmich and Oliver Steinkamp

Summary. Closed-form solutions are the Holy Grail in derivative pricing. We provide such pricing formulas for structured credit derivatives such as first- and k-th-to-default baskets as well as all kinds of tranches of collateralized debt obligations.
First, we employ the conditional independence framework to derive semi-explicit pricing formulas for basket credit default swaps. Second, by introducing a linear factor model for individual hazard rates we obtain pricing formulas in terms of the moment-generating functions of the risk factors. We show how to calibrate this factor model to market data, especially to time series of credit spreads. Thus we distinguish between exogenous and endogenous modelling. For the cases of explicitly given moment-generating functions the pricing formulas become explicit.
This approach is characterized by its great flexibility and easy calibration to observed market data and leads to an efficient pricing by analytical calculation. In particular, our approach eases the calculation of hedge ratios for dynamic hedges of basket credit default swaps (also known as *single-tranche technology* or *correlation trading*) and other complex trading strategies.

Introduction

The scope of this work lies in the intersection of two separate fields: pricing of structured credit derivatives on the one hand and the CreditRisk[+] method for the computation of a credit portfolio loss distribution on the other.

The first field is concerned with the pricing of structured credit derivatives, i.e. credit derivatives that have a portfolio of more than one reference entity as underlying. So far, the market of single-name credit derivatives has matured such that there is a substantial number of reference entities that are tradeable with sufficient liquidity. The tenor of five years has emerged as a market standard, which in general is the most liquid one. Depending on the particular reference entity there might be more liquid contracts with different tenors quoted in the market. Altogether, the single-name credit derivative market is standardized and liquid enough that single-name contracts can serve as underlyings for more complex structures built on those underlyings, e.g. first-, second- or in general k-th-to-default baskets or tranches of synthetic collateralized debt obligations. Nowadays the market and in particular, traders who are running correlation books consisting of such products tend to interpret them as options on the underlying single-name credit default swaps. Hence, the pricing and hedging of these products could be done in an "option trading style" calculating the "greeks", i.e. the partial derivatives of a pricing formula with respect to the relevant input parameter to obtain adequate dynamic hedge ratios.

Though there is no general method in the market for pricing and hedging structured credit derivatives as for equity or interest rate options a kind of standard method became apparent: the method of correlated default time simulation based on a Monte Carlo approach. Within this framework the correlation structure of the portfolio is modelled by a copula function.

The Gaussian copula function enjoys some popularity although its shortcomings in explaining fat-tailed behaviour is well known. Since there is some empirical evidence for fat-tailed phenomena in, e.g. equity returns, which are commonly used as a proxy for asset value returns for calibrating portfolio pricing models, other copula functions are used to incorporate fat tails in the models. One alternative is the t-copula function, which has the desired property, but unfortunately leads to problems in the economic interpretation of the model. Even if asset returns are uncorrelated they show some dependence due to the so-called tail dependence property. This phenomenon still lacks a satisfying explanation in economic terms.

In each case the results of such a model have to be calculated by Monte Carlo simulation, which in general causes no problems as long as only prices are considered. But in calculating hedge ratios (i.e. calculating numerical derivatives of prices obtained by simulation) for dynamic hedging strategies one could face serious numerical difficulties. This directly leads to the demand of analytically tractable solutions for the pricing problem of the products under consideration.

The second field under investigation is the CreditRisk[+] method based on the intensity model approach for calculating the loss distribution of a credit portfolio. The main advantage of this approach is an explicit solution. Nonetheless, this model also has its shortcomings, e.g. the rather abstract definition of background factors resp. the driving risk factors and their pairwise independence. In general, risk factors with meaningful economic interpretation such as sector or country indices are correlated such that they cannot be employed directly for the calibration of CreditRisk[+].

The present work is located at the intersection of both fields and aims at bringing them closer together. For the pricing of structured credit derivatives we present analytically tractable solutions leading to faster and much more accurate algorithms for the determination of prices and dynamic hedge ratios. For the CreditRisk[+] model we develop new methods allowing more general distributions of the risk factors going far beyond the standard model with gamma-distributed risk factors, while preserving the advantages of the model. As a consequence, the CreditRisk$^+$ model can be directly calibrated to observable market data. Although the latter is not our main focus this particular result is obtained as a by-product. Unfortunately, the calculation of the full portfolio loss distribution can be very time consuming if we allow arbitrary distributions for the risk factors.

Up to now, analytically tractable solutions for the pricing of structured credit derivatives are not known. New results of Gregory and Laurent [2, 3] provide semi-analytic solutions, which still need numerical integration with respect to the distribution of the risk factors. To keep this approach tractable Gregory and Laurent restrict themselves to normally distributed risk factors. Unfortunately, the method used in the CreditRisk$^+$ approach cannot be applied directly to the pricing problem although the portfolio loss distribution plays a certain role in the pricing of structured credit derivatives.

In contrast to risk management issues for the pricing problem the term structure of default risk has to be considered. This could immediately be read off the market, since different tenors of a credit default swap contract in general have very different prices. Ignoring this term structure implicitly means to work with constant default swap prices for all tenors which is too restrictive an assumption. Since Black–Scholes it has been known that a fair option price could be obtained by taking expectations of discounted cash flows with respect to an appropriate pricing measure. Since we discount each single cash flow with the riskless curve, timing is a key point, but it turns out that this problem could be solved.

This chapter consists of three sections and is organized as follows. Section 19.1 provides basic notations, definitions and standing assumptions. First of all, single-name credit default swaps serve as a building block for all types of structured credit derivatives. Thus we deduce a well-known pricing formula for single-name default swaps, which, e.g. is used for bootstrapping hazard rates. Next, we turn our attention to the simplest case of first-to-default baskets. Going along the lines of [2, 3] we employ the conditional independence framework to obtain pricing formulas for k-th-to-default baskets and all kinds of tranches of collateralized debt obligations. The results presented in Section 19.1 still depend on the individual conditional default probabilities. Hence we will call the results *semi-analytic*.

Section 19.2 is devoted to the estimation of the conditional default probabilities from observable market data. In the intensity model approach individual default probabilities are described in terms of hazard rates. Therefore, we introduce a linear model for the hazard rates of the reference entities, which can be calibrated to market data, e.g. time series of credit spreads. Thus all individual hazard rates depend on some set of risk factors describing the pairwise default correlations of the reference entities.

Ongoing from our results of Section 19.1 we express the pricing formulas in terms of the moment-generating functions (MGFs) of the set of risk factors. We will show how this can be done in its most general form, which leads to numerically tractable solutions only when the number of reference entities is small. This is the case when we consider the pricing and hedging of first-to-default and k-th-to-default baskets, which in general do not contain more than 15–20 reference entities. For larger baskets like CDOs we face some problems with numerical efficiency if the number of defaults becomes large. In the case of CDO baskets, the essential part of the loss distribution is the one up to, say, 25–30 defaults, which can be calculated with sufficient efficiency if the MGFs of the risk factors are known, or, if we impose the mild condition that the risk factors are elliptically distributed.

Finally, in Section 19.3 we discuss several distributions for the risk factors. In general, any distribution can be considered as long as the MGFs can be determined. We will discuss three cases in more detail: the gamma distribution as in standard CreditRisk$^+$, the lognormal distribution, which is useful when the linear model is calibrated with equity-market-related data, and elliptical distributions. The latter are semi-parametric in the sense that they can be described by a parametric component consisting of a vector of means and a covariance matrix as well as a non-parametric component consisting of the density of the radial component of the elliptical distribution. Among others, one advantage of elliptical distributions is that MGFs can be expressed in terms of the characteristic generator of the distribution, which itself can be estimated from market data.

Finally, we present a kernel density estimation procedure for the density function of the radial component of the considered elliptical distribution out of time series of market data. The parametric component of the distribution can be estimated by standard methods, e.g. Kendall's tau estimation.

19.1 Structured Credit Derivatives

In this section we discuss different types of credit derivatives. We start our considerations with the development of a pricing formula for single-name credit default swaps, which serve as building blocks for structured credit derivatives. Employing the conditional independence framework we are closely following [2, 3] and provide pricing formulas for k-th to default baskets and CDO tranches in terms of conditional individual default probabilities.

19.1.1 Survival functions and hazard rates

We consider a portfolio or a basket of n underlying defaultable issuers with associated default times τ_i, $i = 1, \ldots, n$. For all $t_1, \ldots, t_n$ in $\mathbb{R}_+$ we will denote the joint survival function of default times by U and their joint distribution function by F:

$$U : \mathbb{R}_+^n \to [0,1], \quad U(t_1, \ldots, t_n) = \mathbb{Q}(\tau_1 > t_1, \ldots, \tau_n > t_n), \qquad (19.1)$$

$$F : \mathbb{R}_+^n \to [0,1], \quad F(t_1, \ldots, t_n) = \mathbb{Q}(\tau_1 \leq t_1, \ldots, \tau_n \leq t_n), \qquad (19.2)$$

where $\mathbb{Q}$ denotes the risk-neutral probability. In our context we will use $\mathbb{Q}$ as an arbitrage-free pricing measure, but we will not discuss the existence or uniqueness of $\mathbb{Q}$. We will denote by U_i and F_i the marginal survival function and the marginal distribution function:

$$U_i : \mathbb{R}_+ \to [0,1], \quad U_i(t_i) = \mathbb{Q}(\tau_i > t_i), \tag{19.3}$$

$$F_i : \mathbb{R}_+ \to [0,1], \quad F_i(t_i) = \mathbb{Q}(\tau_i \leq t_i) = 1 - U_i(t_i), \tag{19.4}$$

We denote by

$$\tau^1 = \min(\tau_1, \ldots, \tau_n) = \tau_1 \wedge \ldots \wedge \tau_n$$

the first to default time and analogously with τ^k the time of the k-th default in the portfolio. Thus, the survival function of τ^1 is given by

$$\mathbb{Q}(\tau^1 > t) = U(t, \ldots, t). \tag{19.5}$$

In the following we will make the convenient and simplifying assumption.

Assumption I. *The marginal distributions of default times τ_i, $i = 1, \ldots, n$, as defined in (19.4) are absolutely continuous with right continuous derivatives f_i, i.e.*

$$F_i(t) = \int_0^t f_i(u)du \quad \forall t \geq 0, \quad i = 1, \ldots, n.$$

Definition 1. *The hazard rate $\lambda_i : \mathbb{R}_+ \to \mathbb{R}_+$ of default time i is*

$$\lambda_i(t) = \lim_{\Delta t \to 0^+} \frac{\mathbb{Q}(\tau_i \in [t, t + \Delta t] | \tau_i \geq t)}{\Delta t}.$$

Lemma 1. *By Assumption I we have*

$$\lambda_i(t) = \frac{f_i(t)}{U_i(t)} = -\frac{U_i'(t)}{U_i(t)}. \tag{19.6}$$

Usually the hazard rate λ_i is piecewise constant and of the form

$$\lambda_i(t) = \sum_j h_{i,j} \mathbf{1}_{[a_j, a_{j+1})}(t) \tag{19.7}$$

with a_j, $h_{i,j} \in \mathbb{R}_+$, $0 = a_0 < a_1 < \ldots$, which is common when stripping defaultable bonds or credit default swaps from market quotes. Thus, Assumption I is consistent with common market practice and means no restriction for our following considerations.

Assumption II. *In addition to Assumption I we henceforth will assume that the joint survival function*

$$U : \mathbb{R}_+^n \to [0,1], \quad (t_1, \ldots, t_n) \mapsto U(t_1, \ldots, t_n)$$

is absolutely continuous with right continuous derivatives.

Note that for real-valued functions absolute continuity implies differentiability Lebesgue almost surely. In the following we will use the term *smooth* in this sense.

Solving the linear differential equation (19.6) and imposing initial conditions $U_i(0) = 1$, Lemma 1 allows us to deal with the following formulas:

$$U_i(t) = \exp\left\{ -\int_0^t \lambda_i(u)du \right\}, \qquad F_i(t) = 1 - \exp\left\{ -\int_0^t \lambda_i(u)du \right\}.$$

Definition 2. *The conditional hazard rates* $h_i : \mathbb{R}_+ \to \mathbb{R}_+$, $i = 1, \dots, n$ *before the first-to-default time are given by*

$$h_i(t) = \lim_{\Delta t \to 0^+} \frac{\mathbb{Q}(\tau_i \in [t, t + \Delta t] | \tau_1 \geq t, \dots, \tau_n \geq t)}{\Delta t}.$$

Lemma 2. *We have*

$$h_i(t) = -\frac{1}{U(t, \dots, t)} \frac{\partial U(t, \dots, t)}{\partial t_i}. \tag{19.8}$$

In the case of independent default times $\tau_1, \dots, \tau_n$ *one can easily show that* $h_i(t) = \lambda_i(t)$ *holds for* $i = 1, \dots, n$.

Definition 3. *By Assumption II the hazard rate* $\lambda^1 : \mathbb{R}_+ \to \mathbb{R}_+$ *of the first default time* τ^1 *is given as*

$$\lambda^1(t) = \lim_{\Delta t \to 0^+} \frac{\mathbb{Q}(\tau^1 \in [t, t + \Delta t] | \tau^1 \geq t)}{\Delta t}$$

$$= \lim_{\Delta t \to 0^+} \frac{\mathbb{Q}(\tau^1 \in [t, t + \Delta t] | \tau_1 \geq t, \dots, \tau_n \geq t)}{\Delta t}.$$

Lemma 3. *Let* $\hat{U}(t) = U(t, \dots, t)$. *Then we have*

$$\lambda^1(t) = -\frac{1}{U(t, \dots, t)} \frac{\partial \hat{U}(t)}{dt} = \sum_{i=1}^n h_i(t). \tag{19.9}$$

For independent default times $\tau_1, \dots, \tau_n$ *the hazard rate of the first default time is the sum of the single-name hazard rates, i.e.* $\lambda^1(t) = \sum_{i=1}^n \lambda_i(t)$.

The proof of Lemma 3 relies on the fact that only one default can occur at the time. This is implied by Assumption II.

19.1.2 Single-name credit default swaps

Credit default swaps (CDS) are the main plain vanilla credit derivative products and serve as a building block for all structured credit instruments discussed later. A CDS offers protection against default of a certain underlying reference entity over a specified period of time. A premium (or spread) X is paid on a regular basis on a certain notional amount N as an insurance fee against the losses from default of the risky position N. The payment of X stops at maturity or at default of the underlying credit, whichever happens first. At the time of default before maturity of the trade, the protection buyer receives the payment $(1 - \delta)N$, where δ is the recovery rate of the underlying credit risky instrument.

We denote by τ the random default time of the reference credit and write $(B(0,t))_{t\geq 0}$ for the curve of risk-free discount factors (risk-free zero bond prices). We assume the discount factors to be given, e.g. when $(B(0,t))_{t\geq 0}$ is stripped from an interest swap curve. Of course, there is a vast amount of theory on how to model the dynamics of interest rate behaviour. This is far beyond our scope, since all we need can be deducted from spot interest rates as seen in the market.

By $T_1, T_2, \ldots, T_m = T$ we denote the premium payment dates of a CDS contract. In general, the premium of a CDS is paid quarterly in arrears, i.e. at the end of each period. For notational convenience we set $T_0 = 0$ and define the day-count fraction Δ by

$$\Delta_{j-1,j} = T_j - T_{j-1}, \quad j = 1, \ldots, m.$$

Further, let X_T denote the fair default swap spread on the credit for maturity T as quoted in the market.

The fair premium for a CDS is the one that makes the present value of the trade zero on day one. It can also be defined as the premium where the default leg and the premium leg of the trade have the same value.

Analogously to (19.3) and (19.4) we write $U(t)$ and $F(t)$ for the survival function and the distribution function of defaults, respectively. Then, assuming a deterministic recovery rate δ we get the following pricing equation for a single-name credit derivative:

$$0 = X_T \sum_{j=1}^{m} \Delta_{j-1,j} B(0, T_j) U(T_j) + (1 - \delta) \int_0^T B(0,t) dU(t). \qquad (19.10)$$

The first term in (19.10) is the present value of the premium payments X_T, which is paid at each T_j, $j = 1, \ldots, m$, provided there has been no default. Consequently the payment is valued using the risk-free discount factor $B(0, T_j)$ multiplied with the survival probability $U(T_j)$. The integral describes the present value of the default payment $(1 - \delta)$ at the time of default. This can be seen as follows. Let $\pi_k = \{0 < t_1, \ldots, t_k = T\}$, $k \in \mathbb{N}$, be a sequence of partitions of $[0, T]$ with mesh converging to zero and $t_0 = 0$ for all partitions. Now the default payment is due at $t_l \in \pi_k$ for some $k, l \in \mathbb{N}$, $l \leq k$, if default happens in the interval $(t_{l-1}, t_l]$. Taking the limit of the discrete time model and recalling the assumed absolute continuity of the marginal distributions of default times we have for the default payment leg

$$\lim_{k \to \infty} \sum_{t_l \in \pi_k} B(0, t_{l+1}) \mathbb{Q}(\tau \in (t_l, t_{l+1}]) = \lim_{k \to \infty} \sum_{t_l \in \pi_k} B(0, t_{l+1}) \left(U(t_l) - U(t_{l+1}) \right)$$

$$= -\lim_{k \to \infty} \sum_{t_l \in \pi_k} B(0, t_{l+1}) \frac{U(t_{l+1}) - U(t_l)}{t_{l+1} - t_l} (t_{l+1} - t_l)$$

$$= -\int_0^T B(0, t) U'(t) dt = -\int_0^T B(0, t) dU(t)$$

which yields (19.10).

19.1.3 First-to-default baskets

First-to-default baskets (FTD) can be best analysed with reference to and as an extension of single-name CDS. An FTD basket works in a similar manner to a CDS with a key difference. The protection seller of an FTD provides protection against the first reference entity that experiences a credit event from a basket of more than one reference entity.

As an example let us consider a basket of five credits with basket notional of EUR 10m. If one of these credits experiences a credit event, the basket default swap terminates and the protection seller pays EUR 10m in exchange for the notional amount of a deliverable obligation of the credit which went in default. If the protection seller immediately sells the defaulted asset in the market, the obtained price P could be interpreted as the recovery rate on the defaulted issuer. Thus, having paid EUR 10m for the asset the protection seller suffers a loss of EUR 10m$\times (1 - P)$.

In baskets of this type the protection seller is exposed to the credit risk of all the names in the basket, but in the event of default the seller's maximum loss is limited to the notional amount for only one of the credits. The motivation behind an investment in structures of these types is primarily the leverage. On the other hand, the protection buyer views the basket as a lower-cost method of hedging multiple credits.

Let $U^1(t) = U(t, \ldots, t)$ be the survival function of the first-to-default time and X_{FTD} the fair premium for the FTD basket. For a homogeneous portfolio, i.e. a portfolio of identical and independent reference entities, we get an equation similar to (19.10):

$$0 = X_{FTD} \sum_{j=1}^m \Delta_{j-1,j} B(0, t_j) U^1(T_j) + (1 - \delta) \int_0^T B(0, t) dU^1(t). \tag{19.11}$$

In the general case the formula for the premium leg is equal to the first term of (19.11).

Next we want to discuss the default leg of an FTD basket in the general case. For each reference entity i we denote the nominal by $\hat{N}_i$ and the recovery rate by δ_i. Then the loss given default of name i is given by $M_i = (1 - \delta_i)\hat{N}_i$, which we consider as deterministic throughout this chapter. To sketch the idea we consider the price of the default leg as the limit of the price in a discrete time model. Using the notation from above the time zero price of the default leg can be approximated by

$$\sum_{i=1}^{n} \sum_{t_l \in \pi_k} M_i B(0, t_{l+1}) \mathbb{Q}(\tau_i \in [t_l, t_{l+1}), \tau_j \geq t_l, j \neq i). \qquad (19.12)$$

We can see immediately that the following approximation holds:

$$-\frac{\partial U(t, \ldots, t)}{\partial t_i} dt \approx - \left(U(t, \ldots, \underbrace{t + \Delta t}_{i\text{-th component}}, \ldots, t) - U(t, \ldots, t) \right)$$
$$= \mathbb{Q}(\tau_i \in [t, t + \Delta t), \tau_j \geq t, j \neq i),$$

consequently we get the limit of (19.12) as $-\sum_{i=1}^{n} M_i \int_0^T \frac{\partial U(t,\ldots,t)}{\partial t_i} B(0, t) dt$. If we combine this result with (19.8), the price of the default leg of an FTD basket can be written as $-\sum_{i=1}^{n} M_i \int_0^T h_i(t) U^1(t) B(0, t) dt$, which yields the pricing equation for the general case

$$0 = X_{FTD} \sum_{j=1}^{m} \Delta_{j-1,j} B(0, T_j) U^1(T_j) + \sum_{i=1}^{n} M_i \int_0^T h_i(t) U^1(t) B(0, t) dt. \qquad (19.13)$$

19.1.4 k-th-to-default baskets

Let us now turn our attention to the more general case of k-th-to-default baskets. In contrast to an FTD basket, where the protection seller is exposed to the first default out of a basket of reference entities, an investor in a k-th-to-default basket takes the risk of the k-th default out of a pool of reference entities. The premium is paid to the protection seller as long as the number of defaults among the reference entities in the basket is less than k. At the time of the k-th default the premium payments terminate and a default payment is made to the protection buyer.

In the following we will deal only with the loss given default M_i of each reference entity instead of the notional of the trade or the notionals of the reference entities itself. In the following the notation N will be solely used as follows. We introduce the counting processes

$$N_i(t) = \mathbf{1}_{\{\tau_i \leq t\}}, \quad N(t) = \sum_{i=1}^{n} N_i(t), \quad N^{(-i)}(t) = \sum_{j \neq i} N_j(t).$$

$N_i(t)$ jumps from 0 to 1 at the default time of entity i, while $N(t)$ counts the number of defaults in the portfolio or the basket at time t and $N^{(-i)}(t)$ counts the number of defaults among the reference entities different from entity i. The process $L(t)$ will denote the cumulative loss on the credit portfolio at time t

$$L(t) = \sum_{i=1}^{n} M_i N_i(t). \tag{19.14}$$

Similarly to τ^1 and the survival function of the first to default $U^1(t) = U(t, \ldots, t)$ we denote by τ^k the time of the k-th to default with corresponding survival function

$$U^k(t) = \mathbb{Q}[\tau^k > t] = \mathbb{Q}[N(t) < k] = \sum_{l=0}^{k-1} \mathbb{Q}[N(t) = l]. \tag{19.15}$$

For more advanced considerations we will employ the *conditional independence framework* to obtain further results. Thus we introduce an (abstract) risk factor S, which basically is a random variable $S : \Omega \to X$ taking values in some measure space $(X, \mathcal{B}, \mu)$ with σ-algebra $\mathcal{B}$ and finite measure μ, i.e. $\mu(X) < \infty$. In the next section we will discuss several specifications of the risk factor S, when S will take values in some low-dimensional real vector space. All following pricing formulas of this section will be given in terms of conditional probabilities depending on the risk factor S. Therefore they are called *semi-analytical*. With our results presented in the next section at hand, all these formulas turn into analytical tractable ones, when we provide a methodology to obtain all conditional probabilities from given market data.

We start our considerations of k-th-to-default baskets with the examination of the premium payment leg. Again, we denote the premium payment dates by $T_1, T_2, \ldots, T_m = T$. Then the premium payment leg is given by the sum of discounted payments weighted with the probability that the payment actually takes place, hence

$$\sum_{j=1}^{m} \Delta_{j-1,j} B(0, T_j) U^k(T_j) = \sum_{j=1}^{m} \Delta_{j-1,j} B(0, T_j) \mathbb{Q}[N(T_j) < k]$$

$$= \mathbb{E}\left[\sum_{j=1}^{m} \Delta_{j-1,j} B(0, T_j) \mathbb{Q}[N(T_j) < k | S] \right],$$

where the last equation somehow looks artificial, but it turns out to be beneficial for our purposes, since we are able to use the conditional independence framework. This approach eases the upcoming calculations dramatically, since the individual default times τ_i are independent, if the default probabilities are conditioned on the risk factor S.

Thus it remains to calculate the probabilities $\mathbb{Q}[N(t) = k | S]$, for $k = 0, 1, \ldots, n$ conditional on the risk factor S. To achieve this we consider the conditional PGF of $N(t)$, which is

$$G_{N(t)}(z|S) = \sum_{k=0}^{n} \mathbb{Q}[N(t) = k | S] z^k$$

by definition. On the other hand, conditional independence implies

$$G_{N(t)}(z|S) = \prod_{i=1}^{n} G_{N_i(t)}(z|S) = \prod_{i=1}^{n} \left(1 - p_t^{i|S} + p_t^{i|S} z\right)$$

with $p_t^{i|S} = \mathbb{Q}[\tau_i \leq t|S]$ for notational convenience. Hence the premium payment leg could be evaluated by formal expansion of the polynomial

$$\sum_{k=0}^{n} \mathbb{Q}[N(t) = k|S] z^k = \prod_{i=1}^{n} \left(1 - p_t^{i|S} + p_t^{i|S} z\right). \tag{19.16}$$

Now we turn our attention to the default payment leg. We start our considerations with the evaluation of the probability of name i being the k-th default in the basket and occurring in the interval $(t_0, t_1]$, $t_0 < t_1$. We have[1]

$$\mathbb{Q}[\tau^k = \tau_i, \tau^k \in (t_0, t_1]] = \mathbb{Q}[N(t_0) = k-1, \tau_i \in (t_0, t_1]] =$$
$$= \mathbb{Q}[N(t_0) = k-1, N_i(t_1) - N_i(t_0) = 1]$$
$$= \mathbb{Q}[N^{(-i)}(t_0) = k-1, N_i(t_1) - N_i(t_0) = 1]$$

since $N(t_0) = N^{(-i)}(t_0)$ holds in the case of $\tau_i > t_0$. The following lemma shows how to evaluate the last expression for a slightly more general case.

Lemma 4. *Let $p_t^{i|S} = \mathbb{Q}[\tau_i \leq t|S]$ denote the probability of reference entity i being in default before time t, conditioned on the risk factor S. Then, for each $i, k = 1, \ldots, n$, the probabilities*

$$\mathbb{Q}\left[N_i(t_2) - N_i(t_1) = 1, N^{(-i)}(t_0) = k - 1\right], \quad t_0 \leq t_1 \leq t_2,$$

are given by the formal polynomial in $z \in \mathbb{R}$

$$\sum_{m=0}^{n-1} \mathbb{Q}\left[N^{(-i)}(t_0) = m, N_i(t_2) - N_i(t_1) = 1\right] z^m =$$
$$= \mathbb{E}\left[\left(p_{t_2}^{i|S} - p_{t_1}^{i|S}\right) \cdot \prod_{j \neq i} \left(1 - p_{t_0}^{j|S} + p_{t_0}^{j|S} z\right)\right], \quad z \in \mathbb{R}, \tag{19.17}$$

where the expectation is taken with respect to S. The polynomial within the expectation can be computed by formal expansion.

---

[1] The following equation only holds for the limit $t \to t_0$ with further conditions on the involved jump processes. The considered processes (e.g. Poisson processes) do not cause any problems, as Assumption II assures an absolutely continuous joint distribution function, which implies that only one jump can occur at the time.

Proof. By conditional independence we have

$$\sum_{m=0}^{n-1} \mathbb{Q}\left[N^{(-i)}(t_0) = m, N_i(t_2) - N_i(t_1) = 1|S\right] z^m$$

$$= \mathbb{Q}[N_i(t_2) - N_i(t_1) = 1|S]\sum_{m=0}^{n-1} \mathbb{Q}\left[N^{(-i)}(t_0) = m|S\right] z^m$$

$$= \mathbb{Q}[N_i(t_2) - N_i(t_1) = 1|S]\, \mathbb{E}\left[z^{N^{(-i)}(t_0)}|S\right]$$

$$= \left(p_{t_2}^{i|S} - p_{t_1}^{i|S}\right) \prod_{j\neq i} \left(1 - p_{t_0}^{j|S} + p_{t_0}^{j|S} z\right).$$

Taking expectations with respect to S yields the assertion. $\qquad\square$

Proposition 1. *Assume smooth conditional default probabilities $p_t^{i|S} = \mathbb{Q}[\tau_i \leq t|S]$ and define*

$$Z_m^i(t) = \lim_{t'\to t} \frac{1}{t'-t}\mathbb{Q}[N_i(t') - N_i(t) = 1, N^{(-i)}(t) = m]$$

for $m = 0,\ldots,n-1$ and $i = 1,\ldots,n$. Then the $Z_m^i(t)$ are given by

$$\sum_{m=0}^{n-1} Z_m^i(t)z^m = \mathbb{E}\left[\frac{dp_t^{i|S}}{dt}\prod_{j\neq i}\left(1 - p_t^{j|S} + p_t^{j|S}z\right)\right], \quad z \in \mathbb{R}, \qquad (19.18)$$

where the expectation is taken upon the risk factor S. Finally, the price of the default payment leg of a k-th-to-default basket is

$$\int_0^T B(0,t)\sum_{i=1}^n M_i Z_{k-1}^i(t)dt, \qquad (19.19)$$

which could be evaluated with help of a formal expansion of (19.18).

Proof. In a k-th-to-default basket contract the default payment is triggered at time t, if there have been already $k-1$ defaults in the basket and if reference entity i goes in default at time t, i.e. the process $N_i(t)$ has a jump at time t. Thus, the price of the default payment leg is given by taking expectations with respect to S over the following stochastic integrals with respect to the jump processes $N_i(t)$

$$\mathbb{E}\left[\int_0^T B(0,t)\sum_{i=1}^n M_i \mathbf{1}_{\{N^{(-i)}(t)=k-1\}}dN_i(t)\right].$$

Note that each process $N_i(t)$ can only jump once. Hence the sum of stochastic integrals above is bounded by some deterministic constant depending only on the sizes of individual exposures and time to maturity T. Let $\pi_m = \{t_1, \ldots, t_l, \ldots, t_m = T\}$, $m \in \mathbb{N}$, be a sequence of partitions of the interval $[0, T]$ with mesh converging to zero. We have

$$\mathbb{E}\left[\int_0^T B(0,t)\mathbf{1}_{\{N^{(-i)}(t)=k-1\}}dN_i(t)\right]$$

$$= \mathbb{E}\left[\lim_{m\to\infty}\sum_{t_l\in\pi_m}B(0,t_l)\mathbf{1}_{\{N^{(-i)}(t_l)=k-1\}}\left(N_i(t_{l+1})-N_i(t_l)\right)\right]$$

$$= \lim_{m\to\infty}\mathbb{E}\left[\sum_{t_l\in\pi_m}B(0,t_l)\mathbf{1}_{\{N^{(-i)}(t_l)=k-1\}}\left(N_i(t_{l+1})-N_i(t_l)\right)\right]$$

$$= \lim_{m\to\infty}\sum_{t_l\in\pi_m}B(0,t_l)\frac{\mathbb{Q}[N_i(t_{l+1})-N_i(t_l)=1,N^{(-i)}(t_l)=k-1]}{t_{l+1}-t_l}(t_{l+1}-t_l)$$

$$= \int_0^T B(0,t)Z_{k-1}^i(t)dt,$$

where we could interchange expectation and limit, since the sum over the partition π_m is always bounded by 1. From Lemma 4 we obtain

$$\sum_{k=0}^{n-1}Z_k^i(t)z^k = \sum_{k=0}^{n-1}\lim_{t'\to t}\frac{1}{t'-t}\mathbb{Q}[N_i(t')-N_i(t)=1,N^{(-i)}(t)=k]z^k$$

$$= \lim_{t'\to t}\frac{1}{t'-t}\mathbb{E}\left[\left(p_{t'}^{i|S}-p_t^{i|S}\right)\prod_{j\neq i}\left(1-p_t^{j|S}+p_t^{j|S}z\right)\right]$$

$$= \mathbb{E}\left[\lim_{t'\to t}\frac{p_{t'}^{i|S}-p_t^{i|S}}{t'-t}\prod_{j\neq i}\left(1-p_t^{j|S}+p_t^{j|S}z\right)\right] = \mathbb{E}\left[\frac{dp_t^{i|S}}{dt}\prod_{j\neq i}\left(1-p_t^{j|S}+p_t^{j|S}z\right)\right],$$

which concludes the proof.   □

Using conditional probabilities in the definition of $Z_m^i(t)$ leads to alternative expressions for the default leg of a k-th-to-default basket as the next corollary will show. It turns out that this alternative will be helpful in some situations when it comes to numerical evaluation.

Corollary 1. *Let us assume the conditions of Proposition 1 and define*

$$Z_m^{i|S}(t) = \lim_{t'\to t}\frac{1}{t'-t}\mathbb{Q}[N_i(t')-N_i(t)=1,N^{(-i)}(t)=m|S]$$

for $m = 0, \ldots, n-1$ and $i = 1, \ldots, n$. Then we have

$$\sum_{m=0}^{n-1}Z_m^{i|S}(t)z^m = \frac{dp_t^{i|S}}{dt}\prod_{j\neq i}\left(1-p_t^{j|S}+p_t^{j|S}z\right),$$

and $\mathbb{E}\left[Z_m^{i|S}(t)\right] = Z_m^i(t)$, where the expectation is taken over S. The price of the default payment leg of a k-th-to-default basket can be written as

$$\mathbb{E}\left[\sum_{i=1}^n M_i \int_0^T B(0,t)\mathbb{Q}[N^{(-i)}(t) = k-1|S]dp_t^{i|S}\right], \qquad (19.20)$$

where the expectation is taken over S and $\mathbb{Q}[N^{(-i)}(t) = k|S]$ is obtained from the formal expansion of the polynomial $\prod_{j\neq i}\left(1 - p_t^{j|S} + p_t^{j|S}z\right)$, $z \in \mathbb{R}$.

Proof. From the proof of Lemma 4 we obtain

$$\sum_{k=0}^{n-1} Z_k^{i|S}(t)z^k = \sum_{k=0}^{n-1} \lim_{t'\to t} \frac{1}{t'-t}\mathbb{Q}[N_i(t') - N_i(t) = 1, N^{(-i)}(t) = k|S]z^k$$

$$= \lim_{t'\to t} \frac{1}{t'-t}\sum_{k=0}^{n-1} \mathbb{Q}[N_i(t') - N_i(t) = 1, N^{(-i)}(t) = k|S]z^k$$

$$= \lim_{t'\to t} \frac{p_{t'}^{i|S} - p_t^{i|S}}{t'-t} \prod_{j\neq i}\left(1 - p_t^{j|S} + p_t^{j|S}z\right) = \frac{dp_t^{i|S}}{dt}\prod_{j\neq i}\left(1 - p_t^{j|S} + p_t^{j|S}z\right).$$

Further, conditional independence implies

$$Z_k^{i|S}(t) = \lim_{t'\to t} \frac{1}{t'-t}\mathbb{Q}[N_i(t') - N_i(t) = 1|S]\mathbb{Q}[N^{(-i)}(t) = k|S]$$

$$= \frac{dp_t^{i|S}}{dt}\mathbb{Q}[N^{(-i)}(t) = k|S].$$

Finally, by smoothness of $\mathbb{Q}$ we have $Z_m^i(t) = \mathbb{E}\left[Z_m^{i|S}(t)\right]$ for $m = 0,\ldots,n-1$ and $i = 1,\ldots,n$, which implies

$$\int_0^T B(0,t)\sum_{i=1}^n M_i Z_{k-1}^i(t)dt = \int_0^T B(0,t)\sum_{i=1}^n M_i\mathbb{E}\left[Z_{k-1}^{i|S}(t)\right]dt$$

$$= \mathbb{E}\left[\int_0^T B(0,t)\sum_{i=1}^n M_i Z_{k-1}^{i|S}(t)dt\right]$$

$$= \mathbb{E}\left[\int_0^T B(0,t)\sum_{i=1}^n M_i\mathbb{Q}[N^{(-i)}(t) = k-1|S]dp_t^{i|S}\right].$$

$\square$

Summarizing our results of Proposition 1 and Corollary 1 we have found the following pricing equation for k-th-to-default baskets. Let X_{kTD} be the fair premium. Then we have

$$0 = X_{kTD}\mathbb{E}\left[\sum_{j=1}^{m}\Delta_{j-1,j}B(0,T_j)\mathbb{Q}[N(T_j)<k|S]\right] - \int_0^T B(0,t)\sum_{i=1}^{n}M_i Z_{k-1}^i(t)dt,$$

resp., with the default leg given by (19.20). The quantities $\mathbb{Q}[N(t) = k|S]$, $k = 0, 1, \ldots, n$ and $Z_m^i(t)$, $m = 0, \ldots, n-1$, $i = 1, \ldots, n$ can be evaluated by formal expansion of (19.16) and (19.18).

Thus we have two pricing formulas for the default leg of a k-th-to-default basket. Both could be evaluated semi-analytically by formal expansion of certain polynomials and straightforward numerical integration with respect to time t and the distribution of S. All we have to know are the conditional default time distributions $p_t^{i|S} = \mathbb{Q}[\tau_i \leq t|S]$ of each reference entity i.

In (19.19) the integration over S is done first. Thus, when no closed-form solution for $p_t^{i|S}$ is available, then for each step in the numerical integration procedure with respect to t some integral with respect to S has to be computed, which might lead to severe numerical problems in some cases, e.g. if S is of more than one dimension. If closed-form solutions for $p_t^{i|S}$ are at hand, evaluation of (19.19) would be the method of choice.

Expression (19.20) evaluates the integral with respect to t first and finally the expectation with respect to S. We suggest using (19.20) when the integration with respect to S is more complex and has to be done numerically.

19.1.5 Synthetic Collateralized Debt Obligations

Let us now consider another common basket credit derivative product. Collateralized debt obligations (CDOs) are basket credit derivatives, whose contingent cash flow depends on the cumulative loss of a given portfolio of credit risks. The sources of credit risk in such a structured product can be very different. Here we concentrate on the case where the underlying portfolio consists of n single-name credit default swaps. The risks and revenues of this portfolio are tranched in a so-called capital structure with tranches showing different risk/return profiles.

In general, a capital structure of a CDO consists of three different types of tranches: an *equity* (also called *first-loss*) *tranche*, one *mezzanine tranche*, and a *senior tranche*. An investor, who e.g. sells protection on a mezzanine tranche is exposed to the credit losses on the reference portfolio exceeding a certain threshold during the life of the transaction and pays the buyer of protection an amount equal to the excess of such losses over the threshold, but only up to a pre-specified coverage limit. The lower threshold is also called *attachment point*, *subordination* or *size of first loss*. The coverage limit is usually referred to as the *tranche size*, *tranche notional* or simply *notional*.

To compensate the investor for taking risks, the protection buyer pays the investor a premium. The premium is a percentage of the outstanding notional amount of the transaction and is paid periodically, generally quarterly. The outstanding notional amount is the original tranche size reduced by the losses that have been covered by the tranche. To illustrate this point, assume that the subordination of a tranche is EUR 56m and the tranche size is EUR 10m. If EUR 60m of credit losses have occurred, the premium will be paid on the outstanding amount of EUR 6m (tranche size of EUR 10m minus EUR 4m that represents the amount of losses which exceeds the subordination of EUR 56m).

Equity tranches do not have a subordination, i.e. they absorb, as the name indicates, the first loss of the protfolio no matter which reference entity caused this loss. Thus equity tranches face the highest risks and therefore they receive the highest premium. Conversely, senior tranches face the lowest risk in a CDO structure, since they have the largest subordination of all tranches. In general, senior tranches cover all losses exceeding the respective subordination up to the total portfolio size.

In the following we will denote the lower threshold of a CDO tranche by A and the upper threshold by B with $0 \leq A < B$. Thus, the subordination is A and the notional is $N = B - A$. Further, we denote the cumulative loss on the portfolio at time t by

$$L(t) = \sum_{i=1}^{n} M_i N_i(t),$$

which is a non-decreasing pure jump process. Then the loss $K(t)$ of a CDO tranche at time t is given by

$$K(t) = \int_0^{L(t)} \mathbf{1}_{[A,B)}(x)dx = \min(N, \max(L(t) - A, 0)). \qquad (19.21)$$

Having defined the payout profile of a CDO tranche we are led to the following interpretation. In the same way as a call or put on a certain stock is an option on the underlying stock, a CDO tranche can be viewed as an option, where the cumulative portfolio loss plays the role of the underlying. Similar to the classical Black–Scholes framework, CDO tranches can be evaluated using a risk-neutral probability distribution of the underlying quantity.

To consider the premium payment leg we denote the premium payment dates as in the last section by $T_1, T_2, \ldots, T_m = T$. Then the premium payment leg is given by

$$\mathbb{E}\left[\sum_{j=1}^{m} \Delta_{j-1,j} B(0, T_j)(N - K(T_j))\right] = \sum_{j=1}^{m} \Delta_{j-1,j} B(0, T_j)(N - \mathbb{E}[K(T_j)]).$$

The above discussion shows that the protection seller of a CDO tranche pays the increments of the process $K(t)$. Discounting these cash flows and taking the expectation yields the default payment leg, which is $\mathbb{E}\left[\int_0^T B(0,t)dK(t)\right]$. Using the definition of Stieltjes integrals and the fact that the process $K(t)$ is bounded by the nominal of the tranche, we obtain

$$\mathbb{E}\left[\int_0^T B(0,t)dK(t)\right] = \mathbb{E}\left[\lim_{m\to\infty}\sum_{t_l\in\pi_m} B(0,t_l)\left(K(t_{l+1})-K(t_l)\right)\right]$$

$$= \lim_{m\to\infty}\sum_{t_l\in\pi_m} B(0,t_l)\left(\mathbb{E}[K(t_{l+1})]-\mathbb{E}[K(t_l)]\right) = \int_0^T B(0,t)d\mathbb{E}[K(t)].$$

This leads to the following pricing equation for a CDO tranche. If we denote the fair premium of a tranche by X_{CDO} we have

$$0 = X_{CDO}\sum_{j=1}^m \Delta_{j-1,j}B(0,T_j)(N-\mathbb{E}[K(T_j)]) - \int_0^T B(0,t)d\mathbb{E}[K(t)].$$

Hence, for the evaluation of the premium and default payment leg of a CDO tranche it suffices to calculate the expected loss $\mathbb{E}[K(t)]$ on the tranche for each time t. It follows from (19.21) that this can be done with help of the portfolio loss distribution, i.e. $\mathbb{Q}[L(t)\le x]$.

Now we proceed with the determination of the portfolio loss distribution. As usual in this set-up we work with discretized losses and fix a (sufficiently small) loss unit L_0. Thus for each individual reference entity i we could write the loss given default as $M_i = \tilde{\nu}_i L_0$ for some $\tilde{\nu}_i > 0$. Now choose a positive integer ν_i to obtain a rounded version of M_i/L_0.

Let us consider the PGF of $L(t)$ conditional on the risk factor S. By conditional independence we have

$$G_{L(t)}(z|S) = \mathbb{E}\left[z^{L(t)}|S\right] = \prod_{i=1}^n \left(1 - p_t^{i|S} + p_t^{i|S}z^{\nu_i}\right),$$

where $p_t^{i|S} = \mathbb{Q}(\tau_i \le t|S)$. Formal expansion yields

$$G_{L(t)}(z|S) = c_0(S) + c_1(S)z + \ldots + c_m(S)z^m,$$

where $m = \sum_{i=1}^n \nu_i$. The coefficients $c_j(S)$, $j = 0,1,\ldots,m$, of the formal expansion are used to obtain $\mathbb{Q}[L(t) = j] = \mathbb{E}[c_j(S)]$; e.g., the probability that no loss in the basket occurs is given by $\mathbb{Q}[L(t) = 0] = \mathbb{E}\left[\prod_{i=1}^n \left(1 - p_t^{i|S}\right)\right]$.

19.2 Estimating Conditional Probabilities

So far we have developed semi-explicit pricing formulas for different structured credit derivatives. For all these formulas, in particular for the loss distribution of the whole portfolio, it remains to determine the conditional default probabilities for each reference entity. In the original CreditRisk$^+$ framework a gamma distribution is used to model the stochastic behaviour of the risk factors driving individual default probabilities. Here we want to pursue a more general way of obtaining the conditional default probabilities from empirical market data. The aim is to develop a general framework, which allows us to determine a set of appropriate risk factors S driving the correlation structure as well as the corresponding conditional default probabilities $p_t^{i|S}$.

19.2.1 Estimating hazard rates

Survival and default probabilities corresponding to a reference entity i are determined by its hazard rate function $\lambda_i(t)$, $0 \leq t \leq T$. Each individual hazard rate function can be estimated with the help of (19.10) using observable market quotes and an assumption on the recovery rate of the reference entity under consideration.

If we want to estimate hazard rate functions in their most general form we would have to consider all market information for each single reference entity. Suppose we have found $m \geq 1$ market prices for different tenors $T_1 < T_2 \ldots < T_m$ for reference entity i, i.e. we have the price vector

$$\left[X_{T_1}^{(i)}, \ldots, X_{T_m}^{(i)} \right]^T$$

of CDS quotes for reference entity i. Let $T_0 = 0$. Using well-known bootstrapping methods or more advanced techniques of Martin et al. [4] and the pricing formula (19.10) we obtain a piecewise constant hazard rate function

$$\lambda_i(t) = \sum_{j=1}^{m} \lambda_{i,j} \mathbf{1}_{[T_{j-1}, T_j)}(t)$$

for each reference entity $i = 1, \ldots, n$.

The methods we will present in the following could be considered in this most general situation. For the ease of presentation and the sake of readability we restrict our considerations to the case of one liquid market quote for each single reference entity. This approach is consistent with common market practice and enjoys some popularity because of its easy calibration.

Denote the tenor of the most liquid contract for reference entity i by $T^{(i)}$. In general, the most liquid tenor is $T^{(i)} \equiv 5$ years for almost all traded names. Thus we obtain a flat hazard rate function for each reference entity, i.e.

$$\lambda_i(t) = \lambda_i, \quad 0 \leq t \leq T,$$

which implies the following survival and default probabilities

$$U_i(t) = e^{-\lambda_i t}, \quad F_i(t) = 1 - e^{-\lambda_i t}.$$

Using the latter to simplify (19.10) and assuming a flat riskless interest rate curve we end up with the following approximating equations:

$$\lambda_i = \frac{X_T^{(i)}}{1 - \delta_i}, \quad \Delta \lambda_i = \frac{\Delta X_T^{(i)}}{1 - \delta_i}. \tag{19.22}$$

Both equations are very helpful for estimating default correlations from time series data of CDS quotes without having to resort to stock market data and corresponding asset value approaches.

19.2.2 Risk factors

One basic problem in the valuation of structured credit derivatives is concerned with the fact that one cannot work consequently in the risk-neutral world. In particular, the parameters describing the interdependencies in the evolution of the creditworthiness of two reference entities cannot be estimated from market quotes of traded instruments. Thus it is in general not possible to obtain implicit correlations, and more complex measures of dependence such as tail dependence are not observable from the market. Nevertheless, to provide pricing tools for structured credit derivatives we have to fall back on a mixture of market information, i.e. market prices of traded instruments, and modelling using historical data.

Assuming a given recovery rate for each reference entity we obtain survival and default probabilities from the pricing equation (19.10) out of market quotes. In other words, the marginal distributions describing the stochastic behaviour of each individual evolution of creditworthiness are obtained from market data and hence a part of the risk-neutral measure. As explained above, it is not possible to get the joint survival and default probabilities in the same way. In current approaches so-called copula functions are used to overcome the problem of finding joint survival probabilities. Apart from the fact that the choice of a particular copula function is somewhat arbitrary, the calibration of the copula within a certain class of functions (e.g. the estimation of the free parameters for the Gauss- or t-copula) is based on historical data as well.

In contrast to the asset value models, we will explain the interdependencies between two reference entities by systematic risk factors, which influence individual default rates. Let us denote the set of systematic risk factors by $R = (R_1, \ldots, R_K)$. We describe the credit spread $X_T^{(i)}$ of reference entity i with tenor T by the following linear model:

$$\frac{\lambda_i}{\sigma_i} = \frac{X_T^{(i)}}{\sigma_{X_T^{(i)}}} = a_{i,0}\varepsilon_i + \sum_{k=1}^{K} a_{i,k}R_k, \quad 1 \leq i \leq n, \qquad (19.23)$$

where $\sigma_{X_T^{(i)}}$ is the credit spread volatility and σ_i the volatility of the hazard rate λ_i. Note that $\sigma_{X_T^{(i)}}$ and σ_i are related to each other by the estimation procedure of the hazard rate function. An approximating relationship could be deduced from (19.22).

In general we assume that the systematic risk factors are pairwise uncorrelated and that the individual risks are pairwise independent as well as independent from the systematic factors. Now, it depends on the distributional assumption on the random vector

$$S = (\varepsilon_1, \ldots, \varepsilon_n, R_1, \ldots, R_K)$$

if the assumption is sufficient that the components of $R = (R_1, \ldots, R_K)$ are uncorrelated or if we have to impose the much stronger condition of pairwise independence.

Using the above specification of the survival probabilities the expressions introduced in the last section read as follows:

$$\mathbb{Q}(\tau_i \leq t|S) = p_t^{i|S} = F_i^{|S}(t) = 1 - e^{-\lambda_i t}, \tag{19.24}$$

$$\mathbb{Q}(\tau_i > t|S) = U_i^{|S}(t) = e^{-\lambda_i t}. \tag{19.25}$$

Furthermore, we have

$$U_i(t) = \mathbb{E}\left[U_i^{|S}(t)\right] = \mathbb{E}\left[e^{-\lambda_i t}\right] = \mathbb{E}\left[\exp\left\{-\left(a_{i,0}\sigma_i \varepsilon_i + \sum_{k=1}^{K} a_{i,k}\sigma_i R_k\right)t\right\}\right]$$

$$= \mathbb{E}\left[\exp\left\{-\left(a_{i,0}\sigma_i \varepsilon_i t\right)\right\}\right] \mathbb{E}\left[\exp\left\{-\sum_{k=1}^{K} a_{i,k}\sigma_i R_k t\right\}\right]$$

$$= M_{\varepsilon_i}\left(-a_{i,0}\sigma_i t\right) M_R\left(-a_{i,1}\sigma_i t, \ldots, -a_{i,K}\sigma_i t\right), \tag{19.26}$$

$$U(t_1, \ldots, t_n) = \mathbb{E}\left[U^{|S}(t_1, \ldots, t_n)\right]$$

$$= \prod_{i=1}^{n} M_{\varepsilon_i}\left(-a_{i,0}\sigma_i t\right) M_R\left(-\sum_{i=1}^{n} a_{i,1}\sigma_i t_i, \ldots, -\sum_{i=1}^{n} a_{i,K}\sigma_i t_i\right), \tag{19.27}$$

where

$$M_X(u) = \mathbb{E}\left[e^{Xu}\right], \quad u \in \mathbb{R}^d,$$

denotes the moment-generating function (MGF) of a d-dimensional random variable $X = (X_1, \ldots, X_d)$. Without loss of generality we could assume in (19.23) that the risk factors R_k and the individual risks ε_i have variance 1. This implies the following relationship of CDS spread, resp., hazard rate correlation

$$\rho(\lambda_i, \lambda_j) = \text{cov}\left(\frac{\lambda_i}{\sigma_i}, \frac{\lambda_j}{\sigma_j}\right) = \text{cov}\left(\frac{X_T^{(i)}}{\sigma_{X_T^{(i)}}}, \frac{X_T^{(j)}}{\sigma_{X_T^{(j)}}}\right) \tag{19.28}$$

$$= \text{cov}\left(a_{i,0}\varepsilon_i + \sum_{k=1}^{K} a_{i,k}R_k, a_{j,0}\varepsilon_j + \sum_{k=1}^{K} a_{j,k}R_k\right) = \sum_{k=1}^{K} a_{i,k}a_{j,k}.$$

In particular, for $i = j$ and $i = 1, \ldots, n$

$$\sum_{k=1}^{K} a_{i,k}^2 = 1. \tag{19.29}$$

Again, comparing (19.27) with the copula function approach we see that market prices are reflected in the marginal disributions. We also have to calibrate our model to market data, which is done by estimating the remaining parameters σ_i, $i = 1, \ldots, n$ in (19.26) and (19.27) describing the spread volatilities. We set up (19.23) such that spread correlations are explained by the linear model but name-specific volatilities still have to be calibrated. To do this we plug (19.26) into (19.10) and solve for σ_i, $i = 1, \ldots, n$. Therefore, we could interpret the σ_i, $i = 1, \ldots, n$, as the model-implied credit spread volatilities.

19.2.3 Application to pricing

We have expressed the individual and joint survival probabilities in equations (19.26) and (19.27) in terms of the MGF

$$M_R(u) = \mathbb{E}\left[e^{Ru}\right], \quad u \in \mathbb{R}^K,$$

of the risk factor $R = (R_1, \ldots, R_K)$ and the MGFs

$$M_{\varepsilon_i}(v) = \mathbb{E}[e^{\varepsilon_i v}], \quad v \in \mathbb{R}, \quad i = 1, \ldots, n$$

of the individual risks. Let us reconsider (19.11) and (19.13) and note that it is already possible to price a first-to-default basket. It remains to specify the distribution of the risk factor and the individual risks.

Now we want to go on from our results of the first section, where we obtained semi-explicit pricing formulas for more complex-structured credit derivatives. We deal with a portfolio of n reference entities. Thus the distribution $\mathbb{Q}[N(t) = k|S)]$, $k = 0, 1, \ldots, n$ plays an important role in our considerations. From (19.16) we obtain the PGF of $N(t)$ conditional on the risk factor S. We have

$$G_{N(t)}(z|S) = \prod_{i=1}^{n} G_{N_i(t)}(z|S) = \prod_{i=1}^{n} \left(q_t^{i|S} + \left(1 - q_t^{i|S}\right) z \right) = \sum_{k=0}^{n} g_k^n z^k,$$

where

$$q_t^{i|S} = \mathbb{Q}[\tau_i > t|S] = 1 - p_t^{i|S} = e^{-\lambda_i t}, \quad i = 1, \ldots, n,$$

is the conditional survival probability of reference entity i up to time t and $g_k^n = \mathbb{Q}[N(t) = k|S]$ the conditional probability for exactly k out of n reference entities being in default at time t. By conditional independence we have

$$g_0^n = \prod_{i=1}^{n} q_t^{i|S}, \quad g_n^n = \prod_{i=1}^{n} \left(1 - q_t^{i|S}\right),$$

while for $k = 1, \ldots, n-1$ the following recursion holds:

$$g_k^n = g_k^{n-1} q_t^{n|S} + g_{k-1}^{n-1}\left(1 - q_t^{n|S}\right) = q_t^{n|S}\left(g_k^{n-1} - g_{k-1}^{n-1}\right) + g_{k-1}^{n-1}. \quad (19.30)$$

Equation (19.30) follows immediately from

$$\sum_{k=0}^{n} g_k^n z^k = G_{N_n(t)}(z|S) \prod_{i=1}^{n-1} G_{N_i(t)}(z|S) = \left(q_t^{n|S} + \left(1 - q_t^{n|S}\right)\right) \sum_{k=0}^{n-1} g_k^{n-1} z^k,$$

resp. from the following consideration. Suppose we have a portfolio consisting of $n-1$ reference entities and we are adding a new entity to it. This new entity survives until time t or it does not. To have exactly k defaults in the enlarged portfolio we have to consider two cases. In the first case there are already k defaults among the original $n-1$ reference entities with a surviving new one, while in the second case there are $k-1$ defaults in the original portolio and the new reference entity is in default as well up to time t.

Now we want to express the coefficients g_k^n for $k = 0, \ldots, n$ in terms of survival probabilities. First of all we have for $k = 0, \ldots, n$

$$g_k^n = \sum_{1 \le j_1 < j_2 < \ldots < j_k \le n} \left(1 - q_t^{j_1|S}\right) \cdot \ldots \cdot \left(1 - q_t^{j_k|S}\right) \prod_{\substack{i=1 \\ i \ne j_1, \ldots, j_k}}^{n} q_t^{i|S}. \quad (19.31)$$

To evaluate (19.30) for all $k = 0, \ldots, n$ we prove the following lemma.

Lemma 5. *For arbitrary $n \in \mathbb{N}$ we have*

$$g_0^n = \prod_{i=1}^{n} q_t^{i|S}, \qquad g_n^n = 1 + \sum_{\nu=1}^{n} (-1)^\nu \sum_{1 \le j_1 < \ldots < j_\nu \le n} q_t^{j_1|S} \cdot \ldots \cdot q_t^{j_\nu|S}, \quad (19.32)$$

and for $k = 1, \ldots, n-1$

$$g_k^n = \sum_{\substack{1 \le j_1 < \\ < j_k \le n}} \prod_{\substack{i=1 \\ i \ne j_1, \ldots, j_k}}^{n} q_t^{i|S} + \sum_{\nu=1}^{k-1} (-1)^\nu \binom{n-k+\nu}{\nu} \sum_{\substack{1 \le j_1 < \\ < j_{k-\nu} \le n}} \prod_{\substack{i=1 \\ i \ne j_1, \ldots, j_{k-\nu}}}^{n} q_t^{i|S} + (-1)^k \binom{n}{k} \prod_{i=1}^{n} q_t^{i|S}$$

$$= \sum_{\substack{1 \le j_1 < \\ < j_k \le n}} \prod_{\substack{i=1 \\ i \ne j_1, \ldots, j_k}}^{n} q_t^{i|S} - \binom{n-k+1}{1} \sum_{\substack{1 \le j_1 < \\ < j_{k-1} \le n}} \prod_{\substack{i=1 \\ i \ne j_1, \ldots, j_{k-1}}}^{n} q_t^{i|S} + \binom{n-k+2}{2} \sum_{\substack{1 \le j_1 < \\ < j_{k-2} \le n}} \prod_{\substack{i=1 \\ i \ne j_1, \ldots, j_{k-2}}}^{n} q_t^{i|S}$$

$$- \binom{n-k+3}{3} \sum_{\substack{1 \le j_1 < \\ < j_{k-3} \le n}} \prod_{\substack{i=1 \\ i \ne j_1, \ldots, j_{k-3}}}^{n} q_t^{i|S} + \ldots + (-1)^k \binom{n}{k} \prod_{i=1}^{n} q_t^{i|S}. \quad (19.33)$$

Proof. While the first equality in (19.32) is evident, the second is a direct consequence of the inclusion–exclusion principle, i.e. the probability $\mathbb{Q}[N(t) = n|S]$ could be written as conditional probability of the complementary event. In this case it means that at least one reference entity survives until time t. Note further that with the convention $\prod_{i \in \emptyset} q_t^{i|S} = 1$ the second equation in (19.32) is of the same type as (19.33). We will further use the convention $\sum_{i \in \emptyset} \cdots = 0$, where we interpret, e.g., $\sum_{i=1}^{0} \cdots$ as a sum over the empty set. The proof of (19.33) will be done in three steps by an induction over n.

Step 1: We fix some $k \in \mathbb{N}$ and assume that the assertion holds for g^n_{k-1} and g^n_k. Now we show that this assumption implies the assertion for g^{n+1}_k. Therefore we use recursion (19.30) to calculate

$$
q^{n+1|S}_t \left(g^n_k - g^n_{k-1} \right) = q^{n+1|S}_t \left(\sum_{\substack{1 \leq j_1 < \ldots \\ < j_k \leq n}} \prod_{\substack{i=1 \\ i \neq j_1, \ldots, j_k}}^{n} q^{i|S}_t - \left\{ \binom{n-k+1}{1} + 1 \right\} \right.
$$

$$
\times \sum_{\substack{1 \leq j_1 < \ldots \\ < j_{k-1} \leq n}} \prod_{\substack{i=1 \\ i \neq j_1, \ldots, j_{k-1}}}^{n} q^{i|S}_t + \sum_{\nu=2}^{k-1} (-1)^\nu \left\{ \binom{n-k+\nu}{\nu} + \binom{n-k+\nu}{\nu-1} \right\} \sum_{\substack{1 \leq j_1 < \ldots \\ < j_{k-\nu} \leq n}} \prod_{\substack{i=1 \\ i \neq j_1, \ldots, j_{k-\nu}}}^{n} q^{i|S}_t
$$

$$
\left. + (-1)^k \left\{ \binom{n}{k} + \binom{n}{k-1} \right\} \prod_{j=1}^{n} q^{j|S}_t \right) \tag{19.34}
$$

$$
= \sum_{1 \leq j_1 < \ldots < j_k \leq n} q^{n+1|S}_t \prod_{\substack{i=1 \\ i \neq j_1, \ldots, j_k}}^{n} q^{i|S}_t - ((n+1)-k+1) \sum_{\substack{1 \leq j_1 < \ldots \\ < j_{k-1} \leq n}} q^{n+1|S}_t \prod_{\substack{i=1 \\ i \neq j_1, \ldots, j_{k-1}}}^{n} q^{i|S}_t
$$

$$
+ \sum_{\nu=2}^{k-1} (-1)^\nu \binom{(n+1)-k+\nu}{\nu} \sum_{\substack{1 \leq j_1 < \ldots \\ < j_{k-\nu} \leq n}} q^{n+1|S}_t \prod_{\substack{i=1 \\ i \neq j_1, \ldots, j_{k-\nu}}}^{n} q^{i|S}_t + (-1)^k \binom{n+1}{k} \prod_{i=1}^{n+1} q^{i|S}_t .
$$

We further have

$$
g^n_{k-1} = \sum_{\substack{1 \leq j_1 < \ldots \\ < j_{k-1} \leq n}} \prod_{\substack{i=1 \\ i \neq j_1, \ldots, j_{k-1}}}^{n} q^{i|S}_t + \sum_{\nu=1}^{k-1} (-1)^\nu \binom{n-k+\nu+1}{\nu} \sum_{\substack{1 \leq j_1 < \ldots \\ < j_{k-\nu-1} \leq n}} \prod_{\substack{i=1 \\ i \neq j_1, \ldots, j_{k-\nu-1}}}^{n} q^{i|S}_t
$$

$$
= \sum_{\substack{1 \leq j_1 < \ldots \\ < j_{k-1} \leq n}} \prod_{\substack{i=1 \\ i \neq j_1, \ldots, j_{k-1}}}^{n} q^{i|S}_t - ((n+1)-k+1) \sum_{\substack{1 \leq j_1 < \ldots \\ < j_{k-2} \leq n}} \prod_{\substack{i=1 \\ i \neq j_1, \ldots, j_{k-2}}}^{n} q^{i|S}_t
$$

$$
+ \sum_{\nu=2}^{k-1} (-1)^\nu \binom{(n+1)-k+\nu}{\nu} \sum_{\substack{1 \leq j_1 < \ldots \\ < j_{k-\nu-1} \leq n}} \prod_{\substack{i=1 \\ i \neq j_1, \ldots, j_{k-\nu-1}}}^{n} q^{i|S}_t . \tag{19.35}
$$

Moreover

$$
\sum_{\substack{1 \leq j_1 < \ldots \\ < j_k \leq n}} q^{n+1|S}_t \prod_{\substack{i=1 \\ i \neq j_1, \ldots, j_k}}^{n} q^{i|S}_t + \sum_{\substack{1 \leq j_1 < \ldots \\ < j_{k-1} \leq n}} \prod_{\substack{i=1 \\ i \neq j_1, \ldots, j_{k-1}}}^{n} q^{i|S}_t = \sum_{\substack{1 \leq j_1 < \ldots \\ < j_k \leq n+1}} \prod_{\substack{i=1 \\ i \neq j_1, \ldots, j_k}}^{n+1} q^{i|S}_t , \tag{19.36}
$$

$$
(n+2-k) \sum_{\substack{1 \leq j_1 < \ldots \\ < j_{k-1} \leq n}} q^{n+1|S}_t \prod_{\substack{i=1 \\ i \neq j_1, \ldots, j_{k-1}}}^{n} q^{i|S}_t \tag{19.37}
$$

$$
+ (n+2-k) \sum_{\substack{1 \leq j_1 < \ldots \\ < j_{k-2} \leq n}} \prod_{\substack{i=1 \\ i \neq j_1, \ldots, j_{k-2}}}^{n} q^{i|S}_t = \binom{(n+1)-k+1}{1} \sum_{\substack{1 \leq j_1 < \ldots \\ < j_{k-1} \leq n+1}} \prod_{\substack{i=1 \\ i \neq j_1, \ldots, j_{k-1}}}^{n+1} q^{i|S}_t ,
$$

and

$$
\sum_{\nu=2}^{k-1}(-1)^\nu \binom{(n+1)-k+\nu}{\nu} \sum_{\substack{1\le j_1<\dots \\ <j_{k-\nu}\le n}} q_t^{n+1|S} \prod_{\substack{i=1 \\ i\neq j_1,\dots,j_{k-\nu}}}^{n} q_t^{i|S}
$$

$$
+\sum_{\nu=2}^{k-1}(-1)^\nu \binom{(n+1)-k+\nu}{\nu} \sum_{\substack{1\le j_1<\dots \\ <j_{k-\nu-1}\le n}} \prod_{\substack{i=1 \\ i\neq j_1,\dots,j_{k-\nu-1}}}^{n} q_t^{i|S}
$$

$$
=\sum_{\nu=2}^{k-1}(-1)^\nu \binom{(n+1)-k+\nu}{\nu} \sum_{\substack{1\le j_1<\dots \\ <j_{k-\nu}\le n+1}} \prod_{\substack{i=1 \\ i\neq j_1,\dots,j_{k-\nu}}}^{n+1} q_t^{i|S}. \tag{19.38}
$$

By recursion (19.30) we obtain g_k^{n+1} by addition of (19.34) and (19.35). In combination with (19.36), (19.37) and (19.38) this leads to

$$
g_k^{n+1} = \sum_{\substack{1\le j_1<\dots \\ <j_k\le n+1}} \prod_{\substack{i=1 \\ i\neq j_1,\dots,j_k}}^{n+1} q_t^{i|S}
$$
$$
+\sum_{\nu=1}^{k-1}(-1)^\nu \binom{(n+1)-k+\nu}{\nu} \sum_{\substack{1\le j_1<\dots \\ <j_{k-\nu}\le n+1}} \prod_{\substack{i=1 \\ i\neq j_1,\dots,j_{k-\nu}}}^{n+1} q_t^{i|S} + (-1)^k \binom{n+1}{k} \prod_{i=1}^{n+1} q_t^{i|S},
$$

which finally concludes the proof of step 1.

Step 2: Now we show that if the assertion holds for $g_0^n, g_1^n, \dots, g_n^n$, then it is true for $g_0^{n+1}, g_1^{n+1}, \dots, g_{n+1}^{n+1}$ as well. We distinguish the following cases.

(i) For g_0^{n+1} the assertion follows from the first equation in (19.32).
(ii) The case g_{n+1}^{n+1} is a consequence of the second equation in (19.32).
(iii) The result of step 1 implies the assertion for g_k^{n+1}, $k=1,\dots n$.

Step 3: Finally, it remains to prove that the assertion of the lemma holds for g_0^2, g_1^2 and g_2^2. Again, for g_0^2 and g_2^2 the assertion follows from (19.32), while for g_1^2 we have

$$
g_1^2 = \left(1-q_t^{1|S}\right)q_t^{2|S} + \left(1-q_t^{2|S}\right)q_t^{1|S} = \left(q_t^{1|S}+q_t^{2|S}\right) - 2q_t^{1|S}q_t^{2|S},
$$

which is an expression of type (19.33) for $n=2$ and $k=1$. □

Now we can write the distribution of the number of defaults in closed form in terms of the MGF of $R=(R_1,\dots,R_K)$ and ε_i, $i=1,\dots,n$. Lemma 5 gives

$$
g_0^n = \mathbb{Q}[N(t)=0|S] = \prod_{i=1}^n q_t^{i|S} = \exp\left(-t\sum_{i=1}^n \lambda_i\right)
$$
$$
= \exp\left(\sum_{k=1}^K R_k\left(-t\sum_{i=1}^n a_{i,k}\sigma_i\right)\right)\prod_{i=1}^n \exp\left(\varepsilon_i(-ta_{i,0}\sigma_i)\right),
$$

which implies

$$\mathbb{Q}[N(t)=0] = \mathbb{E}[g_0^n] = M_R\left(-t\sum_{i=1}^{n}a_{i,1}\sigma_i,\ldots,-t\sum_{i=1}^{n}a_{i,K}\sigma_i\right)\prod_{i=1}^{n}M_{\varepsilon_i}(-a_{i,0}\sigma_i t)$$

$$= U(t,\ldots,t) = U^1(t). \tag{19.39}$$

In the same way we obtain

$$\mathbb{E}[g_n^n] = \mathbb{Q}[N(t)=n] = 1 - \sum_{\nu=1}^{n}(-1)^{\nu+1}\sum_{1\le j_1<\ldots<j_\nu\le n}\mathbb{E}\left[q_t^{j_1|S}\cdot\ldots\cdot q_t^{j_\nu|S}\right]$$

$$= 1 - \sum_{\nu=1}^{n}(-1)^{\nu+1}\sum_{1\le j_1<\ldots<j_\nu\le n}M_R\left(-t\sum_{\mu=1}^{\nu}a_{j_\mu,1}\sigma_{j_\mu},\ldots,-t\sum_{\mu=1}^{\nu}a_{j_\mu,K}\sigma_{j_\mu}\right)$$

$$\times\prod_{\mu=1}^{\nu}M_{\varepsilon_{j_\mu}}\left(-ta_{j_\mu,0}\sigma_{j_\mu}\right), \tag{19.40}$$

$$\mathbb{E}[g_k^n] = \mathbb{Q}[N(t)=k] = \sum_{\substack{1\le j_1<\ldots\\<j_k\le n}}\mathbb{E}\left[\prod_{\substack{i=1\\i\ne j_1,\ldots,j_k}}^{n}q_t^{i|S}\right]$$

$$+\sum_{\nu=1}^{k-1}(-1)^\nu\binom{n-k+\nu}{\nu}\sum_{\substack{1\le j_1<\ldots\\<j_{k-\nu}\le n}}\mathbb{E}\left[\prod_{\substack{i=1\\i\ne j_1,\ldots,j_{k-\nu}}}^{n}q_t^{i|S}\right] + (-1)^k\binom{n}{k}\mathbb{E}\left[\prod_{j=1}^{n}q_t^{j|S}\right]$$

$$=\sum_{\substack{1\le j_1<\ldots\\<j_k\le n}}M_R\left(-t\sum_{\substack{i=1\\i\ne j_1,\ldots,j_k}}^{n}a_{i,1}\sigma_i,\ldots,-t\sum_{\substack{i=1\\i\ne j_1,\ldots,j_k}}^{n}a_{i,K}\sigma_i\right)\prod_{\substack{i=1\\i\ne j_1,\ldots,j_k}}^{n}M_{\varepsilon_i}(-a_{i,0}\sigma_i t)$$

$$+\sum_{\nu=1}^{k-1}(-1)^\nu\binom{n-k+\nu}{\nu}\sum_{\substack{1\le j_1<\ldots\\<j_{k-\nu}\le n}}M_R\left(-t\sum_{\substack{i=1\\i\ne j_1,\ldots,j_{k-\nu}}}^{n}a_{i,1}\sigma_i,\ldots,-t\sum_{\substack{i=1\\i\ne j_1,\ldots,j_{k-\nu}}}^{n}a_{i,K}\sigma_i\right)$$

$$\times\prod_{\substack{i=1\\i\ne j_1,\ldots,j_{k-\nu}}}^{n}M_{\varepsilon_i}(-a_{i,0}\sigma_i t) + (-1)^k\binom{n}{k}M_R\left(-t\sum_{i=1}^{n}a_{i,1}\sigma_i,\ldots,-t\sum_{i=1}^{n}a_{i,1}\sigma_i\right)$$

$$\times\prod_{i=1}^{n}M_{\varepsilon_i}(-a_{i,0}\sigma_i t) \tag{19.41}$$

for $k = 1,\ldots,n-1$.

We are now able to evaluate the premium leg of a k-th-to-default basket in terms of the MFG of the risk factors and the individual risks. Note that the coefficients $g_k^n = g_k^n(t)$ actually are time dependent. Thus we have

$$X_{kTD} \, \mathbb{E}\left[\sum_{j=1}^{m} \Delta_{j-1,j} B(0,T_j) \mathbb{Q}[N(T_j) < k|S]\right]$$

$$= X_{kTD} \sum_{j=1}^{m} \Delta_{j-1,j} B(0,T_j) \sum_{l=0}^{k-1} \mathbb{E}[\mathbb{Q}[N(T_j) = l|S]]$$

$$= X_{kTD} \sum_{j=1}^{m} \Delta_{j-1,j} B(0,T_j) \sum_{l=0}^{k-1} \mathbb{E}[g_k^n(t)]. \qquad (19.42)$$

The price of the default leg of a k-th-to-default basket is given by

$$\int_0^T B(0,t) \sum_{i=1}^{n} M_i Z_{k-1}^i(t)dt,$$

where

$$\sum_{m=0}^{n-1} Z_m^i(t) z^m = \mathbb{E}\left[-\frac{dq_t^{i|S}}{dt} \prod_{j\neq i} \left(q_t^{j|S} + \left(1 - q_t^{j|S}\right) z\right)\right], \quad z \in \mathbb{R}.$$

Without loss of generality we let $i = n$ and obtain

$$\sum_{m=0}^{n-1} Z_m^n(t) z^m = \mathbb{E}\left[-\frac{dq_t^{n|S}}{dt} \prod_{j=1}^{n-1} \left(q_t^{j|S} + \left(1 - q_t^{j|S}\right) z\right)\right], \quad z \in \mathbb{R}.$$

Since $q_t^{n|S} = e^{-\lambda_n t}$ we have $\frac{dq_t^{n|S}}{dt} = -\lambda_n e^{-\lambda_n t}$ which yields

$$\mathbb{E}\left[-\frac{dq_t^{n|S}}{dt} \prod_{j=1}^{n-1} \left(q_t^{j|S} + \left(1 - q_t^{j|S}\right) z\right)\right] = \mathbb{E}\left[\lambda_n q_t^{n|S} \sum_{j=0}^{n-1} g_j^{n-1} z^j\right],$$

i.e.

$$Z_m^n(t) = \mathbb{E}\left[\lambda_n q_t^{n|S} g_m^{n-1}\right], \quad m = 1,\ldots,n-1.$$

If we use the explicit expressions for λ_n and g_m^{n-1}, we obtain for $m = 0$, e.g.,

$$Z_0^n(t) = \mathbb{E}\left[\lambda_n \prod_{i=1}^{n} q_t^{i|S}\right] = \mathbb{E}\left[\left(a_{n,0}\sigma_n \varepsilon_n + \sigma_n \sum_{k=1}^{K} a_{n,k} R_k\right)\right.$$

$$\left. \times \exp\left(\sum_{k=1}^{K} R_k \left(-t \sum_{i=1}^{n} a_{i,k}\sigma_i\right)\right) \prod_{i=1}^{n} \exp\left(\varepsilon_i(-ta_{i,0}\sigma_i)\right)\right]$$

and consequently

$$Z_0^n(t) = -M_R\left(-t\sum_{i=1}^{n}a_{i,1}\sigma_i,\ldots,-t\sum_{i=1}^{n}a_{i,K}\sigma_i\right)a_{n,0}\sigma_n M'_{\varepsilon_n}\left(-a_{n,0}\sigma_n t\right)$$

$$\times\prod_{i=1}^{n-1}M_{\varepsilon_i}\left(-a_{i,0}\sigma_i t\right) - \prod_{i=1}^{n}M_{\varepsilon_i}\left(-a_{i,0}\sigma_i t\right)$$

$$\times\sigma_n\sum_{k=1}^{K}a_{n,k}\frac{\partial}{\partial u_k}M_R\left(-t\sum_{i=1}^{n}a_{i,1}\sigma_i,\ldots,-t\sum_{i=1}^{n}a_{i,K}\sigma_i\right).\quad(19.43)$$

Replacing n by $n-1$ and k by m in (19.33) we get for $m=1,\ldots,n-1$

$$Z_m^n(t) = \mathbb{E}\left[\lambda_n q_t^{n|S}g_m^{n-1}\right] = \sum_{\substack{1\le j_1<\cdots\\<j_m\le n-1}}\mathbb{E}\left[\lambda_n\prod_{\substack{i=1\\i\neq j_1,\ldots,j_m}}^{n}q_t^{i|S}\right] + \sum_{\nu=1}^{m-1}(-1)^\nu\binom{n-k+\nu-1}{\nu}$$

$$\times\sum_{\substack{1\le j_1<\cdots\\<j_{m-\nu}\le n-1}}\mathbb{E}\left[\lambda_n\prod_{\substack{i=1\\i\neq j_1,\ldots,j_{m-\nu}}}^{n}q_t^{i|S}\right] + (-1)^m\binom{n-1}{m}\mathbb{E}\left[\lambda_n\prod_{i=1}^{n}q_t^{i|S}\right].(19.44)$$

Let us note that the expectation in the last term of the sum (19.44) has already been calculated in (19.43). Further, the factor $q_t^{n|S}$ occurs in all products of (19.44), since the excluded indices $j_1<\cdots<j_m$, resp., $j_1<\cdots<j_{m-\nu}$ are elements of $\{1,\ldots,n-1\}$. Now, to obtain explicit formulas for $Z_m^n(t)$, $m=1,\ldots,n-1$, from (19.44), it suffices to evaluate the expressions

$$\mathbb{E}\left[\lambda_n\prod_{\substack{i=1\\i\neq j_1,\ldots,j_{m-\nu}}}^{n}q_t^{i|S}\right]$$

for $\nu=1,\ldots,m-1$. Thus, similar to (19.43) we obtain

$$\mathbb{E}\left[\lambda_n\prod_{\substack{i=1\\i\neq j_1,\ldots,j_{m-\nu}}}^{n}q_t^{i|S}\right] = -M_R\left(-t\sum_{\substack{i=1\\i\neq j_1,\ldots,j_{m-\nu}}}^{n}a_{i,1}\sigma_i,\ldots,-t\sum_{\substack{i=1\\i\neq j_1,\ldots,j_{m-\nu}}}^{n}a_{i,K}\sigma_i\right)a_{n,0}\sigma_n$$

$$\times M'_{\varepsilon_n}(-a_{n,0}\sigma_n t)\prod_{\substack{i=1\\i\neq j_1,\ldots,j_{m-\nu}}}^{n-1}M_{\varepsilon_i}(-a_{i,0}\sigma_i t) - \prod_{\substack{i=1\\i\neq j_1,\ldots,j_{m-\nu}}}^{n}M_{\varepsilon_i}(-a_{i,0}\sigma_i t)\sigma_n\sum_{k=1}^{K}a_{n,k}$$

$$\times\frac{\partial}{\partial u_k}M_R\left(-t\sum_{\substack{i=1\\i\neq j_1,\ldots,j_{m-\nu}}}^{n}a_{i,1}\sigma_i,\ldots,-t\sum_{\substack{i=1\\i\neq j_1,\ldots,j_{m-\nu}}}^{n}a_{i,K}\sigma_i\right).\quad(19.45)$$

For $i\neq n$ we get the corresponding formulas for $Z_m^i(t)$ from (19.43) and (19.44) by appropriate re-indexing.

Using the presented closed-form solutions we are able to price basket credit derivatives with a small number of reference entities in a very efficient way. Besides FTD baskets, second-to-default baskets (STD) are a popular product, which usually do not contain more than $n = 10$–15 reference entities. Let us consider an example of an STD basket with tenor of 5 years and quarterly premium payments. To keep things easy we use 30/360 as a day count method, although ACT/360 (with the actual number of days ACT) is the market standard. For the premium payment leg we have, e.g.,

$$
\frac{X_{STD}}{4} \sum_{j=1}^{20} B\left(0, \frac{j}{4}\right) \left(\mathbb{E}\left[g_0^n\left(\frac{j}{4}\right)\right] + \mathbb{E}\left[g_1^n\left(\frac{j}{4}\right)\right] \right)
$$

$$
= \frac{X_{STD}}{4} \sum_{j=1}^{20} B\left(0, \frac{j}{4}\right) \left\{ (1-n) M_R\left(\frac{-j}{4} \sum_{i=1}^{n} a_{i,1}\sigma_i, \ldots, \frac{-j}{4} \sum_{i=1}^{n} a_{i,K}\sigma_i\right) \right.
$$

$$
\times \prod_{i=1}^{n} M_{\varepsilon_i}(-a_{i,0}\sigma_i t)
$$

$$
\left. + M_R\left(-\frac{j}{4} \sum_{\substack{i=1 \\ i \neq j}}^{n} a_{i,1}\sigma_i, \ldots, -\frac{j}{4} \sum_{\substack{i=1 \\ i \neq j}}^{n} a_{i,K}\sigma_i\right) \prod_{\substack{i=1 \\ i \neq j}}^{n} M_{\varepsilon_i}(-a_{i,0}\sigma_i t) \right\}.
$$

Now we generalize the method to calculate the distribution of the number of defaults for the portfolio loss distribution. According to Section 19.1.5 the rounded loss given default of reference entity i is given by $M_i = \nu_i L_0$ for $i = 1, \ldots, n$. The PGF of $L(t)$ conditional on the risk factor S is given by

$$
G_{L(t)}(z|S) = \prod_{i=1}^{n} \left(q_t^{i|S} + (1 - q_t^{i|S}) z^{\nu_i} \right) = c_0^n + c_1^n z + \ldots + c_m^n z^m, \quad (19.46)
$$

where $m = \sum_{i=1}^{n} \nu_i$. Note that the coefficients c_j^n, $j = 1, \ldots, m$ still depend on the risk factors and the individual risks, although the notation used might indicate something else. Without loss of generality we could assume $\nu_1 \leq \nu_2 \leq \ldots \leq \nu_n$. We put all reference entities with identical loss given default in one subportfolio. Having done this we obtain M subportfolios P_i consisting of K_i reference entities, $i = 1, \ldots, M$, i.e. $K_1 + \ldots + K_M = n$. Let $q_t^{(i,j)|S}$ be the conditional survival probability of the j-th reference entity in subportfolio i. Thus, we obtain the probabilities $q_t^{(i,j)|S}$ by an appropriate re-indexing of the conditional probabilities $q_t^{i|S}$.

For $i = 1, \ldots, M$ we get from $\mu_i = \nu_{K_1 + \ldots + K_{i-1} + 1} = \ldots = \nu_{K_1 + \ldots + K_{i-1} + K_i}$

$$
G_{L(t)}(z|S) = \prod_{i=1}^{M} \prod_{j=1}^{K_i} \left(q_t^{(i,j)|S} + (1 - q_t^{(i,j)|S}) z^{\mu_i} \right)
$$

$$
= \prod_{i=1}^{M} \left(g_{0,P_i}^{K_i} + g_{1,P_i}^{K_i} z^{\mu_i} + \ldots + g_{K_i,P_i}^{K_i} z^{\mu_i K_i} \right), \quad (19.47)
$$

where
$$g_{j,P_i}^{K_i} = \mathbb{Q}[N_{P_i}(t) = j|S], \quad j = 0, \ldots, K_i, \tag{19.48}$$

and $N_{P_i}(t)$ is a process counting the number of defaults in subportfolio i, $i = 1, \ldots, M$. Therefore it follows automatically that these coefficients are of type (19.32)–(19.33). Hence we obtain for $k \in \{0, \ldots, m\}$

$$c_k^n = \mathbb{Q}[L(t) = k|S] = \sum_{\substack{(l_1,\ldots,l_M)\in\mathbb{N}_0^M \\ \mu_1 l_1 + \cdots + \mu_M l_M = k}} g_{l_1,P_1}^{K_1} \cdot \cdots \cdot g_{l_M,P_M}^{K_M}.$$

Consequently, we have to evaluate the expressions $\mathbb{E}\left[g_{l_1,P_1}^{K_1} \cdot \cdots \cdot g_{l_M,P_M}^{K_M}\right]$. As mentioned, each coefficient $g_{j,P_i}^{K_i}$ is of one of the types (19.32)–(19.33), i.e. multiplying these coefficients as in the latter equation leads to sums of products of the conditional probabilities $q_t^{i|S}$, which again could be evaluated with help of the MGFs M_R and M_{ε_i}, $i = 1, \ldots, n$. Hence, we obtain closed-form solutions for $\mathbb{Q}[L(t) = k] = \mathbb{E}[c_k^n]$, $k = 0, \ldots, m$ and for the loss distributions of the whole portfolio at time t.

For the expected loss of the portfolio at time t we have

$$\mathbb{E}[L(t)] = \mathbb{E}\left[\sum_{i=1}^n M_i N_i(t)\right] = \sum_{i=1}^n M_i - \sum_{i=1}^n M_i U_i(t)$$

$$= \sum_{i=1}^n M_i - \sum_{i=1}^n M_i M_{\varepsilon_i}(-a_{i,0}\sigma_i t) M_R(-a_{i,1}\sigma_i t, \ldots, -a_{i,K}\sigma_i t). \tag{19.49}$$

The loss $K(t)$ of a CDO tranche with subordination A and tranche notional $B - A$ is given by (19.21). Therefore, we have

$$\mathbb{E}[K(t)] = \sum_{k:A\leq k\leq B} (k - A)\mathbb{Q}[L(t) = k] + (B - A)\mathbb{Q}[L(t) > B]$$

$$= \sum_{k:A\leq k\leq B} (k - A)\mathbb{E}[c_k^n] + (B - A)\left(1 - \sum_{k:0\leq k\leq B} \mathbb{E}[c_k^n]\right),$$

where we note that the $c_k^n = c_k^n(t)$ are functions of t. The fair premium of the tranche is now given by

$$0 = X_{CDO} \sum_{j=1}^m \Delta_{j-1,j} B(0, T_j)(N - \mathbb{E}[K(T_j)]) - \int_0^T B(0, t) d\mathbb{E}[K(t)]. \tag{19.50}$$

The conditional loss distribution of the portfolio, i.e. the c_k^n, could be easily calculated by recursion, while the calculation of the unconditional one is more complicated, since for higher losses, i.e. higher k's, we have to add up a lot of numbers, which might be very time consuming. Since equity- or mezzanine tranches of synthetic CDOs are already wiped out after 25–30 defaults we face such kinds of numerical problems in particular when evaluating senior tranches. This can be fixed as follows. Let $K^{Sub}(t)$ be the expected loss of all tranches that are subordinated to the senior tranche. Then we have

$$\mathbb{E}\left[K^{Senior}(t)\right] = \mathbb{E}[L(t)] - \mathbb{E}\left[K^{Sub}(t)\right],$$

which is numerically much more efficient due to (19.49).

19.3 MGF of Risk Factors

Now we proceed to our last step to turn the semi-explicit pricing formulas into explicit ones. In the previous section we expressed the pricing formulas in terms of MGFs of the risk factors. In this section we present two approaches to obtaining the lacking MGFs. First, if the distribution of the risk factors is given, resp., if the model is calibrated to a given distribution then we could directly read off the MGFs from this distribution. Of course, we have to make sure that the MGF of the chosen distribution is available. As an example we show how this will work in the case of gamma- and lognormal-distributed risk factors.

Second, if we want to deduce risk factors from empirical data, then we proceed with the following recipe. In this case it turns out to be convenient to work in the class of elliptical distributions. First of all we perform a principal component analysis on the given credit spread time series to obtain time series for the resulting risk factors. Due to the assumption of elliptically distributed risk factors we deduce a time series of the radial component, which characterizes the non-parametric part of the elliptical distribution. Employing standard methods we obtain an estimate for the density of the radial component. Applying a Hankel transform to this density yields the corresponding characteristic generator. This directly gives the MGF, which finally enters our pricing formula.

19.3.1 Exogenous modelling

The advantage of our approach is the great flexibility regarding how market information enters the modelling. It turned out that the linear model (19.23) essentially describes the interrelations between the different reference entities. Now, there are several methods to establish this linear model, resp., to choose the systematic risk factors. One possible approach is to explain the systematic component of the credit spread evolution by fundamental or macro-economic factors. The main problem is to find appropriate factors that explain the systematic part of the credit spread sufficiently well such that the remaining individual risks ε_i, $i = 1, \ldots, n$, are independent. We call this approach *exogenous modelling*. In contrast to a Merton-based copula function, which is calibrated using historical data the exogenous modelling has the advantage that only current market information enters the model.

The basic question that automatically arises is, what kind of distributional assumptions on the risk factors $R_1, \ldots, R_K$ and on the individual risks ε_i, $i = 1, \ldots, n$, make sense? In general this can be answered only by testing a certain distribution hypothesis on given empirical data. The distributions of the risk factors and the individual risks enter the formulas for the (joint) survival probabilities (19.26) and (19.27) only by the appearance of its MGFs, which we have not specified up to now.

If we employ, e.g., the classical CreditRisk$^+$ approach in an exogenous modelling, we would estimate the coefficients

$$a_{i,1}, \ldots, a_{i,K}, \quad i = 1, \ldots, n,$$

with help of a least-squares approximation by using the following assumptions on the distribution of the risk factors:

$$\mathbb{E}[\varepsilon_i] = \mu_{i,0}, \quad \mathbb{E}[R_k] = \mu_k,$$

$$\varepsilon_i \sim \Gamma\left(\mu_{i,0}^2, \frac{1}{\mu_{i,0}}\right), \quad R_k \sim \Gamma\left(\mu_k^2, \frac{1}{\mu_k}\right),$$

for $i = 1, \ldots, n$ and $k = 1, \ldots, K$. Therefore, the MGFs are given by

$$M_{\varepsilon_i}(u) = \left(1 - \frac{u}{\mu_{i,0}}\right)^{-\mu_{i,0}^2}, \quad i = 1, \ldots, n,$$

$$M_{R_k}(u) = \left(1 - \frac{u}{\mu_k}\right)^{-\mu_k^2}, \quad k = 1, \ldots, K,$$

$$M_R(u_1, \ldots, u_K) = \prod_{k=1}^{K}\left(1 - \frac{u_k}{\mu_k}\right)^{-\mu_k^2}.$$

Further, assuming independence of the components of $R = (R_1, \ldots, R_K)$, (19.26) and (19.27) simplify to

$$U_i(t) = \left(1 + \frac{a_{i,0}\sigma_i t}{\mu_{i,0}}\right)^{-\mu_{i,0}^2} \prod_{k=1}^{K}\left(1 + \frac{a_{i,k}\sigma_i t}{\mu_k}\right)^{-\mu_k^2}, \tag{19.51}$$

$$U(t_1, \ldots, t_n) = \prod_{i=1}^{n}\left(1 + \frac{a_{i,0}\sigma_i t}{\mu_{i,0}}\right)^{-\mu_{i,0}^2} \prod_{k=1}^{K}\left(1 + \frac{\sum_{i=1}^{n} a_{i,k}\sigma_i t_i}{\mu_k}\right)^{-\mu_k^2}. \tag{19.52}$$

To illustrate the use of the latter we reconsider the calibration of the model to market prices of CDS, which is done with respect to the volatility parameter. With the help of (19.10) we obtain for reference entity i

$$0 = X_T \sum_{j=1}^{m} \Delta_{j-1,j} B(0, T_j) \left(1 + \frac{a_{i,0}\sigma_i t}{\mu_{i,0}}\right)^{-\mu_{i,0}^2} \prod_{k=1}^{K}\left(1 + \frac{a_{i,k}\sigma_i t}{\mu_{i,k}}\right)^{-\mu_k^2}$$

$$+ (1 - \delta) \int_0^T B(0, t)d\left\{\left(1 + \frac{a_{i,0}\sigma_i t}{\mu_{i,0}}\right)^{-\mu_{i,0}^2} \prod_{k=1}^{K}\left(1 + \frac{a_{i,k}\sigma_i t}{\mu_{i,k}}\right)^{-\mu_k^2}\right\}.$$

Using (19.52) we could easily price FTD baskets, while for k-th-to-default baskets we obtain the relevant quantities $Z_m^i(t)$ in a similar way.

A widely used approach for an exogenous modelling is to split up the systematic risk component of a reference entity into different risk factors by using a sector analysis. Common approaches use several industrial or country risk-related sectors. The sector analysis finally implies factor weights $a_{i,1}, \ldots, a_{i,K}$, $i = 1, \ldots, n$, $k = 1, \ldots, K$, while the coefficient describing the individual risk of a reference entity follows from (19.29).

If one wants to work in the Merton-based firm value framework, hence calibrating the linear model (19.23) with equity market information such as equity price and equity volatility data, it would make sense to consider lognormal-distributed risk factors. For example, if industry or country equity indices or, more general, macroeconomic indices are used as market data, we cannot expect them to be uncorrelated. Though the problem of correlated risk factors could be solved by orthogonalizing the factors with a principal component analysis (see Chapter 15) we still face the problem that we do not have a closed-form solution of (19.26) and (19.27) due to the lognormal distribution of the risk factors. Thus we have to fall back to some Monte Carlo simulation in this case, hence we could drop the assumption of uncorrelated risk factors.

Let $R_1, \ldots, R_K$ and ε_i, $i = 1, \ldots, n$, be lognormal-distributed random variables, where $\varepsilon_1, \ldots, \varepsilon_n$ are pairwise independent and independent from $R_1, \ldots, R_K$. Further, we assume that

$$\text{var}(\varepsilon_i) = \text{var}(R_k) = 1, \quad i = 1, \ldots, n, \quad k = 1, \ldots, K,$$

$$\rho(R_k, R_l) = \rho_{k,l}, \quad 1 \leq k, l \leq K.$$

Based on the latter we could evaluate (19.26) and (19.27) with help of the following expressions.

The components that are necessary for the pricing of structured credit derivatives such as (19.26), (19.27) resp. (19.39), (19.40) and (19.41) as well as (19.43) and (19.44) are always of the following type:

$$\mathbb{E}\left[\exp\left(\sum_{k=1}^{K} R_k \xi_k(t)\right)\right],$$

where $\xi_k(t)$ is a linear function of t and $R_1, \ldots, R_K$ are lognormal-distributed random variables. The evaluation of such expressions with the help of Monte Carlo simulations is described in Section 13.3 of Chapter 13 by Reiß.

19.3.2 Elliptical distributions

A k-dimensional distribution is spherically symmetric if it is invariant under the action of the orthogonal group $\mathcal{O}(k)$. The density $f_Y(y)$ of a spherically distributed random vector $Y : \Omega \to \mathbb{R}^k$ is a function of $\|Y\|$, i.e.

$$f_Y(y) = g(\|y\|^2) = g(y^T y) \tag{19.53}$$

for some scalar function $g : \mathbb{R}_+ \to \mathbb{R}_+$, the so-called *density generator*. Similary, the characteristic function $\Phi_Y(t) = \mathbb{E}[\exp(it^T Y)]$ is a function of $\|t\|$, i.e.

$$\Phi_Y(t) = \psi(\|t\|^2) = \psi(t^T t). \tag{19.54}$$

The scalar function $\psi : \mathbb{R} \to \mathbb{R}_+$ is called the *characteristic generator*. For the random variable Y we have the stochastic representation

$$Y = RW, \tag{19.55}$$

where W is uniformly distributed on the unit sphere in $\mathbb{R}^k$ and $R > 0$ is a scalar random variable with $Q = R^2 = \|Y\|^2$. The density of Q is given in terms of the density generator g by

$$f_Q(q) = \frac{\pi^{\frac{k}{2}}}{\Gamma\left(\frac{k}{2}\right)} q^{\frac{k}{2}-1} g(q). \tag{19.56}$$

A k-dimensional distribution is elliptically contoured if it is the image of a spherically symmetric distribution under an affine transformation. We shall confine our attention to the absolutely continuous, full rank, $\mathbf{L}^2$ case. Then an elliptically distributed random vector X has the following representation:

$$X = \mu + A^T Y, \tag{19.57}$$

with Y spherically distributed, mean vector μ and covariance matrix $\Sigma = A^T A$. With (19.55) we have the stochastic representation

$$X = \mu + RA^T W. \tag{19.58}$$

For the density of X we get with (19.53) the following formula:

$$f_X(x) = |\Sigma|^{-\frac{1}{2}} g\left((x - \mu)^T \Sigma^{-1}(x - \mu)\right). \tag{19.59}$$

The characteristic function of X is of the form

$$\Phi_X(t) = \mathbb{E}[\exp(it^T X)] = \exp(it^T \mu)\psi(t^T \Sigma t). \tag{19.60}$$

We write $f \sim EC_k(\mu, \Sigma, g)$. Then $\theta := (\mu, \Sigma)$ is the parametric part of the model and the function g is the non-parametric part.

The link between the density generator part g and the characteristic generator ψ is given by

$$\psi(t^T t) = \int_{\mathbb{R}^k} \exp(it^T y) g(y^T y) dy. \tag{19.61}$$

Examples

1. The density generator of the multivariate normal distribution is given by

$$g(u) = (2\pi)^{-k/2} \exp\left(-\frac{1}{2}u\right) \tag{19.62}$$

and the characteristic generator is

$$\psi(u) = \exp\left(-\frac{1}{2}u\right).$$

2. Consider a multivariate t-distribution with m degrees of freedom. For $m > k$ the density generator is defined by

$$g(u) = (m\pi)^{-k/2} \frac{\Gamma\left(\frac{k+m}{2}\right)}{\Gamma\left(\frac{m}{2}\right)} \left(1 + \frac{u}{m}\right)^{-\frac{k+m}{2}}.$$

19.3.3 Endogenous modelling and elliptical models

Suppose we have historical data of credit spread for the reference entities in the considered portfolio. If we perform a dimension reduction employing a principal component analysis we will obtain the linear factor model

$$\frac{\lambda_i}{\sigma_i} = \frac{X_T^{(i)}}{\sigma_{X_T^{(i)}}} = \mu_i + a_{i,0}\varepsilon_i + \sum_{k=1}^{K} a_{i,k}R_k, \quad 1 \le i \le n, \tag{19.63}$$

where $R = (R_1, \ldots, R_K)$ is a vector of uncorrelated standardized risk factors, i.e. $\mathbb{E}[R_k] = 0$, $\mathrm{var}(R_k) = 1$, $k = 1, \ldots, K$. The vector R is the systematic part of the risk factor that results from the dimension reduction.

If we assume a spherical distribution for $(R_1, \ldots, R_K)$ with characteristic generator ψ, resp., with corresponding density generator g, i.e.

$$R = (R_1, \ldots, R_K) \sim \mathrm{EC}_K(0, \mathrm{I}_K, g, \psi),$$

then we obtain

$$M_R(u) = \mathbb{E}\left[e^{u^T R}\right] = \psi(-u^T u)$$

for the MGF of $(R_1, \ldots, R_K)$. Using the latter, (19.26) and (19.27) imply

$$U_i(t) = e^{-\sigma_i \mu_i t} M_{\varepsilon_i}\left(-a_{i,0}\sigma_i t\right) \psi\left(-\sigma_i^2 t^2 \sum_{k=1}^{K} a_{i,k}^2\right), \tag{19.64}$$

$$U(t_1, \ldots, t_n) = \exp\left(-\sum_{i=1}^{n} \mu_i \sigma_i t_i\right) \prod_{i=1}^{n} M_{\varepsilon_i}\left(-a_{i,0}\sigma_i t_i\right)$$

$$\times \psi\left(-\sum_{k=1}^{K}\left(\sum_{i=1}^{n} a_{i,k}\sigma_i t_i\right)^2\right). \tag{19.65}$$

In the same way we could evaluate (19.39)–(19.41) and (19.43)–(19.44). To clarify the principle we present the result of (19.41). We have for $k = 1, \ldots, n{-}1$

$$
\mathbb{E}[g_k^n] = \sum_{\substack{1 \le j_1 < \cdots \\ < j_k \le n}} \psi\left(-t^2 \sum_{l=1}^{K}\left(\sum_{\substack{i=1 \\ i \ne j_1, \ldots, j_k}}^{n} a_{i,l}\sigma_i\right)^2\right) \prod_{\substack{i=1 \\ i \ne j_1, \ldots, j_k}}^{n} M_{\varepsilon_i}\left(-a_{i,0}\sigma_i t\right)
$$

$$
+ \sum_{\nu=1}^{k-1}(-1)^\nu \binom{n-k+\nu}{\nu} \sum_{\substack{1 \le j_1 < \cdots \\ < j_{k-\nu} \le n}} \psi\left(-t^2 \sum_{l=1}^{K}\left(\sum_{\substack{i=1 \\ i \ne j_1, \ldots, j_{k-\nu}}}^{n} a_{i,l}\sigma_i\right)^2\right)
$$

$$
\times \prod_{\substack{i=1 \\ i \ne j_1, \ldots, j_{k-\nu}}}^{n} M_{\varepsilon_i}\left(-a_{i,0}\sigma_i t\right) + (-1)^k \binom{n}{k}
$$

$$
\times \psi\left(-t^2 \sum_{l=1}^{K}\left(\sum_{i=1}^{n} a_{i,l}\sigma_i\right)^2\right) \prod_{i=1}^{n} M_{\varepsilon_i}\left(-a_{i,0}\sigma_i t\right). \tag{19.66}
$$

19.3.4 Model calibration

Let us now assume that the mean vector μ and the covariance matrix Σ are already estimated from the data and that we have calibrated the model parameters of (19.63) using a principal component analysis. Equations (19.65) to (19.66) reveal that it remains to determine the characteristic generator ψ of $R = (R_1, \ldots, R_K)$ coming from the dimension reduction.

Since R is a radially symmetric random vector the characteristic function is given by

$$
\Phi_R(t) = \psi(t^T t) = \mathbb{E}\left[e^{it^T R}\right] = \int_0^\infty \frac{J_{\frac{K-2}{2}}(\|t\|r)\Gamma(\frac{K}{2})}{\left(\frac{1}{2}\|t\|r\right)^{\frac{K-2}{2}}} dH_\rho(r)
$$

$$
= \int_0^\infty \frac{J_{\frac{K-2}{2}}(\|t\|r)\Gamma(\frac{K}{2})}{\left(\frac{1}{2}\|t\|r\right)^{\frac{K-2}{2}}} h_\rho(r)dr, \tag{19.67}
$$

where $J_n(x)$ is the Bessel function

$$
J_n(x) = \sum_{k=0}^{\infty} \frac{(-1)^k}{k!\,\Gamma(n+k+1)}\left(\frac{x}{2}\right)^{n+2k}
$$

and $h_\rho(r)$ the density function of the radial component of R with respect to the representation $R = \rho u$ (cf. (19.55)).

Equation (19.67) yields the decomposition of the Fourier transform of a radially symmetric density function into the spherical part and the radial part, where the latter contains all the relevant information of the distribution. Therefore we could estimate the characteristic generator ψ, resp., the characteristic function Φ_R by estimating the densitiy function h_ρ of ρ.

For a positive random variable ρ and $x > 0$ we have

$$H_\rho(x) = P(\rho \leq x) = P(\rho^2 \leq x^2) = H_{\rho^2}(x^2)$$

and thus

$$h_\rho(x) = \frac{d}{dx} H_\rho(x) = \frac{d}{dx} H_{\rho^2}(x^2) = 2x\, h_{\rho^2}(x^2). \tag{19.68}$$

We remark that (19.55) leads to an estimation of the density generator g of R as well. Hence we have to estimate the density h_{ρ^2}.

Let $R^{(1)}, \ldots, R^{(\varpi)}$ be a random sample of R resulting from the time series data that entered the principal component analysis (19.63). Denote

$$R^{(u)} = \left(R_1^{(u)}, \ldots, R_K^{(u)} \right), \quad u = 1, \ldots, \varpi,$$

and define $\rho_u^2 = \|R^{(u)}\|^2$, $u = 1, \ldots, \varpi$. Now we proceed with a non-parametric estimation based on standard methods.

We are led by the following motivation. The random vector R is elliptically distributed according to our assumptions. Starting with the normal distribution as reference distribution, i.e. $f_R^{(\text{ref})} = n(0, I_K)$, we obtain from (19.56) and (19.62) that $\rho^2 = \|R\|^2$ is χ^2-distributed with K degrees of freedom. Thus,

$$h_{\rho^2}^{(\text{ref})}(x) = \begin{cases} 0, & x < 0, \\ \frac{1}{2^{K/2}\Gamma(\frac{K}{2})} x^{\frac{K}{2}-1} e^{-\frac{x}{2}}, & x \geq 0. \end{cases} \tag{19.69}$$

In general the distribution of R differs from the normal distribution. Hence, the distribution of ρ^2 differs from the χ_K^2 distribution. Nevertheless, we will use the reference distribution as a starting point for our density estimation by using χ_k^2 as the kernel, i.e. we choose the following approach:

$$\hat{h}_{\rho^2}(x) = \frac{1}{\varpi h} \sum_{u=1}^{\varpi} \chi_K^2 \left(\frac{x - \rho_u^2}{h} \right). \tag{19.70}$$

According to [1, Section 4.2] we choose the optimal band width $\hat{h}$ based on

$$\hat{h} = \left(\frac{\|K\|_2^2}{\|h''\|_2^2 \, (\mu_2(K))^2 \, \varpi} \right)^{\frac{1}{5}},$$

where k is the kernel and h is the density, which is actually unknown. We replace h by the reference density to prevent a vicious circle. Therefore

$$\hat{h} = \varpi^{-\frac{1}{5}} c(K),$$

where $c(K)$ is a constant given by

$$c(K) = \left(\frac{\Gamma(K-1)}{(2K + K^2) \sum_{i=3}^{7} m_i \Gamma(K-i)} \right)^{\frac{1}{5}}$$

depending only on the dimension K of the random vector R and

$$m_3 = \frac{1}{16}, \quad m_4 = -\frac{1}{2}\left(\frac{K}{2} - 1\right), \quad m_5 = -m_4\left(\frac{3K}{2} - 1\right),$$

$$m_6 = -2m_4^2\left(\frac{K}{2} - 2\right), \quad m_7 = -2m_6\left(\frac{K}{2} - 2\right).$$

Now we employ the latter for the estimation $\hat{h}_\rho$ of the density of R. From (19.68) and (19.70) we obtain

$$\hat{h}_\rho(x) = 2x\hat{h}_{\rho^2}(x^2) = \frac{2x}{\varpi h} \sum_{u=1}^{\varpi} \chi_K^2\left(\frac{x^2 - \rho_u^2}{h}\right)$$

$$= \frac{2x}{\varpi h} \frac{1}{2^{K/2}\Gamma\left(\frac{K}{2}\right)} \sum_{u=1}^{\varpi} \left(\frac{x^2 - \rho_u^2}{h}\right)^{\frac{K}{2}-1} \exp\left(-\frac{1}{2}\frac{x^2 - \rho_u^2}{h}\right). \quad (19.71)$$

Using (19.71) we get from (19.67)

$$\hat{\Phi}_R(t) = \hat{\psi}(t^T t) = \int_0^\infty \frac{J_{\frac{K-2}{2}}(\|t\|r)\Gamma\left(\frac{K}{2}\right)}{\left(\frac{1}{2}\|t\|r\right)^{\frac{K-2}{2}}} \hat{h}_\rho(r)dr$$

$$= \sum_{v=0}^{\infty} \sum_{u=1}^{\varpi} \frac{(-1)^v\|t\|^{2v}}{v!\,\Gamma\left(\frac{K-2}{2} + v + 1\right)\varpi h\, 2^{K/2}}$$

$$\times \int_0^\infty \left(\frac{r^2}{4}\right)^v \left(\frac{r^2 - \rho_u^2}{h}\right)^{\frac{K}{2}-1} 2r \exp\left(-\frac{1}{2}\frac{r^2 - \rho_u^2}{h}\right) dr$$

$$= \frac{1}{U}\sum_{u=1}^{U} \exp\left(\frac{1}{2}\rho_u^2\right) \sum_{v=0}^{\infty} \frac{\left(-\frac{h}{2}\|t\|^2\right)^v}{v!\,\Gamma\left(\frac{K-2}{2} + v + 1\right)}$$

$$\times \int_0^\infty r^v \left(\frac{r - \rho_u^2}{2}\right)^{\frac{K}{2}-1} \exp(-r)\,dr. \quad (19.72)$$

For even K we could explicitly calculate the latter integral

$$\int_0^\infty r^v \left(\frac{r - \rho_u^2}{2}\right)^{\frac{K}{2}-1} e^{-r}\,dr = \sum_{m=0}^{\frac{K}{2}-1} (m+v)!\binom{\frac{K}{2}-1}{m}\left(-\frac{\rho_u^2}{2}\right)^{\frac{K}{2}-1-m}$$

while for odd K we have to be content with the power series expansion

$$\int_0^\infty r^v \left(\frac{r - \rho_u^2}{2}\right)^{\frac{K}{2}-1} e^{-r}\,dr = \left(-\frac{\rho_u^2}{2}\right)^{\frac{K}{2}-1} \sum_{m=0}^{\infty} (m+v)!\binom{\frac{K}{2}-1}{m}\left(-\frac{2}{\rho_u^2}\right)^m.$$

References

1. N.H. Bingham, R. Kiesel, and R. Schmidt. A semi-parametric approach to risk management. Working paper, submitted, 2003.
2. J. Gregory and J. P. Laurent. Basket default swaps, CDO's and factor copulas. Working paper, Oct. 2002.
3. J. Gregory and J. P. Laurent. I will survive. *Risk*, 16(6):103–107, 2003.
4. R. Martin, K. Thompson, and C. Browne. Price and probability. *Risk*, 14(1):115–117, 2001.

Index

actuarial model, 34
  collective, 34
  individual, 34
aggregated downside risk, 262, 276
aggregated expected net return, 276
aggregated expected return, 262
aggregated fitness variable, 57
aggregated net return, 270
aggregated risk, 270
allocation, 31, 32
  Euler, 32
  full, 31
asset-backed securities, 168, 311–315,
    319
  pricing, 321

basket credit default swap, 325
basket credit derivative, 339
  pricing, 352
Bernoulli variable, 9
  generating function, 9
beta distribution, 140, 292
binary model, 52, 57
binomial distribution
  compound negative, 8
  negative, 132
bottom-up approach, 26
Brownian motion, 222, 224–226, 228
business cycle indicator, 206

capital allocation, 25, 92
capital budget, 264
Carleman's condition, 114
Cauchy distribution, 114, 115

characteristic function, 114
central moment, 93
characteristic function, 111–126, 130,
    147, 149, 150, 158, 216, 218–221,
    223–225, 228, 357, 359, 360
characteristic generator, 357–360
Chebyshev's inequality, 27, 291
Cholesky decomposition, 136, 137, 224,
    227
coefficient of variation, 300
coefficient operator, 83
coherent risk measure, 26, 28, 29, 105
coherent risk spectrum, 295, 296, 303,
    306, 307
collateral risk, 139
collateralized debt obligation, 279, 280,
    292, 293, 325, 328, 339, 340
  synthetic, 326, 339
  tranche, 325, 327, 328, 339–341, 353
collateralized loan obligation, 314
compound gamma distribution, 153,
    156–163
compound negative binomial distribu-
    tion, 8
compound Poisson distribution, 33
computation time, 75, 91, 93, 97, 118,
    125, 126, 189, 196, 210, 213, 220,
    275, 286
computational complexity, 75, 93, 193,
    263, 268
computational effort, 85, 119, 125, 215,
    216, 220, 226, 229, 275, 281, 291
concentration risk, 246

Printing: Strauss GmbH, Mörlenbach
Binding: Schäffer, Grünstadt